"Schulte has studied and been on the cutting edge of the Fintech revolution in China for more than five years. Schulte's book is an enlightening and impressive look at the future of technological developments in Asia as the political winds between the US and China blow cooler."

Rob Citrone
Founder and CIO, Discovery Capital

"With a keen understanding of the high stakes competition between the US and China, the authors offer a timely analysis of how and where emerging technologies impact the two countries and the companies leading the charge."

Daniel Tu
Group CIO, Ping An (2013–2017), Hong Kong

"Paul Schulte is at the forefront of research in Fintech and AI developments. His insights and predictions are ignored at your own peril!"

Frank Wang
Deputy CEO and Executive Director, OCBC Wing Hang Bank (1999–2018), Hong Kong

"Paul has a remarkable talent for seeing into the future, observing trends that most people ignore and putting things together several years ahead of most analysts. Read this book."

David Halpert
Founder and CIO, Prince Street Capital, Singapore

"As a director in a publicly traded Chinese company, I can honestly say there is nobody in the banking industry who understands the PRC Fintech landscape better than Paul Schulte. Schulte predicted the rise of Alibaba, Tencent, & Ant long before the talking heads in the West knew they existed."

Austin Groves
Director of Revenue, Acorn International, Shanghai

"Schulte reminds us that AI and its rapidly emerging enabling technologies, like quantum computing, are unlocking new ways for people to work by allowing humans to focus on what humans do best — creative and critical thinking."

Michael Brett
CEO, QxBranch, Washington, D.C.

"Once again, Paul frames the dynamically changing intersection of technology and finance, and helps us focus on where they are heading, rather than where they have been."

Russell Kopp
Managing Partner, Options Group, Hong Kong

AI & Quantum Computing for Finance & Insurance

Fortunes and Challenges for China and America

Singapore University of Social Sciences - World Scientific Future Economy Series

ISSN: 2661-3905

Series Editor
David Lee Kuo Chuen *(Singapore University of Social Sciences, Singapore)*

Subject Editors
Guan Chong *(Singapore University of Social Sciences, Singapore)*
Ding Ding *(Singapore University of Social Sciences, Singapore)*

Singapore University of Social Sciences - World Scientific Future Economy Series introduces the new technology trends and challenges that businesses today face, financial management in the digital economy, blockchain technology, smart contract and cryptography. The authors describe current issues that the business leaders and finance professionals are facing, as well as developments in digitalisation. The series covers several increasingly important new areas such as the fourth industrial revolution, Internet of Things (IoT), blockchain technology, artificial intelligence (AI) and many other forces of disruption and breakthroughs that shape today's realities of the economy. A better understanding of the changing environment in the future economy can enable business professionals and leaders to recognise realities, embrace changes, and create new opportunities — locally and globally — in this inevitable digital age.

Published*

Vol. 1   *AI & Quantum Computing for Finance & Insurance: Fortunes and Challenges for China and America*
by Paul Schulte and David Lee Kuo Chuen

Forthcoming Titles

The Emerging Business Models
Guan Chong, Jiang Zhiying and Ding Ding

Financial Management in the Digital Economy
Ding Ding, Guan Chong and David Lee Kuo Chuen

Blockchain Foundational Issues and Beyond
Lo Swee Won and David Lee Kuo Chuen

Inclusive Disruption: Digital Capitalism, Deep Technology and Trade Disputes
David Lee Kuo Chuen and Linda Low

*More information on this series can also be found at
https://www.worldscientific.com/series/susswsfes

Singapore University of Social Sciences - World Scientific
Future Economy Series : 1

AI & Quantum Computing for Finance & Insurance

Fortunes and Challenges for China and America

Paul SCHULTE
Schulte Research, Singapore
Singapore University of Social Sciences (SUSS)

David LEE Kuo Chuen
Singapore University of Social Sciences (SUSS)

SINGAPORE UNIVERSITY
OF SOCIAL SCIENCES

 World Scientific

Published by

World Scientific Publishing Co. Pte. Ltd.

5 Toh Tuck Link, Singapore 596224

USA office: 27 Warren Street, Suite 401-402, Hackensack, NJ 07601

UK office: 57 Shelton Street, Covent Garden, London WC2H 9HE

Library of Congress Cataloging-in-Publication Data
Names: Schulte, Paul, 1963–   author. | Lee, David (David Kuo Chen) author.
Title: AI & quantum computing for finance & insurance : fortunes and challenges for
    China and America / by Paul Schulte (Schulte Research, Singapore &
    Singapore University of Social Sciences, Singapore) and
    David Kuo Chuen Lee (Singapore University of Social Sciences, Singapore).
Other titles: AI and quantum computing for finance and insurance
Description: New Jersey : World Scientific, [2019] | Series: Singapore University of
    Social Sciences - World Scientific future economy series ; Volume 1
Identifiers: LCCN 2019012738 | ISBN 9789811203893 (hc : alk. paper)
Subjects: LCSH: Financial services industry--Information technology. | Insurance--Accounting. |
    Internet banking. | Money--Technological innovations.
Classification: LCC HG173 .S328 2019 | DDC 332.0285/63--dc23
LC record available at https://lccn.loc.gov/2019012738

British Library Cataloguing-in-Publication Data
A catalogue record for this book is available from the British Library.

First published 2019 (Hardcover)
Reprinted 2019 (in paperback edition)
ISBN 978-981-120-918-5 (pbk)

For any available supplementary material, please visit
https://www.worldscientific.com/worldscibooks/10.1142/11371#t=suppl

Desk Editors: Aanand Jayaraman/Yulin Jiang

Typeset by Stallion Press
Email: enquiries@stallionpress.com

Printed in Singapore

To the Five Golden Rings: Boris, Austin, Suraj, Chris and Simon.

— Paul Schulte

To the Five that my True Love sent to me:
Glee, Dlee1, Dlee2, Dlee3 and Shawnlwk.

— David Lee

Acknowledgments

The authors are indebted to many people who shared their views and provided much information for the book. We are especially thankful to Gavin Liu, who did much of the PowerPoint Slides and the primary research, Diana An Xinxin, who provided research assistance, Jiang Yulin, who encouraged us to publish, Lee Hooi Yean, who marketed the title, and many more that have helped us in one way or another. Jason Kang from Harvard University did much work on the content as well as the chapter on quantum computing. Matt Zayco was very helpful in the content as well. Thanks are also due to Benedict Printz, Will Stuart, Austin Groves, Yimeng Sun, Kenta Iwasaki, Houchen Li, William Drake, JY Phuang, Nik Lemesko, Philip Hultsch, James Naylor, Lee Chor Pharn, Veronica Tan, Koh Wee Siong, Su Tay, Mohanty Supnendu, Damien Pang, Gerald Nah, Roy Teo, Prof Yan Li, Andy Wang, Prof Pei Sai Fan, Dr Xiao Feng, Bo Shen, Roland Sun, Remington Ong, Renqi Shen, Emma Cui, Wei Shi Khai, Chen Yan Feng, Sun Lilin, Prof Cheong Hee Kiat, Prof Tsui Kai Chong, Prof Lee Pui Man, Prof Allan Chia Beng Hock, Prof Euston Quah, Yeo Guat Kwang, Tome Oh, Selina Lin, Caroline Lim, Lo Swee Won, Cheryl Wang Yu, Sherry Li, Addy Crezee, Anson Zealle, Chia Hock Lai, Zann Kwan, Daphne Ng, Dr Ernie Teo, Dr Naseem Naqvi, Rev. Martin Ignacio Diaz Velasquez, Dr Choi Gongpil, Paul Neo, Jessie Wang Xiao Meng, Ulan Quqige, and Davina Tham. Special thanks also to Harry Smorenberg and Joyce Ang from currency research. Charles Liu in Beijing was a central and vital interlocutor in this. Others who are outstanding thinkers in this area who helped to clarify our views are investor titans, such as Adam Levinson, Rob Citrone, Eric Bushell, John Burbank, Matthew Williams, Amit Rajpal, David Halpert, David Dredge, Todd Tibbetts, David Courtney, Kyu Ho, Gary Ang, Melissa Guzy, and Danny Lee. Others who helped form ideas include the mercurial Fred Feldkamp, ever-present Andrew Work, global lawyer Brian Ganson, and digital media guru Dede Nickerson. Veterans of the Asia scene with great insight are Gao Xi Qing, Kenneth Ng, Daniel Tu, Roy Wu, Yang Yang, Kevin Tang, Jim Stent, David Halperin, Matthew Williams, Scott Sleyster and Jorge Sebastiao and Ryan Thall.

Paul SCHULTE and David LEE Kuo Chuen

Contents

Introduction

In the thick of this fourth industrial revolution, navigating the technology landscape has never been a more complex task. Early idealism that the Internet would free us from oppression as well as economic inequality and move the world toward a grand utopian future has proven elusive. As a human race, our speed to innovate overwhelms our ability to civilize technology. Our impulse to share overwhelms our need for privacy. Our need for progress as nations eclipses those countries which have failed to innovate, causing jealousy and resentment. Having the world's information available to billions of people at the touch of a button has morphed from convenience and diversity into the pervasive monitoring of ordinary citizens and encroachment of individual privacy. The results are claims of fake news by cynical governments, manipulation of national elections by bad actors, wide-scale electronic theft of money and identity, and increasing censorship. A growing chorus sings the vices rather than virtues of technology, and there is a creeping sense that humans may not be the ones in control anymore.

We believe that attempts to understand the powers of technology must be grounded in evidence-based research and take into consideration ethics, culture, politics, philosophy, human behavior, and psychology. As Kevin Kelly put it in *The Inevitable*: *A utopia has no problems to solve, but therefore no opportunities either.* The opportunity afforded by Blockchain and its potential for privacy protection, distribution of trust and coded governance was what intrigued us and initially brought us, the two authors, together.

Now, the question at the back of our minds is whether the same "utopian promises of freedom, fairness and equality" of the Internet are back to haunt us, and whether we have all placed too much faith in blockchain technology. We do not have all the answers. However, as we watched China, we both believed that the "heyday" of financial institutions was passing as Alibaba and Tencent introduced one new blockbuster financial product after the other. Our shared experience of investment, research, and board advisory in the China orbit gave us insight into the implications of convergence between disruptive technology and the speed of this innovation, particularly with regard to how the US has come to perceive China as a threat and not an ally. Blockchain is no utopia, but we want to show that it offers great promise nonetheless.

We realize everyone is overwhelmed with data, so we went heavy on PowerPoint slides and light on text. In this way, important ideas can be conveyed in a few seconds of concise data rather than endless text which no one has time to read anymore. PowerPoint is short hand and is, in many ways, more difficult than writing as it forces brevity. It truly represents the spirit of Mark Twain: "I did not have time to write a short letter, so I wrote a long one instead." This is a series of short "Powerpoint letters" on the challenges and opportunities in this brave new world which is a fascinating subset of finance, AI, realpolitik, and behavioral economics.

Money follows great ideas. In China and the US, the marketplace of ideas was awash with money for the next technology startup that had the potential to become a unicorn. The rules seemed simple. These companies had to be good at servicing every aspect of their customers' lives and were often aided by artificial intelligence in anticipating and meeting people's basic needs (衣食住行), or clothing, food, accommodation, and transport). Technology giants were poised to take over the services of incumbents and, therefore, commanded steep valuations. However, financial market professionals soon found that these unicorns were a great challenge to comprehend, monitor, and regulate.

Furthermore, investors have come to see both Chinese and American technology companies as pawns of the national security apparatus. Given the rapid innovation of emerging technology, the speed of acquisitions and increasingly bellicose rhetoric, it was difficult to fathom corporate activity. Without personal experience of the technology as a user and comprehension of the Chinese language, familiarity with the technology, business models and internal politics, it was almost impossible to achieve.

We both gave talks on this subject independently at universities around the world, but continued to meet over conferences, especially at the Singapore University of Social Science (SUSS). We taught courses together at SUSS and soon realized that our teaching materials converged from different viewpoints. That was the moment when both of us — who have been analyzing and investing in the financial markets since the 1980s — decided that we should compile and share some of our work with PowerPoint presentation slides in this book. These slides are self-explanatory and the most efficient way to get the latest systematic coverage to a wider readership.

The central theme of this book is Deep Technology (DT) and its commercialization. The speed of acquisition, research, and development by DT giants is incredible, and few have the time to fully grasp this area of research, which has powerful geopolitical and financial implications. Tellingly, there is still no consensus on the definition of DT, though some define it as a set of cutting-edge and disruptive technologies based on scientific discoveries, engineering, mathematics, physics, and medicine with applications that can have a profound impact on society. DT will certainly include hi-tech and span artificial intelligence, deep learning, machine learning, quantum computing, blockchain, data technology, robotics, autonomous systems, clean tech, energy efficient, food tech, and biotechnology.

The Chinese word for Deep Technology is *shēn kējì* (深科技). The word *shēn* can mean *shēn ào* (深奥), indicating something profound, recondite or difficult to understand, or *shēn yōu* (深忧), which signals anxiety. In the present climate, anxiety is perhaps the more appropriate phrase to describe DT. There is a nervousness within the US about technology being used to build a digital empire to spy on anyone and everyone. That anxiety is specifically directed at

Chinese DT possibly allowing Chinese security services to intercept the US military, government, and corporate communications. This tension has erupted into the 5G race between the US' allies and China's Huawei. The Chinese government continues to claim that there is no evidence of digital empires. However, William R. Evanina, the director of America's National Counterintelligence and Security Center[1] has articulated the US' perspective: *Chinese company relationships with the Chinese government aren't like private sector company relationships with governments in the West. China's 2017 National Intelligence Law requires Chinese companies to support, provide assistance and cooperate in China's national intelligence work, wherever they operate.*

While many criticize China for what they see as a monopoly on snooping into its citizens' lives, Timothy Nee, President of Dorman Consulting Associates, makes a similar point with regard to the US intel community: *Most Americans realize that there are two groups of people who are monitored regularly as they move about the country. The first group is monitored involuntarily by a court order requiring that tracking device be attached to their ankle. The second group includes everyone else, almost all of whom volunteer each day to be monitored by placing a mobile phone in a purse or pocket.*[2]

NSA employee and whistleblower Edward Snowden gives yet another view: *Privacy isn't about something to hide. Privacy is about something to protect. And that's who you are. Privacy is the right to the self. Privacy is what gives you the ability to share with the world who you are on your own terms.*[3] This grand conviction presumably motivated him to disclose the NSA's vast global surveillance network.

Both of us firmly believe that there is a disparity in the quantity as well as the quality of discussion about the information being harvested by both the US and Chinese governments. When Shoshana Zuboff coined the phrase "surveillance capitalism", she describes its

---

[1]https://www.dni.gov/index.php/ncsc-newsroom/item/1889-2018-foreign-econ omic-espionage-in-cyberspace
[2]https://www.insurancejournal.com/news/national/2013/10/02/307073.htm
[3]https://www.businessinsider.sg/edward-snowden-privacy-argument-2016-9/?r =US&IR=T

basis in *The rapid buildup of institutionalized facts — brokerage, analytics, data mining, ... powerful network effects, state collaboration ... and unprecedented concentrations of information — ... In this new regime, global architecture of computer mediation turns the electronic text of the bounded organization into an intelligent world-spanning organism that I call Big Other.* Her point is that surveillance capitalism spans both sides of the Pacific Ocean — we are all playing on one chessboard, with the NSA as the US' Queen on one side, and the MSS as China's Queen on the other.

Geopolitics, national security and fund management in the tech industry are only becoming more deeply intertwined. In late January 2019, National Security Advisor John Bolton himself inadvertently revealed that the trade war is not about trade but about 5G and naval technological superiority. The trade negotiations are a tactic to stall for time — a diplomatic throwaway line to accommodate Trump's campaign promises. Moreover, when Bolton goes on to say that "China needs to play by the rules that everyone else plays by", he presumably means that the PLA and MSS need to stop spying and stealing, but not so their US counterparts.

John Le Carré, who worked for the MI5 and MI6 before perfecting the spy novel, says the world of spying "is a world of constant deceit and shades of grey". It is dotted with highly dysfunctional, duplicitous, amoral and extremely selfish practices largely driven by blackmail. Unfortunately, this state activity has infiltrated part of the investment process. Tech giants are already wrapped up in the activities of political actors and national intelligence agencies, at times to their great advantage, and at times not. Huawei has become the new site of an Industry 4.0 proxy war between the US and China.

This new cold war is about cryptography and privacy. Perhaps, Satoshi Nakamoto anticipated this potential monopoly on technological power in his 2008 white paper advocating a design that imagines every node playing a role, and a network being decentralized with a distributed trust to prevent contraction of information power. That Bitcoin network went live in January 2009 after the Global Financial Crisis. A movement was started to design different blockchain payments systems to replace fiat currency. The cypherpunks believe coded governance can overcome irresponsible government and central banks, becoming distributed and autonomous, but perhaps overlooked that off chain would remain centralized in governance or mining.

One country which is grappling with these issues and has chosen to embrace many of the principles of the cyberpunk economy is Singapore. The country has renamed its National Trade Platform to the Networked Trade Platform, perhaps recognizing that the new world is about Digitalization, Disintermediation, Democratization, Decentralization, and Diminishing Oneself. The questions Singapore faces are as follows: (1) Is one central role of government to provide a blockchain ecosystem with privacy protection? (2) Is a worthy goal of government to be one node among many with no centralized ruler. In other words, is it possible for no one ruling party to be in charge while every party is an equal node — a technological 'primes inter pares'? (3) Does Singapore need to consider being part of a "connected G0 world" that runs on its own? At Davos 2019, Singapore's Minister of Finance Heng Swee Keat said, "One important area... for Singapore is how do we reposition Singapore to this new world of industry 4.0 globalization. I believe Singapore can be a global Asia node for technology, innovation, and enterprise ... to increase the cross-cultural literacy of Singaporeans so that we can connect with people from all over the world."

We leave it to the readers to conclude for themselves whether the clash of the US and China will leave some open, agile, teachable, smart, and digital-democratic countries better off. This book is designed as a deep dive into the new relationships and functions of financial institutions in the context of (1) society's demand for better services but also privacy, (2) learning new AI skills but also alleviating fear of chronic unemployment, (3) appreciating the profound changes happening in the health insurance sector but also highlighting the dangers to universal exposure of personal health data, and (4) seeing China as a global innovator but also an actor to challenge the US and not undermine it.

Part 1 of the book is all about the lizard brain and how the frailties of the human mind can often trick us or undermine good judgment. AI tools are precisely designed to help us overcome these 'blind spots' in all areas of life and NOT to eliminate the employee in the workplace. Part 2 of the book looks at who is implementing these tools the fastest and most efficiently in financial services. We examine the leaders and laggards. Part 3 of the book looks at corporate case studies globally with a specific look at China, since it contains many of the global leaders. Part 4 looks at specific industries and countries which are at the forefront of this golden age of technological innovation: quantum computing, cloud services, and insurtech. Lastly, we ask about the sustainability of smart choices made by Singapore in its quest to be a global tech city and what else can it do to remain relevant.

We hope that this book will be useful to anyone interested in investing their capital for impact and to anyone who is drafting policies to ensure we are in a safer world. We want to make the disclaimers that under no circumstances should readers treat the information and written comments as financial advice. As the second author is an independent director of a subsidiary of Ping An, he was not involved in the research nor commentary of the company and its subsidiaries in this book.

We are optimistic that most companies do not subscribe to the idea of surveillance capitalism, but are working towards the idea of inclusive FinTech.[4] Indeed, many of these companies have lifted millions of people out of poverty and have strengthened the financial status of micro, small and medium enterprises.

If we have no faith in technology to liberate us, nor do we place our hope in humanity to protect us, then who else is there to trust? In a world where the winner takes all, perhaps it is wiser to petition our Creator to empower those who preserve human dignity to step up. We must first do our part in understanding the technology and its complex landscape. This is precisely what we both set out to do in this book.

---

[4]https://www.worldscientific.com/worldscibooks/10.1142/10949

PART 1

How to Escape the Corporate Lizard Brain: The True Role of AI

Chapter 1

Purpose of AI: Naming and Shaming the Corporate Lizard Brain — One Meeting at a Time

1.1 How Us Frail Humans Can Make Peace with AI

There is a part of the brain known as the limbic cortex that neuroanatomists have attributed as the seat of emotion, addiction, mood and lots of other metal and emotional processes. Amygdala, the part of the limbic system that is responsible for primitive survival instincts such as fear and aggression, is also called "The Lizard Brain" or "The Reptilian Brain". This is because the limbic system is about all a lizard has for brain function[1] (see Figure 1.1).

> *The lizard brain is the reason you're afraid, the reason you don't do all the art you can, the reason you don't ship when you can. The lizard brain is the source of the resistance.*
> *The lizard brain is hungry, scared, angry, and horny.*
> *The lizard brain only wants to eat and be safe.*
> *The lizard brain cares what everyone else thinks, because status in the tribe is essential to its survival.*

— Seth Godin[2]

---

[1]https://www.psychologytoday.com/us/blog/where-addiction-meets-your-brain/201404/your-lizard-brain

[2]https://www.goodreads.com/quotes/304213-the-lizard-brain-is-hungry-scared-angry-and-horny-the https://seths.blog/2010/01/quieting-the-lizard-brain/

Figure 1.1: How us frail humans can make peace with AI: Ditch the lizard brain and embrace algorithms.

In Barbara Tuchman's 1961 book titled, *The Guns of August*, the best sentence in summarising the failure to plan for a long World War I among other wrong moves was as follows:

> "The disposition of everyone on all sides was not to prepare for the harder alternative, not to act upon what they suspected to be true."[3]

She has also identified a historical phenomenon which she terms "Tuchman's law", which has become known as a psychological principle of "perceptual readiness" or subjective probability"[4]:

---

[3] https://books.google.com.sg/books?id=fnVy4v5pZPMC&pg=PA27&lpg=PA2 7&dq=The+disposition+of+everyone+on+all+sides+was+not+to+prepare+fo r+the+harder+alternative,+not+to+act+upon+what+they+suspected+to+be +true&source=bl&ots=oth8HT0G4N&sig=o_S5bYWEXb9gdKwMehNBTW4T ZTs&hl=en&sa=X&ved=2ahUKEwiVq8Oi_dDfAhUOAogKHTw6Cf4Q6AEwA HoECAkQAQ#v=onepage&q=The%20disposition%20of%20everyone%20on%2 0all%20sides%20was%20not%20to%20prepare%20for%20the%20harder%20alter native%2C%20not%20to%20act%20upon%20what%20they%20suspected%20to %20be%20true&f=false

[4] Texas Research Institute of Mental Sciences, *Violence and the Violent Individual: Proceedings of the Twelfth Annual Symposium*, Texas Research Institute of Mental Sciences, Houston, Texas, November 1–3, 1979. Spectrum Publications, p. 412.

Disaster is rarely as pervasive as it seems from recorded accounts. The fact of being on the record makes it appear continuous and ubiquitous whereas it is more likely to have been sporadic both in time and place. Besides, persistence of the normal is usually greater than the effect of the disturbance, as we know from our own times. After absorbing the news of today, one expects to face a world consisting entirely of strikes, crimes, power failures, broken water mains, stalled trains, school shutdowns, muggers, drug addicts, neo-Nazis, and rapists. The fact is that one can come home in the evening — on a lucky day — without having encountered more than one or two of these phenomena. This has led me to formulate Tuchman's Law, as follows: "The fact of being reported multiplies the apparent extent of any deplorable development by five- to tenfold" (or any figure the reader would care to supply).

— Barbara W. Tuchman[5]

In other words, sensational and exaggerated events are what people like to read, and therefore events are depicted as general and pervasive. Negative aspects of things are often played up in history and news while chroniclers tend to forget the positive aspects of crucial events. For example, the failures of startups are a blast over the media while the successes of the impactful few are rarely reported or chronicled in small print.

Psychologists have also studied many similar cognitive bias situations, such as groupthink,[6] fear of authority, lack of imagination and hyper-rational (Figure 1.2). The term "groupthink" was first mentioned by William Whyte,[7] and later Irving Janis[8] developed the theory of groupthink to describe the faulty decision-making that can occur in groups as a result of forces that bring a group together. Mental health professions consider the severe fear of authority as a form of social phobia. Albert Einstein once said that the unthinking respect for authority is the greatest enemy of truth.

A failure of imagination is when something seems obvious to those in the know, predictable (particularly from hindsight) and yet

---

[5]Tuchman, B. (1978). *A Distant Mirror: The Calamitous 14th Century.* New York: Alfred A. Knopf, p. xviii.

[6]https://www.psychologytoday.com/intl/basics/groupthink

[7]Whyte, W. H., Jr. (March 1952). "Groupthink". *Fortune*, 114–117: 142, 146.

[8]Janis, I. L. (November 1971). "Groupthink". *Psychology Today*, 5(6): 43–46, 74–76. Archived from the original on April 1, 2010.

no provision is made for the undesirable outcome. If the individual does not have the ability or refuses to extract the details of previous experiences and assemble them to create an imaginary event, then there is a lack of imagination. Such a lack of constructive-episodic simulation has linkages to old age and the involvement of recalling other types of memory. Hyper-rationality is a defense mechanism against all that is threatening and causes unease. It describes the situations that rationality carried to the extreme and beyond rational limits.

On the other spectrum, Artificial Intelligence (AI) represents a different view of independence, logical, transparent, ever-changing, relentless, and emotionless. With carefully designed self-learning and AI tools, a variety and complete set of methodologies minimize or overcome cognitive biases when applied to real-world situations.

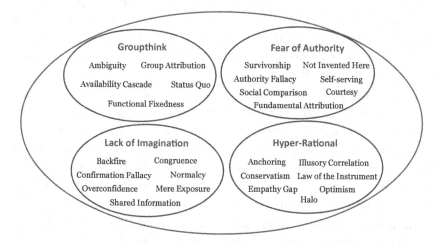

Figure 1.2: Cognitive bias: I MUST BELONG. I can't fail. Polite is good. Boss knows better. Yesterday predicts tomorrow. Data confirm my view.

Technology can do good and also do harm depending on ethical standards driven by people. AI has a good chance of being a foundation for technology to overcome human frailty if governed by ethical rules and regulations.

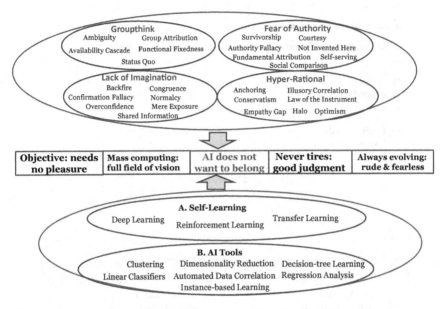

Figure 1.3: Foundation for human frailty: Immediate desire creates perception — desire for \$, pleasing, pleasure, pals, perspective, predictability.

Different types of biases (see Figures 1.3–1.5) will lead to different effects but mostly negative. What followed from the four biases is fear of judgement, fear of failure, fear of the unknown and fear of the irrational. This leads to quitting, hiding, procrastinating and freezing — hardly the outcomes that any corporates or individuals are looking towards. This seems to be the most common observation of modern-day management, especially with the emphasis on conflict of interests and fiduciary duties. It is almost a foregone conclusion that without any technical expertise in the board or management, the "right" thing to do is not to do anything. However, AI has much to contribute.

Types of bias	Negative Effect	Outcome
1. **Group**think is dominant	Fear of Judgement	We quit
2. **A**uthority is dominant	Fear of Failure	We hide
3. **I**magination is dominant	Fear of Unknown	We defer
4. **R**ationality is dominant	Fear of the Irrational	We freeze

Why? We
1) depend on the group;
2) tire easily;
3) want to be liked;
4) frighten easily;
5) love pleasure.

Figure 1.4:   Types of biases/defects.

I. Groupthink is dominant — AI Crowd Sourced Data		II. Fear of Authority is dominant — Mass Data to Check Assumptions	
1	Ambiguity Fallacy	6	Authority Fallacy
2	Availability Cascade	7	Courtesy Fallacy
3	Functional Fixedness	8	Survivorship
4	Group Attribution	9	Fundamental Attribution
5	*Status Quo*	10	Not Invented Here
		11	Self-serving
		12	Social Comparison
III. Lack of Imagination is dominant — AI for Alternative Scenarios		IV. Excess Rationality is dominant — Sorting of Mass Data	
13	Backfire Fallacy	20	Anchoring Fallacy
14	Confirmation Fallacy	21	Conservatism Fallacy
15	Congruence Fallacy	22	Empathy Gap
16	Overconfidence	23	Halo Fallacy
17	Mere Exposure	24	Illusory Correlation
18	Normalcy Problem	25	Law of the Instrument
19	Shared Information	26	Optimism
		27	Risk/Compensation Bias

Figure 1.5:   Summary.

1.2 Cognitive Bias (Human Blind Spots): Humans Need AI as a Recall for Our Defects

By crowdsourcing data using AI, one may overcome the dominance of groupthink (Figures 1.6–1.8). But generally, groupthink suffers from the following:

1. **Ambiguity Fallacy:** Logical fallacies are errors in reasoning that render an argument invalid. The effect of logical fallacies is generally to complicate or blur the true issue in the argument by adding extraneous elements without presenting real evidence.

2. **Availability Cascade Fallacy:** An availability cascade is a self-reinforcing cycle that explains the development of certain kinds of collective beliefs.[9]

---

1. Ambiguity Fallacy
Belief: I won't invest because there is no way for me to know the outcome.

Solution: Take many small bets to test uncertain assumptions.

2. Availability Cascade Fallacy
Belief: I believe the group because the group is wise in its decision.

Solution: Do a quarterly questioning of strategic direction. Be prepared to change that direction. Encourage crazy ideas. Seek out introverted outsiders.

Figure 1.6: Groupthink is dominant (1).

3. **Functional Fixedness Fallacy:** Functional fixedness is a cognitive bias that limits a person to use an object only in the way it is traditionally used.[10]

4. **Group Attribution Fallacy:** The group attribution error refers to people's tendency to believe either (1) that the characteristics of an individual group member are reflective of the group as a whole,

---

[9]"Availability Cascade — Liquisearch.com." n.p., n.d. Web. January 30, 2019, https://www.liquisearch.com/availability_cascade
[10]Duncker, K. (1945). "On problem solving". *Psychological Monographs*, 58(5) (Whole No. 270).

or (2) that a group's decision outcome must reflect the preferences of individual group members, even when external information is available suggesting otherwise.[11,12]

---

3. Functional Fixedness Fallacy
Belief: I believe in traditional ways to solve problems.

Solution: Breaking people out of the mold and encouraging imagination.

4. Group Attribution Fallacy
Belief: I believe that individuals are reflective of the group or that group decisions reflect the preferences of the group.

Solution: Make sure you've got enough data to reliably conclude that the group represents the members. Ensure that the attributes you identify are reflective of the broader group and not just a handful of people you've interviewed from that group.

Figure 1.7:　Groupthink is dominant (2).

5. ***Status Quo* Problem:** *Status quo* is a Latin phrase meaning the existing state of affairs, and generally used to describe inaction in a sense to maintain existing social structure and values.

---

5. Status Quo Problem
Belief: I believe that the *status quo* is ideal.

Solution: When putting together a team, evaluate them against the 15 personality attributes of innovators and entrepreneurs.

Figure 1.8:　Groupthink is dominant (3).

---

[11]Hamill, R.; Wilson, T. D.; Nisbett, R. E. (1980). "Insensitivity to sample bias: Generalizing from atypical cases". *Journal of Personality and Social Psychology*, 39(4): 578–589. doi:10.1037/0022-3514.39.4.578. Archived from the original on May 11, 2016.

[12]Allison, S. T.; Messick, David M (1985). "The group attribution error". *Journal of Experimental Social Psychology*, 21(6): 563–579. doi:10.1016/0022-1031(85)90025-3.

Mass data can be used to check assumptions when there is a fear of authority bias (see Figures 1.9–1.11). But generally, the biases are reinforced by the following (well documented in Wikipedia and reference throughout this chapter, see Figure 1.19):

6. **Authority Fallacy:** Insisting that a claim is true simply because a valid authority or expert on the issue said it was true without any other supporting evidence offered.

7. **Courtesy Fallacy:** As the bias or conflict of interest becomes more relevant to the argument, usually signified by a lack of other evidence, the argument is seen as less of a fallacy and more as a courtesy.[13]

8. **Survivorship Problem:** Survivorship bias or survival bias is the logical error of concentrating on the people or things that made it past some selection process and overlooking those that did not, typically because of their lack of visibility.

---

6. Authority Fallacy
Belief: The big man (woman) knows better.

Solution: Use suggestion box and take it seriously.

7. Courtesy Fallacy
Belief: I believe it is more important to be polite than to be right.

Solution: Create a *safe environment* for giving honest feedback that supports the creative process.

8. Survivorship Problem
Belief: I believe that failure should not be investigated.

Solution: Celebrate failures.

Figure 1.9: Authority is dominant (1).

---

[13] "Availability Cascade — Liquisearch.com." n.p., n.d. Web. January 30, 2019, https://www.liquisearch.com/availability_cascade

9. **Fundamental Attribution Fallacy:** In contrast to interpretations of their own behavior, people tend to (unduly) emphasize the agent's internal characteristics (character or intention), rather than external factors, in explaining other people's behavior. This effect has been described as "the tendency to believe that what people do reflects who they are".[14]

10. **Not Invented Here (NIH):** NIH is a stance adopted by social, corporate, or institutional cultures that avoid using or buying already existing products, research, standards, or knowledge because of their external origins and costs, such as royalties. Research illustrates a strong bias against ideas from the outside.[15]

---

9. Fundamental Attribution Fallacy
Belief: I believe personality is more important than situations and notes.

Solution: Ask "Is environment + context more important than personality?"

10. Not Invented Here
Belief: I believe that in-house solutions are the best.

Solution: Perform a cost–benefit analysis of buy vs build and hold people accountable for cost and schedule blowouts.

Figure 1.10:   Authority is dominant (2).

---

[14]Bicchieri, C. "Scripts and schemas". *Coursera — Social Norms, Social Change II*. Retrieved June 15, 2017.

[15]Piezunka, H.; Dahlander, L. (2014). "Distant search, narrow attention: How crowding alters organizations' filtering of suggestions in crowdsourcing". *Academy of Management Journal*, 58: 856–880.

11. **Self-Serving Problem:** A self-serving bias is any cognitive or perceptual process that is distorted by the need to maintain and enhance self-esteem, or the tendency to perceive oneself in an overly favorable manner.[16]

12. **Social Comparison Problem:** Social comparison bias can be defined as having feelings of dislike and competitiveness with someone who is seen as physically or mentally better than yourself.[17]

---

11. Self-Serving Problem
Belief: I believe I learn more from success than from failure.

Solution: Celebrate and share failures which are ultimately learnings in disguise.

12. Social Comparison Problem
Belief: I believe that people who are smarter than me are a threat.

Solution: Do as Richard Branson and Mark Zuckerberg do: (1) surround yourself with smart people; (2) inspire them with a worthy mission; (3) give them the environment and tools they need to succeed; (4) get the hell out of the way.

Figure 1.11: Authority is dominant (3).

[16]Myers, D.G. (2015). *Exploring Social Psychology*, 7th ed. New York: McGraw Hill Education.

[17]Garcia, S.; Song, H.; Tesser, A. (2010). "Tainted recommendations: The social comparison bias". *Organizational Behavior and Human Decision Processes*, 113(2): 97–101. doi:10.1016/j.obhdp.2010.06.002. Archived from the original on April 18, 2016.

AI can simulate or generate alternative scenarios, thereby alleviating the lack of imagination biases that include (well documented in Wikipedia and reference throughout this chapter, see Figures 1.12–1.15):

13. **Backfire Fallacy:** The backfire effect is a name for the finding that given evidence against their beliefs, people can reject the evidence and believe even more strongly.[18]

14. **Confirmation Fallacy:** Confirmation bias, also called confirmatory bias or myside bias, is the tendency to search for, interpret, favor, and recall information in a way that confirms one's pre-existing beliefs or hypotheses.[19]

---

13. Backfire Fallacy
Belief: Contradictory evidence of my view which causes confusion reinforces my assumption.

Solution: Do not seek out devil's advocate. Find a person who genuinely disagrees.

14. Confirmation Fallacy
Belief: I believe that my time is better spent seeking out data that confirm my assumption.

Solution: When building experiments to test your assumptions, always ensure that your hypotheses is SMART (specific, measurable, actionable, realistic, time-bound) and gains actual market insights and answers what, why, where, who.

Figure 1.12:   Lack of imagination is dominant (1).

[18]Nyhan, B.; Reifler, J. (2010). "When corrections fail: the persistence of political misperceptions". *Political Behavior*, 32(2): 303–330. doi:10.1007/s11109-010-9112-2.

[19]Plous, S. (1993). *The Psychology of Judgment and Decision Making*. New York: McGraw-Hill, p. 233.

15. **Congruence Fallacy:** Congruence bias is a type of cognitive bias similar to confirmation bias. Congruence bias occurs due to people's over-reliance on directly testing a given hypothesis as well as neglecting indirect testing.[20] In an experiment, a subject will test his own usually naive hypothesis again and again instead of trying to disprove it.

16. **Overconfidence Fallacy:** The overconfidence effect is a well-established bias in which a person's subjective confidence in his or her judgments is reliably greater than the objective accuracy of those judgments, especially when confidence is relatively high.[21]

---

15. Congruence Fallacy

Belief: I believe I should only test my current assumption and not examine other assumptions.

Solution: Identify all key and high-risk assumptions, underpinning your idea and test all of these assumptions before forming any conclusions on the best path forward.

16. Overconfidence Fallacy

Belief: I believe that my own mind will offer the best solutions.

Solution: Do as Amazon's Jeff Bezos does and "be stubborn on vision but flexible on the details". Be outcome-focused.

Figure 1.13: **Lack of imagination is dominant (2).**

---

[20]Wason, P. C. (1960). "On the failure to eliminate hypotheses in a conceptual task". *Quarterly Journal of Experimental Psychology*, 12(3): 129–140. doi:10.1080/17470216008416717.

[21]Pallier, G.; Wilkinson, R.; Danthiir, V.; Kleitman, S.; Knezevic, G.; Stankov, L.; Roberts, R. D. (2002). "The role of individual differences in the accuracy of confidence judgments". *The Journal of General Psychology*, 129(3): 257–299. doi:10.1080/00221300209602099. PMID 12224810.

17. **Mere Exposure Fallacy:** The mere-exposure effect is a psychological phenomenon by which people tend to develop a preference for things merely because they are familiar with them. In social psychology, this effect is sometimes called the familiarity principle.[22]

18. **Normalcy Problem:** The normalcy bias, or normality bias, is a belief people hold when facing a disaster. It causes people to underestimate both the likelihood of a disaster and its possible effects because people believe that things will always function the way things normally have functioned.[23]

---

17. Mere Exposure Fallacy
Belief: I believe that what is familiar to me is the best solution.

Solution: Explore the pros and cons of different options in order to double down on the most relevant.

18. Normalcy Problem
Belief: I believe I should not plan for things that have not happened.

Solution: Having people on board who keep abreast of the latest trends and/or engaging future thinkers to advise you on what to look out for is a start.

Figure 1.14:   Lack of imagination is dominant (3).

---

[22]Zajonc, R.B. (2001). "Mere exposure: A gateway to the subliminal". *Current Directions in Psychological Science*, 10(6): 224. doi:10.1111/1467-8721.00154.
[23]Inglis-Arkell, E. (2013). "The frozen calm of normalcy bias". *Gizmodo*. Retrieved May 23, 2017.

19. **Shared Information Problem:** Shared information bias (also known as the collective information sampling bias) is known as the tendency for group members to spend more time and energy discussing information that all members are already familiar with (i.e., shared information), and less time and energy discussing information that only some members are aware of (i.e., unshared information).[24]

---

19. Shared Information Problem
Belief: I believe more time should be spent on easier topics.

Solution: Do work in a way that is time-boxed. For example, spend 30 minutes or so on such tasks before switching to something else.

Figure 1.15: Lack of imagination is dominant (4).

AI is a good technology to sort out massive data to counter the excess rationality biases that include (well documented in Wikipedia and reference throughout this chapter; see Figures 1.16–1.19):

20. **Anchoring Fallacy:** Anchoring or focalism is a cognitive bias for an individual to rely too heavily on an initial piece of information offered (known as the "anchor") when making decisions.

21. **Conservatism Fallacy:** It is an argument in which a thesis is deemed correct on the basis that it is correlated with some past or present tradition. The appeal takes the form of "this is right because we have always done it this way."

---

[24]Forsyth, D.R. (2009). *Group Dynamics*, 5th ed. Pacific Grove, CA: Brooks/Cole.
  Stasser, G.; Stewart, D. (1992). "Discovery of hidden profiles by decision-making groups: Solving a problem versus making a judgment". *Journal of Personality and Social Psychology*, 63(3): 426–434.

---

20. Anchoring Fallacy

Belief: I believe the first piece of information that comes along will inform my opinion.

Solution: Ensure people 'work alone together': write down ideas individually. Group similar ideas. Then vote on them in secret before engaging in a wider discussion.

21. Conservatism Fallacy

Belief: I believe minor changes in the equation will work even when the overall idea is wrong.

Solution: Remember sunk cost theory. Be prepared to end a project no matter how much money has gone into it if your assumptions are proven false.

Figure 1.16:   **Excess rationality is dominant (1).**

22. **Empathy Gap Fallacy:** A hot–cold empathy gap is a cognitive bias in which people underestimate the influences of visceral drives on their own attitudes, preferences, and behaviours.[25,26]

23. **Halo Fallacy:** The halo effect fallacy is the fallacy of concluding from a perceived single positive trait of a person to the conclusion of a generally positive assessment of that person.[27]

24. **Illusory Correlation Fallacy:** Illusory correlation is created when two separate variables are paired together, which leads to an overestimation of how often they co-occur.[28]

---

[25] Van Boven, L.; Loewenstein, G.; Dunning, D.; Nordgren, L.F. (2013). "Changing places: A dual judgment model of empathy gaps in emotional perspective taking". In Zanna, M.P.; Olson, J.M. (eds.), *Advances in Experimental Social Psychology*, Vol. 48. Academic Press, pp. 117–171. doi:10.1016/B978-0-12-407188-9.00003-X. ISBN 9780124071889. Archived from the original (PDF) on May 28, 2016.

[26] Loewenstein, G. (2005). "Hot–cold empathy gaps and medical decision making". *Health Psychology*, 24(4 Suppl): S49–S56. doi:10.1037/0278-6133.24.4.S49. Archived from the original on April 13, 2016.

[27] https://www.researchgate.net/publication/265303656_The_Halo_Effect_Fallacy

[28] Mullen, B.; Johnson, C. (1990). "Distinctiveness-based illusory correlations and stereotyping: A meta-analytic integration". *British Journal of Social Psychology*

---

22. Empathy Gap Fallacy
Belief: I believe people are rational.

Solution: Ask "should emotional considerations trump rational ones?"

23. Halo Fallacy
Belief: I believe that guy because a person has had a good track record that this will continue.

Solution: Define key metrics of success or performance that matter to you and evaluate based on defined merits. A healthy dose of professional judgment paired with hard data usually works best.

Figure 1.17: Excess rationality is dominant (2).

25. **Law of the Instrument:** Maslow's hammer (or gavel), or the golden hammer, is a cognitive bias that involves an over-reliance on a familiar tool.

---

24. Illusory Correlation Fallacy
Belief: I believe two correlated events prove a trend.

Solution: Multivariate testing assumptions can help see if correlation also amounts to causation.

25. Law of the Instrument
Belief: I believe the tools I have are the only tools of need.

Solution: Take many small bets across lots of potential solutions early. To do this quickly and cheaply, you might run solution interviews, ads on social media, testing your offering and directing interested users to a landing page where they can sign up to 'early access' once the product becomes live. We want to test real -world customer behavior.

Figure 1.18: Excess rationality is dominant (3).

(Wiley-Blackwell on behalf of the British Psychological Society), 29(1): 11–28. doi:10.1111/j.2044-8309.1990.tb00883.x.

26. **Optimism Fallacy:** Optimism bias (also known as unrealistic or comparative optimism) is a cognitive bias that causes a person to believe that they are at a lesser risk of experiencing a negative event compared to others.

27. **Risk Compensation Bias:** People typically adjust their behavior in response to the perceived level of risk, becoming more careful where they sense greater risk and less careful if they feel more protected.

---

26. Optimism Fallacy
Belief: I believe that optimism always pays.

Solution: Always test your assumptions by placing lots of small bets. This way, we'll be far more likely not to let overconfidence or optimism about the future unduly result in our doubling down and overinvesting in the wrong thing.

27. Risk Compensation Bias
Belief: I believe you should be more optimistic when things are going well.

Solution: Revisit assumptions, especially in good times, asking about worst case scenarios.

Figure 1.19:   Excess rationality is dominant (4).

1.3 Artificial Intelligence

Figures 1.20–1.30 cover some AI algorithms and methodology in simple language.

---

1. What is it?
A subset of machine learning which aims to store and apply knowledge gained from solving one problem to other related problems.

2. Example
Image recognition knowledge gained for recognizing cars can be applied to recognize trucks.

3. How does it work?
Leverages features extracted from one part to another.

4. How can it be used?
Save time: Retrain the algorithm for other similar tasks.

Figure 1.20: Transfer learning.

Source: Mckinsey, Wikipedia.

---

1. What is it?
A type of machine learning to allow the software to automatically determine the ideal behavior to get optimal result within a specific context.

2. Example
AlphaGo: reinforcement learning under the rule of Go to train the model with rewards for each step based on whether it increased the probability of win and time used.

AlphaGo

3. How does it work?
Algorithms are trained by trial and error (rewards and punishments) using dynamic programming.

4. How can it be used?
Games, Inventory Management, Delivery Management, Trading Strategy, Power System, etc.

Figure 1.21: Reinforcement learning.

Source: Mckinsey, Wikipedia.

1. What is it?

A subfield of machine learning using brain simulations to learn from very large numbers of examples.

2. Example

Natural language processing uses deep learning to make educated guesses for intent, sentiment, topic and relevance.

amazon alexa

3. How does it work?

Using artificial neural networks to deal with large amount of data by data labeling and substantial computing power.

4. How can it be used?

Perform classification for large amount of data (image, voice, video, text, etc.).

Figure 1.22:   Deep learning.

Source: Mckinsey, Wikipedia.

1. What is it?

The process of reducing the number of random variables by obtaining a set of principal variables.

2. Example

A 3D image data can be projected to a 2D dataset which requires much less computing power to process.

Self-Driving Car Project

3. How does it work?

Using feature selection to find subset of original variables and then to transform the data in high-dimension space to lower dimension.

4. How can it be used?

It substantially reduces the size of data and computing power and time needed, widely used in large data processing.

Figure 1.23:   Dimensionality reduction.

Source: Mckinsey, Wikipedia.

1. What is it?
Using multiple learning algorithms to solve the same problem in order to get better performance.

2. Example
Netflix and Amazon uses ensemble learning to get better results for recommendations.

3. How does it work?
Construct a set of hypotheses and use multiple learners to solve the same problem.

4. How can it be used?
Easy way to improve results by combining multiple model outputs.

Figure 1.24:   Ensemble learning.

Source: Mckinsey, Wikipedia.

1. What is it?
A predictive model using a decision tree to predict target variable based on several input variables.

2. Example
Data analytics for better marketing campaigns based on past campaign records.

3. How does it work?
A flowchart-like structure where each node denotes a test and each branch represents the outcome.

4. How can it be used?
Easy-to-understand classification method.

Figure 1.25:   Decision-tree learning.

Source: Mckinsey, Wikipedia.

1. What is it?
A learning algorithm which compares new problem instances with training problems, instead of explicit generalization.

2. Example
Medical field as second-opinion diagnostic tools.

平安好医生

Ping An Good Doctor

3. How does it work?
It constructs hypotheses directly from the training set.

4. How can it be used?
Easy to store a new instance or delete an old instance.

Figure 1.26: Instance-based learning.

Source: Mckinsey, Wikipedia.

1. What is it?
Grouping a set of data which are similar — data can be in the hundreds of millions.

2. Example
Image recognition use clustering to group targets by features (color, size, category, etc.).

3. How does it work?
Grouping can be based on distance, density, intervals of distributions.

4. How can it be used?
Separate out a large dataset.

Figure 1.27: Clustering.

Source: Mckinsey, Wikipedia.

1. What is it?
Classification based on the value of a linear combination.

2. Example
Sentiment analysis to detect positive or negative.

3. How does it work?
Using linear function to achieve classification.

4. How can it be used?
Less time to train and easy to use.

Figure 1.28:   Linear classifiers.

Source: Mckinsey, Wikipedia.

1. What is it?
A statistical process to estimate relationship among variables.

2. Example
Performance correlation analysis among different financial assets.

3. How does it work?
Using regression function to see how the dependent variable changes in response to the change of independent variable.

4. How can it be used?
Easy to use for finding the correlations among variables.

Figure 1.29:   Regression analysis.

Source: Mckinsey, Wikipedia.

1. What is it?
Automatically analyze correlations among related elements.

2. Example
Detect cyber security issue analyzing correlated patterns of threat events.

Security 🛡

3. How does it work?
Detect, extract & analyze dynamic parameters.

4. How can it be used?
Multiple element correlation analysis.

Figure 1.30:   Automated data correlation.

Source: Mckinsey, Wikipedia.

Chapter 2

Case Study: US vs PRC in the Banking Lizard Brain Shift to AI FinTech Machines

2.1 Introduction

In 1956, John McCarthy invited a group of researchers from different disciplines such as language simulation, neuron nets, complexity theory and others to a summer workshop called the Dartmouth Summer Research Project on Artificial Intelligence (AI) to discuss what would ultimately become the field of AI.[1] The term "AI" was chosen at that time to differentiate it from "thinking machines", a field that included cybernetics, automata theory, and complex information processing.

There are several definitions for AI and following are some of them:

AI refers to the study and use of intelligent machines to mimic human action and thought. With the availability of Big Data, advances in computing, and the invention of new algorithms, AI has risen as a disruptive technology in recent years.[2]

---

[1]https://www.forbes.com/sites/bernardmarr/2018/02/14/the-key-definitions-of-artificial-intelligence-ai-that-explain-its-importance/#601f57bb4f5d
[2]https://www.imda.gov.sg/sgdigital/tech-pillars/artificial-intelligence

AI is a branch of computer science dealing with the simulation of intelligent behavior in the computer and the capability of a machine to imitate intelligent human behavior.[3]

AI refers to the theory and development of computer systems able to perform tasks normally requiring human intelligence, such as visual perception, speech recognition, decision-making, and translation between languages.[4]

Generally, most people will think of AI as a system that thinks like a human, or a system that works like a human without figuring out how human reasoning works, or a human reasoning model that performs tasks or services. For those who are interested in more detailed discussions, there are several textbooks worth consulting.[5]

According to Neapolitan and Jiang (2018), AI approaches can be defined as follows: (i) Logical Intelligence (LI), (ii) Probabilistic Intelligence (PI), (iii) Emergent Intelligence (EI), (iv) Neural Intelligence (NI) and (v) Language Understanding (LU). The last three being the most interesting as evolution computation and swarm intelligence under EI are drawing attention as these researches are inspired by biological evolution of decentralized and self-organized systems, NI is modeled loosely after the human brain with Neural networks and deep learning, and LU deals with machine reading comprehension.

Another way is to split approaches into (i) Symbolic, (ii) Deep Learning, (iii) Bayesian Networks and (iv) Evolutionary Algorithms. Symbolic AI's are based on high-level "symbolic" (human-readable) representations of problems, logic, and search.[6] Deep learning (also known as deep structured learning or hierarchical learning) is part of a broader family of machine learning methods based on learning data representations, as opposed to task-specific algorithms. Learning can

---

[3]https://www.merriam-webster.com/dictionary/artificial%20intelligence
[4]https://en.oxforddictionaries.com/definition/artificial_intelligence
[5]https://courses.csail.mit.edu/6.034f/ai3/rest.pdf, Russel, S.; Norvig, P. (2010). *Artificial Intelligence: A Modern Approach*, 3rd ed. Prentice Hall. Neapolitan, R.E.; Jiang, X. (2018). *Artificial Intelligence: With an Introduction to Machine Learning*, 2nd ed. CRC Press.
[6]https://en.wikipedia.org/wiki/Symbolic_artificial_intelligence

be supervised, semi-supervised or unsupervised.[7] Bayesian networks are a type of probabilistic graphical model that uses Bayesian inference for probability computations.[8] An evolutionary algorithm (EA) is a subset of evolutionary computation, a generic population-based metaheuristic optimization algorithm. An EA uses mechanisms inspired by biological evolution, such as reproduction, mutation, recombination, and selection.[9]

In this chapter, we go beyond what is being described in traditional textbooks and look at applications. In order to illustrate the use in practice and the issues of interest to practitioners and investors, we look at specific issues using a two-country study. We will focus on the issue of "Who is Winning the Artificial Intelligence Race between PRC and US: Alibaba, Tencent, Ping An, Baidu and Zhong An vs Alphabet, Amazon, Apple, Facebook and Microsoft." Figure 2.1 presents the race among the technology giants.

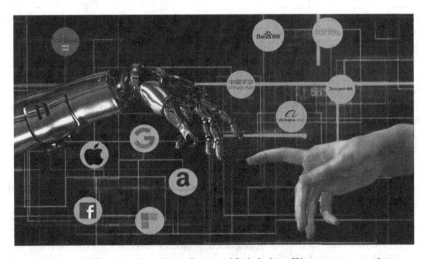

Figure 2.1: **Who's winning the artificial intelligence race between PRC and US: Alibaba, Tencent, Ping An, Baidu and Zhong An vs Alphabet, Amazon, Apple, Facebook and Microsoft.**

---

[7]https://en.wikipedia.org/wiki/Deep_learning

[8]https://towardsdatascience.com/introduction-to-bayesian-networks-81031eeed94e

[9]https://en.wikipedia.org/wiki/Evolutionary_algorithm

2.2 Summary

1. **PRC:** The ecosystem of firms in China has a more complete suite of "finance to lifestyle" AI products than US firms. Alibaba launched the PAI 2.0,[10,11] and it has become a market leader in AI and Cloud, setting a new standard. Ping An seems to be under-researched and is the leader in technology in China for many financial, consumer, health and insurance businesses. Tencent is another very interesting company that is not understood in the English-speaking world and has some very fabulous services that will be described in this chapter.

2. **US:** Amazon is considered to be the undisputed leader for technology in the world. However, PRC AI is as good as Amazon. Facebook, Microsoft, Apple, and Google are leaders in technology with excellent operation capabilities.

---

[10]On March 29, 2017, Alibaba announced that its machine learning platform PAI had been upgraded to PAI 2.0 to help customers easily deploy large-scale data mining and modeling. Its advanced and diversified AI algorithms are aimed at empowering innovations in the life science and manufacturing sectors. https://www.alibabacloud.com/press-room/alibaba-cloud-announces-machine-learning-platform-pai

[11]According to Alibaba, Alibaba Cloud's PAI 2.0 significantly reduces development costs and entry barriers of AI for users.

Key features of PAI 2.0 include the following:

- **Diversified and innovative algorithms:** PAI has more than 100 sets of algorithms covering data preprocessing, neural network, regression, classification, predictions, evaluations, statistical analysis, feature engineering and deep-learning architecture.
- **Deep-learning architecture:** The entire computing architecture is optimized for different deep learning frameworks. It also supports a one-click function to deploy application program interface (API), solving the challenge of modeling and service integration.
- **Large-scale computing power:** Powered by Apsara, Alibaba Cloud's large-scale computing engine, PAI provides ultra-large-scale distributed computing capabilities that allow organizations to handle petabyte-level computing tasks daily.
- **User-friendly interface:** PAI's data visualization capabilities allow developers to quickly include components into workflows by easy drag-and-drop features, greatly enhancing the efficiency of model building and debugging.

3. Both US and PRC firms face regulatory issues and are the core of the debate on the future of workers. Most of all these technology firms wish to operate without government intervention. However, most of them have benefitted from either government funding or lax regulation and unlikely to be outside the influence of their governments in the future.

2.3 AI and Finance: China Gets It While the US is Years Behind

AI is the integration of finance with data from unintentioned (text, chat, images) and intentioned (merchandise receipts) in order to offer anticipatory choices of goods and services to people in the areas of lifestyle, business, investment, cognitive services, and SME activity. In the pursuit of these goals using data and cloud, China has decidedly pulled ahead of the US in general.

With its launch of PAI, Alibaba has pulled ahead of Tencent in the combination of finance with intentioned as well as unintentioned data. Combining finance data and cloud requires a horizontal conglomerate style where new rings of businesses are placed outward from a center dominated by data analysis — not a vertical pyramid structure.

With PAI, Alibaba is the new global leader. Ping An deserves rerating as it grows into a diversified financial healthcare conglomerate. Tencent has superlative technology, of course, but is very expensive (and overhyped). Baidu has lost its edge and seems to be wandering — is it jettisoning old businesses or building new ones? Or both? It is jettisoning old businesses while jumping into AI through acquisitions. Both have serious risks. Moreover, all of Baidu's credit health is systemically deteriorating.

A. US companies — China has also pulled ahead because US companies have had to deal with the following:

   (1) entrenched lobbyists (in DC) who are protecting cartels and vice versa;
   (2) state/federal regulators on the warpath after GFC ($360 billion in fines);

(3) a laundry list of state/federal regulations which prevented the US Big 5 from flexibility in fintech;

(4) a case of conglomerate — it is after the collapse of GE financial services;

(5) retardation of relationships between Big 5 and DC military complex and universities;

(6) an absence of any national strategy;

(7) it seems there is a great board postmortem whenever any product fails.

B. Chinese firms adapt quickly after failure and keep on trying anew. Our rankings for the US companies which take into consideration the suite of AI products, multiple valuations, and several operational and credit criteria show Apple as the best and Amazon as the worst. Amazon and Google have the best AI platforms, but it is a myth that these are superior to China.

C. If we combine operational metrics and valuations to the AI mix, Amazon is least attractive. Microsoft has failed to do anything meaningful after a 15-year headstart. Facebook also fails to impress. Apple is the best overall, but Huawei's new technology is at least as good as Apple's.

2.4   Gaping Holes in the Landscape

AI has been commoditized. China's overall AI push is as good as that of the US — arguably better. All of the companies in this report can now translate thousands of Bibles per second. Moreover, they can all process hundreds of thousands of photos per second. They all have talking personal butlers, staggering video offerings, dazzling facial recognition statistics, and can anticipate our wants, needs, and desires for food, travel, health, education, books, news, wellness, and mates.

So what? AI is now a commodity. It is not enough. We are also economic creatures. Only Alibaba and Ping An offer these AI tools AND a full suite of services in payments, insurance, loans, credit ratings, money market, wealth management, crowdfunding, and FX.

This is the secret sauce. The US firms all have a gaping hole in this regard.

Alibaba's gaping hole is in entertainment/content. They need to overhaul Alibaba Pictures.

The gaping hole in Amazon is financial services (and almost all of the US firms), and entertainment is a big hole for Alibaba. Why the gaping hole in financial services? Aside from the seven reasons mentioned above, ALL of these firms DID embark on financial products, but they were mostly duds. Whoever has heard of Microsoft Wallet? Google Hangout? Facebook Messenger Pay? Even Apple Pay is 80% smaller than Alipay.

Amazon has a small SME loan book, but the other services in wealth management, crowdfunding, ratings, insurance, and any financial products are virtually absent. Amazon should be the largest bank in the world — and it does NOT need a balance sheet. Instead, Facebook and Google are glorified advertising companies which are wrecking the economics of online newspapers nationally.

Lastly, the other gaping hole in the landscape is the absence of other global players from Europe. Europe has none. Japan has none (SoftBank is a holding company). Korea has none. There is no dominant regional player in India, Brazil, Indonesia, Eastern Europe, Scandinavia or Russia. This must make us conclude that what Tencent, Ping An, and Alibaba have done is a unique accomplishment. Few others globally have managed to accomplish anything close to them.

2.5   Other Issues: Politics, Policy and Regulation

A. Every political risk in China is just as valid in the US. The dominance of the US companies in advertising will have eventual political consequences and force a political anti-trust showdown at some point (A GREAT NEW BOOK ON THIS IS "Run Fast, Break Stuff: How Facebook, Google, and Amazon Cornered Culture and Undermined Democracy" — Jonathan Taplin of the USC School of Journalism). These companies are heavily involved in the current strange political streams. This can boomerang. Facebook has received a subpoena from Mueller's team on its activity in the 2016 election.

B. PRC government is intentionally creating competition between Alibaba and Tencent. Baidu began to fall apart the day Google left China in 2010. The Chinese government runs by forcing competition — with provincial governors and corporates. Alibaba, Tencent, and Baidu were the big three, and now Baidu has fallen out. With PAI 2.0, Alibaba has arguably pulled ahead of Tencent. The competition has consequences as upcoming players need to "pick sides." Only regulation can stop the first movers in China and the US. Otherwise, Alibaba, Tencent, Google, and Amazon are unstoppable for now.

C. China aggressively supports the internet economy — it is the foundation of the 5-year plan. The foundation of the current 5-year plan is the "internet Plus economy". Alibaba is likely in the center of this. Furthermore, it is entirely conceivable that Alibaba's new platform PAI 2.0 can be a cornerstone of SOE reform, tax reform and a conduit for millions of jobs in the SME sector. Companies everywhere are expected to play a role in political and social stability — in the US and China. Alibaba portrays itself as a company for the "small potato". All these firms are political creatures, and this does play a role in the stock multiple. So far, Alibaba is playing its cards right. So is Ping An.

2.6 Master Summary: AI Analysis

Internet of Things (IoT) is always an essential combined component of AI (Figure 2.2) in practice as IoT collects the physical data from sensors, mobile, and other digital devices. With the infrastructure linking the Neural Network with Machine Learning, data stored in the cloud platform from financial services, cognitive services, lifestyle, health, autonomous vehicles, robotics, and advertising can be processed and analyzed.

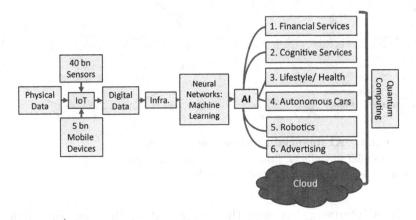

Figure 2.2: Artificial intelligence described on a single chart.
Source: Schulte Research Estimates.

The AI financial services (Figure 2.3) are in the area of payments, insurance, personal loans, small, and medium enterprise (SME) loans, credit rating, money market, wealth management, crowdfunding, and currency exchanges. We can make a comparison of ATPA (Alibaba, Tencent, Ping An, Baidu) and AMGAF (Amazon, Microsoft, Google, Apple, Facebook). These financial services are dominated by Chinese technology companies, such as Alibaba, Tencent, Ping An and to a lesser extent Baidu. The US Technology firms, Amazon, Microsoft, Google, Apple, Facebook, are missing out on most of these opportunities.

Company	Payment	Insurance	Personal Loans	SME Loans	Credit Rating	Money Market	Wealth Mgmt	Crowd-funding	Currency Exchange
Alibaba	AliPay (400 mn) (51.8% ms)	Ant Zhong An	Personal loans - Ant	SME loans – Ant SME Service	Zhima Credit	Yue'bao (CNY1.2tn)	Ant Financial, AliPay	ANTSDAQ	AliPay
Tencent	WePay (300 mn) (38.3% ms)	Zhong An	Weilidai	Weilidai	WeBank	/	WeBank	JD.com	/
Ping An	Ping An Bank	Ping An Insurance	Ping An Bank	Ping An Bank	Ping An Insurance	Ping An Asset Mgmt	Ping An Asset Mgmt	/	Ping An Bank
Baidu	Baidu Wallet (100 mn)	/	/	/	Yes	/	Yes	/	/
Amazon	/	Electronic Damage Insurance (UK)	/	USD$3bn SME loans (Amazon Lending)	/	/	/	/	/
Microsoft	MSFT Wallet	/	/	/	/	/	/	/	/
Google	Google Wallet, Android Pay	/	/	/	/	/	/	/	/
Apple	Apply Pay (85 mn, 450% YOY)	/	/	/	/	/	/	/	/
Facebook	Messenger Pay	/	/	/	/	/	/	/	/

MISSED OPPORTUNITY

Figure 2.3: AI financial services comparison.
Note: Alibaba way ahead. Ping An and Tencent also dominate.
Source: Schulte Research Estimates, Zenith.

It is interesting to note that eight out of the nine companies (Figure 2.4) have their own cloud for storage of data in the form of Intentional, Unintentional, IoT/Car, and Cognitive Services data. The data collection capabilities of most of the companies are very strong, and one success factor is the prevalence of weak privacy protection agreement, and until recently, such data usage and purchase of data are almost without constraints in China.[12] In May 2018, China quietly released the final version of a new data privacy standard that goes even further than the European General Data Protection Regulation (GDPR). This places EU and Chinese data legislation on a far more level footing than American data law.[13] Chinese law places attention to privacy protection with standards on

---

[12]https://www.networksasia.net/article/chinas-data-privacy-law-came-effect-may-and-it-was-inspired-gdpr.1529296999
[13]https://assets.kpmg.com/content/dam/kpmg/cn/pdf/en/2017/02/overview-of-cybersecurity-law.pdf

the collection and usage of personal information. The China National Information Security Standards Technical Committee (SAC/TC260) has developed more than 240 national standards related to cyber-security (i.e., Cloud, Industrial Control Systems, and Big Data) since 2010.[14] The Cybersecurity Law requires personal data collected or generated in China to be stored domestically. Enterprises and organizations that violate the Cybersecurity Law may be penalised up to RMB 1,000,000. Since the law presents clear definitions of network operators and security requirements, it is not surprising that most large financial institutions and technology companies will and have already become network operators.

Company	Intentional Data	Un-intentional Data	IoT/Car	Cognitive Service	Cloud
Alibaba	AliPay, Taobao, Tmall, Alibaba.com, Alibaba Express, Yue'bao, Tmall, Alibaba Cloud (750 mn)	Youku, Weibo, UCWeb, Cainiao Logistics, Yahoo! China, SCMP, AliWangWang, LaiWang, PAI, Ding Talk	"Connected Car" with SAIC, AutoNavi, Ali Health, KFC	Platform for Artificial Intelligence (PAI 2.0), Tmall Genie	Ali Cloud
Tencent	WeChat Pay, 3rd Party Providers (JD.com, Didi, etc.)	WeChat (938 mn), QQ (700 mn), Qzone, WeChat Ecosystem, Gaming	Didi, Dianping review site	WeChat Voice/Image, Tencent Video	Tencent Cloud
Ping An	Ping An Bank, Ping An Insurance, Ping An Asset Mgmt (350 mn)	Ping An Health, Ping An Securities	Ping An Auto Owner, Wanjia Clinics	Facial recognition, Voice print	Ping An Health Cloud
Baidu	Baidu Search, Baidu Wallet	Baidu Search	Food Delivery Service, Project Apollo	Little Fish	Baidu Cloud
Amazon	E-commerce (B2C, C2C)	Shopping search	Echo, Kindle, Whole Foods, Amazon Books, Logistics	Alexa, Rekognition, Polly, Lex, Amazon Video	AWS
Google	Google Play Store, Google Search	Google Search, Android OS, G-mail, Maps, Chrome, Snapchat (166 mn MAU), Youtube, Waymo	Android OS, Waymo	Health, Translation, Google Assistant, Google Face, Deep Mind	Google Cloud
Apple	iTunes (800 mn), Apple Music, Apple Pay (85 mn)	iOS, Safari	iPhone (1 bn), iPad, iPod, Mac, Apple Watch	Siri, Face Recognition	iCloud
Microsoft	Xbox, Microsoft Wallet (small)	LinkedIn, Office, Skype, Bing, IE	Kinect, Microsoft Surface, Windows Phone	Zo, Computer vision/ Speech/ Language API	Azure
Facebook	Messenger Pay	Facebook, Facebook Messenger (1.96 bn), Whatsapp(1.3 bn)	Oculus, Project Titan	Deep face, Deep text, Translation	/

(left margin label: same technology)

Figure 2.4: AI data source comparison.
Note: Alibaba may leapfrog all with PAI, Ding Talk, and Tmall Genie. This puts Alibaba in its own league with Tencent very close.
Source: Schulte Research Estimates.

According to Gartner,[15] the global leader in cloud services is Amazon with 51.8% of the global market share compared to

---

[14]https://www.tc260.org.cn/upload/2018-01-24/1516799764389090333.pdf 信息安全技术个人信息安全规范, December 29, 2017
[15]https://www.gartner.com/newsroom/id/3884500

Microsoft with 13.3%, Alibaba with 4.6%, and Google with 3.3% in 2017 (Figure 2.5). IBM has a global market share of 1.9% with Microsoft growing close to three digits. In China, Alibaba leads with Meituan and Tencent ranking second and third, respectively. Baidu is ranked 10th in China. Alibaba is leading the Asia market. Alibaba Cloud's international operations are registered and headquartered in Singapore. It is expanding its service into countries and regions involving the Belt and Road Initiative. Alibaba Cloud will have an integrated cloud technology and innovation platform for the Olympics 2022. In March, it opened its first data center in Indonesia. In 2017, it launched 316 products and features, 60 were focussed on high-value fields, including artificial intelligence.[16] Its reported revenue was 4.39 billion yuan (US$664.96 million) in 2018Q1, up 103% from the same period last year. Total revenue for the 2017 fiscal year reached 13.39 billion yuan, up 101%.[17]

Company	Market Share (in respective markets)	Comments
Alibaba	41%	Leader in Asia, >100 newly developed AI services. Services in storage, networking, healthcare, logistics, lifestyle, media, business enhancements, and etc.
Tencent	7%	#3 in Chinese Market, services in public/private storage for personal/business use
Baidu	1%	Small cloud service
Amazon	47%	Amazon AWS: Global & US leader
Microsoft	10%	Fastest Growth Rate, 97% YOY. Azure & Office 365
Google	4%	Fast growing, 45 teraflops of data (45 trillion bytes/second). Services include AI APIs', storage, search history, e-mail, etc. 71,000 searches/second. New quantum computer
Apple	1%	Small cloud service, primarily strong in private data collection. Services in location, storage, and security

(left bracket label: same technology)

Figure 2.5: AI cloud comparison: Amazon big winner.
Note: Alibaba leader in Asia.
Source: Schulte Research Estimates, HostUCan, Skyhighnetworks.

---

[16]https://technode.com/2018/05/10/alibaba-cloud-q1-2018-results/
[17]http://www.chinadaily.com.cn/a/201806/27/WS5b3337f5a3103349141df403.html

AI Cognitive Services is a set of machine learning initially developed by Microsoft to solve the problems in the field of Artificial Intelligence (Figure 2.6). The goal was to democratize AI by packaging it into discrete components that are easy for developers to use in their own apps.[18] These services include recognition of the images, facial, voice, natural language learning, and Video. So, it is not surprising to note that Microsoft leads in this field, allowing Web and Universal Windows Platform developers to use the algorithms. While Microsoft has made a wide and deep mark, Alibaba PAI has made even more progress than Microsoft.

Company	Image Recognition	Facial Recognition	Voice	Natural Language Understanding	Video
Alibaba	Document recognition, image search+, PAI	Identity authentication, Alipay	Customer service AI, voice->text service, etc., PAI	Real-time translation services, PAI	Video analysis, broadcast service, PAI
Tencent	Fashion trend analysis	Identity authentication	WeChat Voice/Image	Translation	Tencent Video
Baidu	xPerception, Pixlab API, WICG Shape Detection API	Baidu Facial Recognition (99.7% accuracy)	Voice Search, Text–Speech Converter, Deep Voice (97% accuracy)	Translation, Speech Recog., Kitt.AI, RavenTech	/
Amazon	Amazon Rekognition	Emotion recog., Face Comparison	Alexa, Lex	Alexa, Echo, Polly	Amazon Video
Microsoft	Image understanding, Celebrities/Landmark recog.	Face API, Emotion, Verification, Detection	Speech Verification, Text-Speech Converter	Translation, Text analytics, LUIS	Video analysis, Video Indexer
Google	Image searching	Google Face	Google assistant	Translation AI, Text analytics	YouTube
Apple	Classification/ Detection/ Checking	Facial Rec.	Siri	Siri	/
Facebook	Text recognition, translation	Deep Face	Oculus VR Voice Recog.	Translation, Deep Text	/

(same technology)

Figure 2.6: AI cognitive services comparison: Microsoft winner.
Note: There is a "me too" attitude. Microsoft has made a wide and deep mark. But, Alibaba PAI is deeper and broader than anyone else.
Source: Schulte Research Estimates, Zenith.

As for AI Lifestyle such as media, food, travel, entertainment, interaction, search, education and health, the Chinese technology

---

[18]https://blogs.windows.com/buildingapps/2017/02/13/cognitive-services-apis-vision/

companies are ahead with Tencent and Alibaba taking the lead
(Figure 2.7). The Chinese Ministry of Science and Technology has
identified BAT (Baidu, Alibaba Group Holding, Tencent Holdings) as
the champions to partner the government to spur China's AI strategy
to accelerate the country towards global technology leadership.[19]
"Baidu's focus will be on autonomous driving; the cloud computing
division of Alibaba is tasked with a project called 'city brains', a
set of AI solutions to improve urban life, including smart transport;
Tencent will focus on computer vision for medical diagnosis; while
Shenzhen-listed iFlytek, a dominant player in voice recognition, will
specialize in voice intelligence."

Company	Media	Food	Travel	Entertainment	Interaction	Search	Education	Health
Alibaba	Live media, news production, media interaction, etc.	KFC China, Koubei, Ele.me	Air/train tickets, hotel booking	Audio/video solutions, video game services, e-commerce, Youku, AGTech	Online shopping support	Personalized search, direct marketing service, big data analytics	Media education services	Utilities payment, Hospital-patient comm., smart diagnosis
Tencent	QQ music, Joox, Tencent Video/ News	Meituan Dianping	LY.com	E-commerce, video games	WeChat	WeChat Search, Sogou	Koo Learn, Ke.qq	WeChat Intelligent Hospital
Baidu	Book recomm.	/	Ctrip	/	/	Search engine (76% market share), personalized search, data marketing	Baidu Education (Jiaoyu)	Health Search
Amazon	Books, music, TV streaming	Whole Foods, Amazon Fresh	/	TV streaming, Amazon Studio	/	Search recomm.	Amazon Inspire	/
Microsoft	/	/	/	X-box/gaming	Skype, LinkedIn	Bing	MSFT Education	MSFT Health
Google	Google Videos	/	Google Flights	YouTube	Snapchat, Gmail, Google+	Search engine, personalized search, data marketing	Google for Education	Google Health, Google Fit
Apple	iTunes	/	/	Apps, App Store, Game development kits	Messages	/	/	Health apps
Facebook	Facebook newsfeed	/	/	Facebook games	Facebook, Whatsapp, Instagram	Internal Search Function	/	/

China is way ahead

Figure 2.7: AI lifestyle comparison.
Note: Tencent + Alibaba winner. Microsoft + Baidu falling behind.
Source: Schulte Research Estimates, Zenith.

---

[19]https://www.scmp.com/tech/china-tech/article/2120913/china-recruits-baidu
-alibaba-and-tencent-ai-national-team

2.7 Master Summary: How the Ecosystems are Evolving

Alibaba has expanded its e-commerce business Taobao and Tmall to Finance (Ants Financial) and Tech Service Provider (Alibaba Cloud). Figure 2.8 shows the fullest AI ecosystem that is unmatched by any other, and it has achieved all these innovations while being the most profitable of any tech companies. Alibaba will produce its own AI chip as well as develop quantum processors expanding into a semiconductor business in 2019Q2. AliNPU, the new AI chip had the potential to support technologies used in autonomous driving, smart cities, and smart logistics and driven by its R&D arm Damo Academy. Pingtouge, a new semiconductor subsidiary will focus on customized AI chips and embedded processors. Alibaba acquired IC (Integrated Design) vendor Hangzhou C-Sky Microsystems in April 2018.[20]

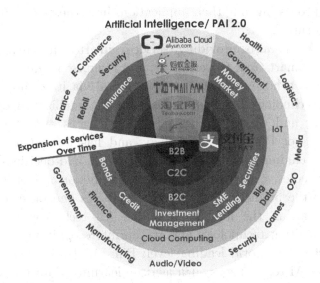

Figure 2.8: Alibaba: e-commerce + the fullest AI ecosystem of any company on earth.
Note: It has achieved this while being the most profitable of any company.
Source: Alibaba Website, Schulte Research.

---

[20]https://www.zdnet.com/article/alibaba-to-launch-own-ai-chip-next-year/

With data from Banking Services, Government Services, Lifestyle Services and Business Services using AI, Alibaba has accessibility to a large pool of data in China (see Figure 2.9). Figures 2.10–2.12 show the different types of deep tech for various Chinese giants.

Tencent also has access to social media data generated by QQ, Wechat, Wechat Pay that are stored in Tencent Cloud, possibly a reach of a billion people's postings, chats, file transfer, photos, locations, and other personal information. Tencent is China's largest social network company with 1 billion users on its app WeChat and 632 million monthly user accounts on social networking platform Qzone. Tencent has extended beyond instant messaging QQ and social networking to gaming, digital assistants, mobile payments, cloud storage, education, live streaming, sports, movies, and artificial intelligence.[21] Tencent AI Lab was set up in 2016 with the mission to "Make AI Everywhere", and it focuses on fundamental research in machine learning, computer vision, speech recognition, and natural language processing, and their applications in Game, Social, Content, and Platform AI.

Ping An is also providing services in Health, Real Estate, Transport, Smart City, Government, and Finance besides insurance. Thus, it has a huge database at its disposal via its cloud services. Ping An Technology is set up to focus on AI, Intelligent Cognition, Blockchain, and Cloud. "Artificial intelligence is one of the core technologies of Ping An Technology, and has formed a series of solutions including predictive AI, cognitive AI, and decision-making AI. The predictive AI based on disease prediction model has been applied to the prediction of influenza, diabetes and other diseases. In particular, in the field of cognitive AI, technologies such as facial recognition technology and voiceprint recognition technology have reached the world leading level; Ping An Brain smart engine integrating AI technologies such as deep learning, data mining, and biometrics can provide six integrated modules including decision-making."[22]

---

[21] https://www.forbes.com/sites/bernardmarr/2018/06/04/artificial-intelligence-ai-in-china-the-amazing-ways-tencent-is-driving-its-adoption/#151b245f479a
[22] https://tech.pingan.com/en/

Zhong An, with Tencent, Alibaba, and Ping An as its largest shareholders, provides pure online insurance services and has slowly expanded into many areas beyond e-commerce insurance that started with the Singles' Day online sales of Alibaba. ZhongAn Technology AI research service covers image recognition, NLP, complex transaction processing and others to (i) help customers understand the users by data analysis, and (ii) to set up intelligent applications in less time with mature AI technology and rich experience.[23] It focuses on Natural Language Processing, Image Processing, Complex Transaction Processing and Machine Learning. Its cloud service platform, Anlink, provides blockchain-based BaaS and AI-based AIaaS and platform security to offer new solutions under different business scenarios such as finance, medical and health care, supply chain, shared R&D, culture, government affairs, and public welfare.[24]

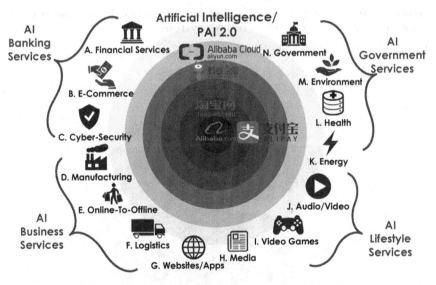

Figure 2.9: Alibaba AI: A truly common man's bible — access to virtually any and all data dumps of all of China.

Source: Alibaba Cloud Website, Schulte Research.

---

[23]https://www.zhongan.io/en/technology.html
[24]https://www.zhongan.io/en/anlink

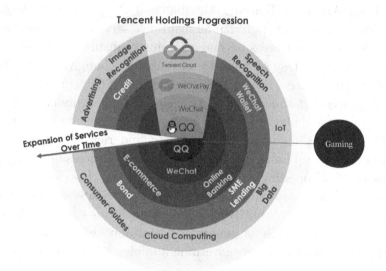

Figure 2.10: Tencent: From social media network to the use of unintentional data to create a full ecosystem.

Source: WeChat, Schulte Research.

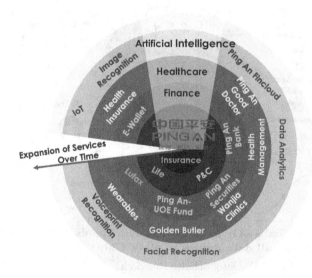

Figure 2.11: Ping An has made greater strides into becoming an autonomous financial ecosystem than any other company.

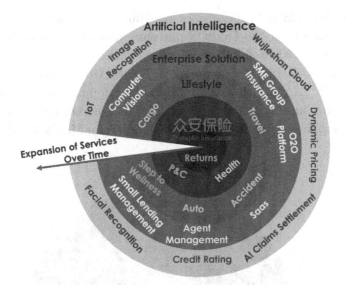

Figure 2.12: Zhong An: Only 3 years old, the largest pure online, cloud-based insurance company globally.

Baidu is another company to be watched as they transformed their search engine to AI services, regaining its expansion momentum after a widescale management shakeup (Figure 2.13). It is now pursuing a riskier strategy of acquisition and expanding its Baidu Cloud business. Baidu has recently launched China's first cloud-to-edge artificial intelligence (AI) chip — Kunlun. With Kunlun, Baidu not only offers an AI platform to help enterprises deploy AI-infused solutions but also have their own hardware to maximise AI processing. The chip can accommodate high-performance requirements of a wide variety of AI scenarios and can be used to provide AI capabilities such as speech and text analytics, natural language processing, and visual recognition.[25] Baidu's program called Apollo is to make Baidu's artificial intelligence technologies available to car makers for free as a brain for their cars in exchange for the access

---

[25]https://cio.economictimes.indiatimes.com/news/business-analytics/will-baidu s-cloud-to-edge-ai-chip-kunlun-change-the-face-of-ai-market-beyond-china/650 66114

to data. DuerOS is Baidu's voice assistant, and Baidu has teamed up with more than 130 DuerOS partners, and the voice assistant is in more than 100 brands of appliances, such as refrigerators, TVs, and speakers. Baidu is also partnering with Huawei to develop an open mobile AI platform to support the development of AI-powered smartphones and Qualcomm to optimize its DuerOS for IoT devices and smartphones using Qualcomm's Snapdragon Mobile Platform.[26]

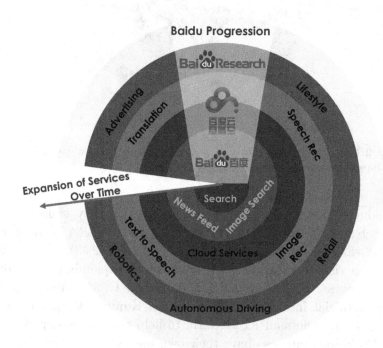

Figure 2.13: **Baidu: Stumbled in its transition from search engine to AI; regaining momentum after widescale management shakeup, and are pursuing risky strategy of acquisition.**

Source: Baidu Website, Schulte Research.

Meanwhile, in the US, the tech giants are transforming fast too. Amazon, originally an online bookstore, has transformed itself into

[26]https://www.forbes.com/sites/bernardmarr/2018/07/06/how-chinese-internet
-giant-baidu-uses-artificial-intelligence-and-machine-learning/#6a5d71f2d557

the number one AI player in almost everything including cloud and machine learning. Efforts have been spent in building the capabilities in robotics, data center business such as the Amazon Web Services (AWS), health and pharmacy with the acquisition of PillPack, voice technology such as Alexa, and voice-based home appliance business such as the Echo[27] (Figure 2.14). Acquisition and expansion of Whole Foods, physical bookstores, and AI-powered cashless Go Stores are signaling that there are combining online and offline businesses using AI.[28] Four major AI products are (i) Amazon Lex — a service for building conversational interfaces into any application using voice

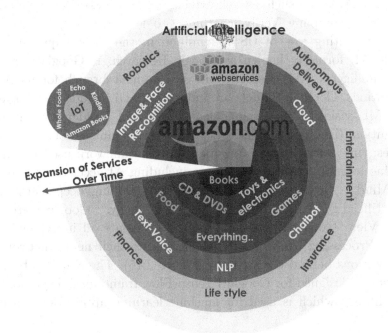

Figure 2.14: Amazon: Online bookstore for 3 years to #1 global player in almost everything, including Cloud, enviable execution + impressive global ambition.

---

[27] https://www.wired.com/story/amazon-artificial-intelligence-flywheel/
[28] https://www.cnbc.com/2018/09/04/inside-amazons-big-plans-to-get-to-2-trill ion-club-health-ai-retail--more.html

and text; (ii) Amazon Polly — a service that turns text into lifelike speech; (iii) Amazon Rekognition — a service that makes it easy to add image analysis to applications; (iv) Amazon Machine Learning — a service that makes it easy for developers of all skill levels to use machine learning technology.[29]

Google's advertising revenue grew from a mere US$70 million in 2001 to a staggering US$95.38 billion in 2017. The recent results showed that Google accounted for 86% of its parent Alphabet's 2018Q2 revenue of US$26.24 billion. Google's other revenues, including cloud services, hardware, and app sales, grew 37% to US$4.4 billion over the same quarter a year earlier. However, Alphabet's other investments, such as self-driving car business Waymo and health-tech company Verily, continued to lose money, accounting for an operating loss of US$732 million in the second quarter.[30] Google AI, formerly known as Google Research, is Google's artificial intelligence (AI) research and development branch for its AI applications. Google products are interesting and include (i) Google Auto ML vision — a machine learning model builder for image recognition; (ii) Google Assistant — a voice assistant AI for Android devices; (iii) TensorFlow — an open source framework used to run machine learning and deep learning, including AI accelerators; (iv) DeepMind — a division responsible for developing deep learning and artificial general intelligence (AGI) technology.[31] Google Search, Street View, Google Photos and Google Translate all use Google's Tensor Processing Unit, or TPU, to accelerate their neural network computations behind the scenes[32] (Figure 2.15). The chip has been specifically designed for Google's TensorFlow framework, a symbolic math library which is used for machine learning applications such

---

[29]https://docs.aws.amazon.com/aws-technical-content/latest/aws-overview/artificial-intelligence-services.html

[30]https://qz.com/1334369/alphabet-q2-2018-earnings-google-is-more-than-just-advertising-now/

[31]https://whatis.techtarget.com/definition/Google-AI

[32]https://cloud.google.com/blog/products/gcp/an-in-depth-look-at-googles-first-tensor-processing-unit-tpu

as neural networks.[33] The third-generation TPU was announced on May 8, 2018 and Google "would allow other companies to buy access to those chips through its cloud-computing service".[34]

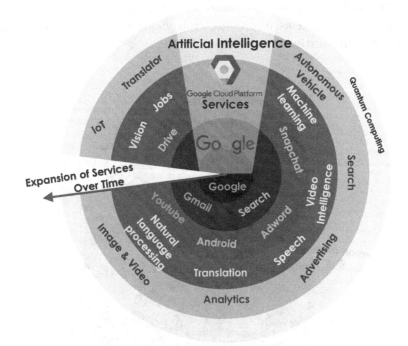

Figure 2.15: Google: From #16 in search engine, Google has used AI to become advertising behemoth. But, it is not on the radar screen in finance.

Microsoft's recent growth has been spurred by its cloud services that include Azure that has its revenue growth in excess of 70% over the previous year (Figure 2.16). Microsoft's focus on fast-growing cloud applications and platforms is helping it beat slowing demand for personal computers that has hurt sales of its popular

---

[33]https://en.wikipedia.org/wiki/Tensor_processing_unit
[34]https://www.nytimes.com/2018/02/12/technology/google-artificial-intelligence-chips.html

Windows operating system.[35] Its AI principles are exciting and most important of all, it is designing AI to be trustworthy that are creating solutions reflecting ethical principles, such as Fairness, Reliability and Safety, Privacy and Security, Inclusiveness, Transparency, and Accountability. The commercialization of some of the projects on mobile devices such as improving crop yields has yet to contribute directly/significantly to revenue.

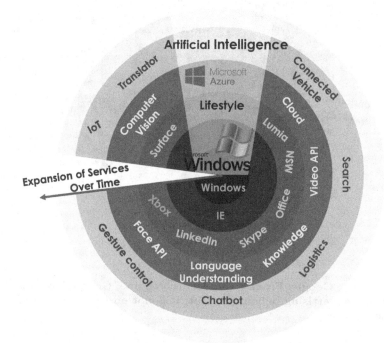

Figure 2.16: **Microsoft: Started as dominant PC operating sustem but failed in the all-important transition to mobile, trying to recover by its Azure Cloud. Why is Microsoft not a dominant leader in SME lending?**

Unlike Microsoft, Apple's AI strategy continues to focus on running workloads locally on devices, rather than relying heavily

[35]https://www.reuters.com/article/us-microsoft-results/microsoft-sales-and-pr ofit-beat-estimates-on-cloud-growth-idUSKCN1MY2T8

on cloud-based resources, as competitors Google, Amazon, and Microsoft do[36] (Figure 2.17). With more privacy protection going forward, the company emphasis is on user privacy and selling devices. The Create ML framework is the app maker to train AI models on Mac. Xcode is Apple's own app for coding programs for its devices. Swift, rather than Python, is Apple's programming language and the advantage is the dragging and dropping when training models with a bunch of data. The updated Core ML software is also a smaller model and takes up less space on devices once they are embedded into apps.[37]

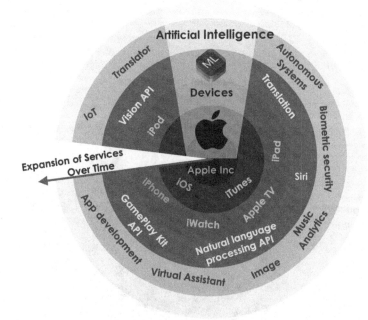

Figure 2.17: Apple: **This is a hardware company with awesome marketing, unimpressive AI and seems to have stumbled in the progression. Too late to the game?**

---

[36]https://www.cnbc.com/2018/06/13/apples-ai-strategy-devices-not-cloud.html

Facebook created Facebook AI Research (FAIR) group in 2013 to advance the understanding of the nature of intelligence in order to create intelligent machines (Figure 2.18). The research code, data sets, and tools like PyTorch, fastText, FAISS, and Detectron are open source.[38] However, given the issues on privacy intrusions and misuse of data, the good efforts are being overshadowed. However, it is still interesting to observe how it expands their research efforts into areas such as developing machines able to acquire models of the world through self-supervised learning, training machines to reason, and training machines to plan and conceive complex sequences of actions. Their concentration is on robotics, visual reasoning, and dialogue systems.

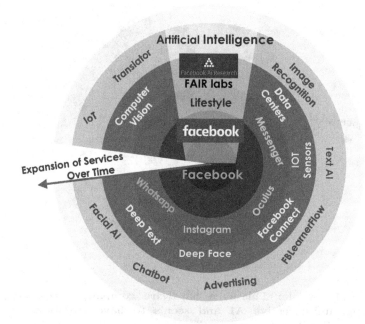

Figure 2.18:   Facebook: The greatest one-trick pony ever. They control the social network plus people think this is important — but for its own sake? One of the worst scores.

---

[38]https://code.fb.com/ai-research/fair-fifth-anniversary/

2.8 Project Conclusions: Why China Has Pulled Ahead

It is interesting to observe that when it comes to commercialization, few countries can beat China, especially the technological companies. When a Chinese project or initiative fails to impress or is otherwise unsatisfactory or with low market acceptance, successful replacements in a different direction are created to take over (Figure 2.19). When Alibaba's Lai Wang, started in 2011, failed to counter the challenge of Wechat of Tencent, it refocused on other businesses rather than competing head-on. Similarly, Tencent's Pai Pai failed to mimic the success of Taobao and closed in April 2016, 2 years after working with JD.com in which it acquired 15% and 11 years after Pai Pai was set up.[39] US firms, on the contrary, have few successful replacements.

Company	Duds	Successful Replacements
Alibaba	Lai Wang, Alibaba.com (HK)	Ant Financial, AliPay, Taobao, Tmall, Alibaba Cloud, PAI 2.0, Zhima Credit, ANTSDAQ, Yue'bao, Youku, Weibo, etc.
Tencent	Pai Pai, E-commerce	WePay, Weilidai, WeBank, JD.com, WeChat, Didi, Tencent Cloud, etc.
Ping An	Ping An Good Car	Ping An Bank, Ping An Insurance, Ping An Asset Mgmt, Ping An Health, etc.
Baidu	O2O Wai Mai	Baidu Wallet
Microsoft	Microsoft Wallet	Skype
Google	Google Hangout	Android Pay
Facebook	Messenger Pay	/
Apple	Apple Pay	/
Amazon	Amazon Lending Amazon Insurance	/

Duds with no replacement

Figure 2.19: **Chinese firms have implemented and jettisoned quicker. China dropped duds and created financial empires. US firms stopped their efforts.**

Note: China had duds but morphed quickly and pivoted effectively.

---

[39]https://walkthechat.com/tencents-product-failure-history/

The conclusions are that while manufacturing of chips is important, cloud is the foundation of AI strategy at this stage of the competition (Figure 2.20). Alibaba dominates in Asia while Amazon dominates in the US. Cognitive services are the foundation of AI development especially in regard to voice, image, face, video, language recognition, and learning. While there is differentiation in the deepness of the AI technologies, the "me too" development strategy seems to have created a feeling that AI services are already commoditized. Small and Medium Enterprises (SMEs) and corporates have access to cheap AI tools that allow them to access new forms of credit and offer new products. This new age of the SME will empower the smaller enterprises to level the playing field, provided the incumbents do not exert pressure on the regulators via their power of sheer employment numbers. Given that regulators are wary of job losses, the incumbents have a certain grip over the policymakers and legislators in implementing anti-competition rules in the current climate. In China, firms are leap-frogging banks because of the failure of the banks to serve the underserved. Furthermore, with weak privacy protection in the earlier years, data technology combined with AI allowed firms to offer new financial services to millions of unbanked or underbanked population and SMEs.

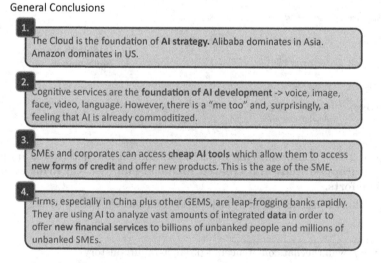

General Conclusions

1. The Cloud is the foundation of **AI strategy.** Alibaba dominates in Asia. Amazon dominates in US.

2. Cognitive services are the **foundation of AI development** -> voice, image, face, video, language. However, there is a "me too" and, surprisingly, a feeling that AI is already commoditized.

3. SMEs and corporates can access **cheap AI tools** which allow them to access **new forms of credit** and offer new products. This is the age of the SME.

4. Firms, especially in China plus other GEMS, are leap-frogging banks rapidly. They are using AI to analyze vast amounts of integrated **data** in order to offer **new financial services** to billions of unbanked people and millions of unbanked SMEs.

Figure 2.20: Project conclusions (1).

The observations point to the fact that Chinese firms are better at monetizing technology for mass adoption. With President Xi emphasizing on technology transformation with AI as leading technology, China has a long-term, coherent plan while the US is still struggling to come to grip with how AI can be commercialized for mass adoption besides military uses. It is also a known fact that even with tighter privacy protection law in China, the Chinese are more willing to surrender their private data to vendors. While we focus on Alibaba's achievements, Ping An has moved further into new territory than any other competitor. The Chinese government encourages the integration of finance and lifestyle with a clear national policy of proliferating credit to individuals and SMEs (see Figure 2.21).

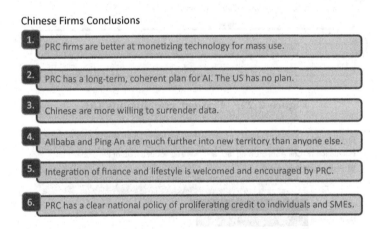

Chinese Firms Conclusions

1. PRC firms are better at monetizing technology for mass use.
2. PRC has a long-term, coherent plan for AI. The US has no plan.
3. Chinese are more willing to surrender data.
4. Alibaba and Ping An are much further into new territory than anyone else.
5. Integration of finance and lifestyle is welcomed and encouraged by PRC.
6. PRC has a clear national policy of proliferating credit to individuals and SMEs.

Figure 2.21: Project conclusions (2).

There are good reasons why the Chinese firms are having more successes over the US firms (Figure 2.22). Firstly, the incumbents and lobbying groups in the US have impeded progress by their sheer determination to protect their own turfs. Secondly, multiple regulators who were on the warpath after the devastation of the GFC with US$300 billion of finds, thus discouraging new consumer credit products. Thirdly, many of the Chinese companies learned to "eat dirt" and were toughened by a highly competitive environment. QR codes were banned for a short period due to the lobby by the incumbents, but the ban was eventually lifted due to the reasoning on the ground of social benefits. Finally, China succeeded since there was "there–there" with the only path to success being "new–new". Banking and regulatory structure were absent and monetary infrastructure were very immature, resulting in regulation not being able to front-run innovation. Financial innovation is more like manufacturing innovation as the Chinese lead in robot installations (Figures 2.22 and 2.23).

Reasons for China's Success over US

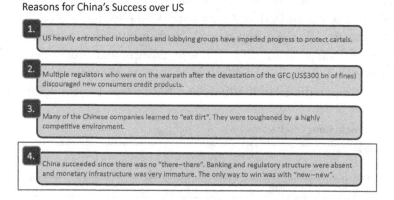

Figure 2.22: Project conclusions (3).

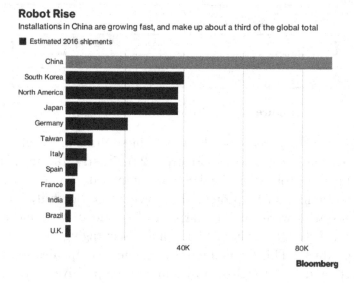

Figure 2.23: Robot installations in China are 1/3 of global total.
Source: International Federation of Robotics, National Robot associations, Bloomberg Intelligence.

2.9 Master Summary: Ranking for AI, Financials, and Valuations

Following are a few ranking tables based on specific metrics. These analyses are not meant for investment purposes, and one of the authors is an independent director of Lu International, a subsidiary of Ping An. The analyses are based on financial, operational, and AI performance. Figures 2.24–2.28 show the summary and various rankings.

		First	Second	Third	Laggard
Business Performance	Operational Health	*Facebook*	*Alibaba*	*Tencent*	*Amazon*
	Revenue Diversification	*Microsoft*	*Alibaba*	*Tencent*	*Facebook*
	Valuation	*Ping An*	*Apple*	*Google*	*Tencent*
	Credit Health	*Google*	*Apple*	*Tencent*	*Baidu*
Artificial Intelligence Performance	Fintech	*Alibaba*	*Ping An*	*Tencent*	*Facebook*
	Insurtech	*Ping An*	*Alibaba*	*Tencent*	*Facebook*
	Healthtech	*Ping An*	*Alibaba*	*Amazon*	*Baidu*
	Cloud	*Amazon*	*Microsoft*	*Alibaba*	*Facebook*
	Data Source	*Alibaba*	*Ping An*	*Tencent*	*Baidu*
	Cog Services	*Google*	*Alibaba*	*Microsoft*	*Facebook*
	Lifestyle	*Tencent*	*Alibaba*	*Amazon*	*Baidu*
	Securities	*Ping An*	*Alibaba*	*Tencent*	*Microsoft*
Overall Winners/Losers		Ping An	Alibaba	Tencent	Baidu/Facebook

Figure 2.24: Master summary: Financial, operational, artificial intelligence analysis.

	AI Scorecard	Financials	Valuation	Average
1. Ping An	2	4	1	2.3
2. Alibaba	1	2	7	3.3
3. Google	5	4	3	4.0
4. Apple	6	5	2	4.3
5. Tencent	3	3	8	4.7
6. Facebook	8	1	6	5.0
7. Microsoft	7	6	4	5.7
8. Baidu	9	7	5	7.0
9. Amazon	4	8	9	7.0

Figure 2.25: Overall rankings: Ping An and Alibaba dominate in all categories.

Note: Ranking of each company's performance across all three areas: Artificial Intelligence Services, Financials, and Valuation. "1" is best, "9" is worst.

	1 Alibaba	2 Ping An	3 Tencent	4 Amazon	5 Google	6 Apple	7 Microsoft	8 Facebook	9 Baidu
FinTech	10	10	9	7	6	7	4	4	5
InsureTech	9	10	9	6	4	4	4	4	4
HealthTech	9	10	8	8	7	7	7	5	4
Cloud	8	5	7	10	8	6	9	4	5
Data source	9	8	9	8	8	8	7	7	6
Cognitive Services	9	8	7	8	10	7	9	7	8
Lifestyle	9	8	10	9	9	9	7	9	6
Securities	9	10	9	4	5	6	4	5	5
Average	9.0	8.6	8.5	7.5	7.1	6.8	6.4	5.6	5.4

Figure 2.26: Artificial intelligence scorecard.

Notes: Chinese firms have diversity and consist of a full menu. US firms tend to specialize through one brand. This age of AI requires diversified ecosystem. Specialist firms lose out.

*Scoring from 1 (worst) to 10 (best).

Source: Schulte Research Estimates.

	1 Facebook	2 Alibaba	3 Tencent	4 Alphabet	5 Apple	6 Microsoft	7 Baidu	8 Amazon
Retained Earnings (US$mn)	3	5	4	1	2	8	6	7
Gross Margin (%)	1	2	5	4	7	3	6	8
Net Margin (%)	1	2	3	6	5	4	7	8
Revenue CAGR (5yrs) (%)	2	1	3	6	7	8	4	5
Cash/Total Asset	4	1	3	5	6	8	7	2
Tangible Leverage	1	3	5	2	6	8	4	7
ROE (%)	2	5	3	6	1	4	7	8
ROC (%)	3	6	2	5	1	4	7	8
ROA (%)	3	2	1	4	5	6	7	8
Average	2.3	3.1	3.3	4.6	4.6	6.2	6.3	7.2

Figure 2.27: Financial rankings.

Notes: Facebook/Alibaba best in operations; Baidu/Amazon worst. Microsoft also mediocre.

*Ranking from 1st (best) to 9th (worst).

*ROE/ROC/ROA are next 4Q estimates.

	1 Ping An	2 Apple	3 Alphabet	4 Microsoft	5 Baidu	6 Facebook	7 Alibaba	8 Tencent	9 Amazon
P/E	1	2	4	3	5	6	7	8	9
P/B	1	4	2	6	3	5	7	8	9
PEG	2	7	6	8	3	1	4	5	9
Price/ Free Cash Flow	1	2	4	3	5	7	6	8	9
EV/EBITDA	NA	1	2	3	5	4	7	8	6
EV/EBIT	NA	1	2	3	5	4	6	7	8
Average	2.4	2.8	3.3	4.3	4.3	4.5	6.2	7.5	8.5

Figure 2.28: Valuation rankings.

Notes: Ping An cheapest; Amazon most expensive.

*Ranking from 1st (best) to 9th (worst).

2.10 Master Summary: Strategic Direction

Successful technology companies tend to focus on building the ecosystem. By that, we mean that they build the capabilities of the entire spectrum of technologies by integrating the high PE rather than those with low PE. A combination of IOT plus AI with an integrated ecosystem presents excellent business opportunities. The Chinese firms are more diversified in revenue than the US firms. In terms of advertising as a ratio of revenue, the US firms such as Facebook and Alphabet/Google have a higher ratio. Data-rich firms have the opportunities to increase their revenue in this area (Figures 2.29–2.33).

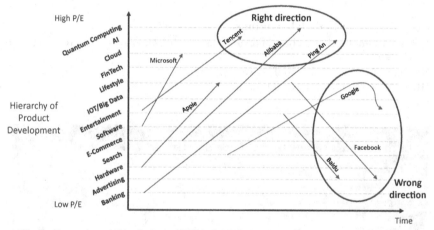

Alibaba, Tencent, and Ping An have IOT/AI *plus* fully integrated systems. Baidu dropped the ball.

Figure 2.29: Strategy directions: The key is integrated ecosystems.
Note: Specialization in IOT + AI is a *liability*. Google/Facebook have thrown away spectacular opportunities.
Source: Schulte Research.

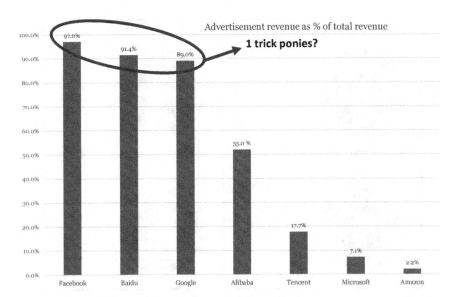

Figure 2.30: Advertisement: Facebook is leading, other data-rich firms have the potential to expand revenue source.

Source: Schulte Research Estimates, Bloomberg.

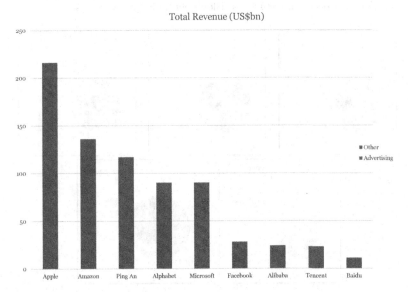

Figure 2.31: Advertisement: Facebook is leading, other data-rich firms have the potential to expand revenue sources.

Source: Schulte Research Estimates, Bloomberg.

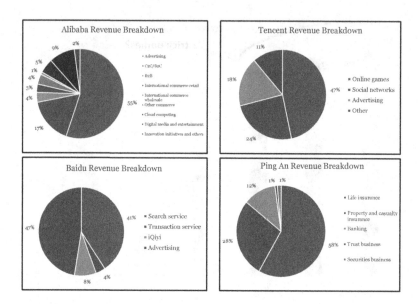

Figure 2.32:   Chinese companies revenue breakdown.
Note: China much more diversified than the US: PRC has a deeper, more integrated ecosystem.
Source: Schulte Research Estimates, Bloomberg, Company filings.

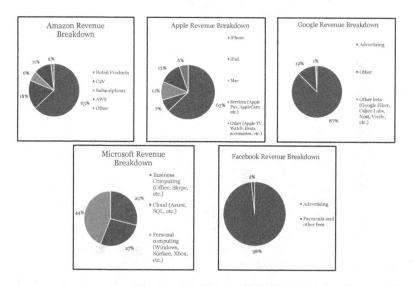

Figure 2.33:   US companies revenue breakdown.
Note: These five companies are essentially specializing. I think the US companies have moved away from the conglomerate model at the wrong time?
Source: Schulte Research Estimates, Bloomberg, Company filings.

2.11 Master Summary: Research and Development — Strategic Development

Most of the US technology companies have a similar vision for the future which is based on AI, Big Data and Language (Figure 2.34). However, each of them has their niche, with Facebook focusing on Virtual Reality, Microsoft on Human–Computer Interaction, Apple on Chips, Amazon on Drones and Robotic and Google on Quantum Computing. Chinese firms are similar but more diversified.

Companies	Research Direction	Unique Point
Facebook	1.Facebook AI Research (FAIR) – semantic analysis, sentient analysis, etc. 2.Natural Language Processing and Speech – translation services, word sense disambiguation, etc. 3.Applied Machine Learning – streamline content delivery, connect users with desired content 4.Data Science – human interaction analysis, market intelligence, etc. 5.Virtual Reality – augmented reality experiences	Virtual Reality
Microsoft	1.Artificial Intelligence 2.Computer Vision 3.Human – Computer Interaction 4.Human Language Technologies 5.Search and Information Retrieval	Human – Computer Interaction
Apple	1.Apple Car/Autonomous Driving 2.Augmented Reality 3.HealthTech – Health Functions within Apple Watch 4.Chips – Hardware improvement	Chips
Amazon	1.Image/Face Recognition 2.Voice Recognition 3.ChatBot AI 4.Drone Delivery 5.Warehouse Robotics	Drone, Robotics
Google	1.Machine Translation System 2.Deep Learning Algorithms – Detection of Diabetic Retinopathy 3.Quantum Computing 4.Big Data Analytics 5.Machine Intelligence/Perception	Quantum Computing

Figure 2.34:   Research directions: US firms.

Note: All companies have a similar vision of the future: AI, Big Data, and Language!

Source: Schulte Research.

Alibaba's unique point is Securities Exchange and Intelligent Manufacturing. Tencent has portfolio management as its unique point. Ping An has focused on Wealth Management Products (WMPs) as its unique point among many other initiatives. Zhong An has blockchain as its unique point with Baidu focusing on high-performance computing as its unique point (Figure 2.35). However, given the dynamic market, this unique point of focus may switch with time and market trend. Acquisitions will also reveal the direction of the companies, and Figures 2.36 and 2.37 show the trend of acquisitions by these companies.

Companies	Research Direction	Unique Point
Alibaba	1.Financial Securities Exchange Core 2.Customer Service AI 3.Intelligent Manufacturing – Cloud computing/big data analytics 4.Health Data Platform – Integration of patient data, image analysis, diagnosis 5.Commercial Vehicle Networks – logistics enhancement and optimization	Securities Exchange, Intelligent Manufacturing
Tencent	1.Artificial Intelligence 2.Machine Learning – Stock Picking, HFT, Portfolio Management 3.Big Data – Trend Analysis 4.Cloud Computing	Portfolio Management
Ping An	1.Data Mining – Social Media User Behavior, Engagement Rates 2.Computational Algorithms 3.Big Data Analytics – Purchasing Patterns, WMPs	WMPs
Zhong An	1.Financial Platforms/Cores – E-commerce Platforms, Insurance Cores, Finance Cores 2.Data Analytics/ Risk Management Modelling 3.AI – Transaction Decision-making 4.Blockchain – Digitalize Off-chain assets, Data Storage, Secure ID 5.AI – Customer Service ChatBot	Blockchain
Baidu	1.Artificial Intelligence – image/speech recognition 2.High-Performance Computing/Big Data Analytics 3.Natural Language Processing 4.Deep Learning 5.Augmented Reality	High-performance computing

Figure 2.35: Research directions: Chinese firms.
Note: All companies have a similar vision of the future. Chinese firms show higher degree of diversification.
Source: Synergy Research, Schulte Research.

Company	Acquisition	Amount (US$)	Year	Description	Comment
Facebook	WhatsApp	19bn	2014	Messaging	Consolidation of social network industry
	Oculus	2bn	2014	VR	
	Instagram	1.01bn	2012	Social Network	
	LiveRail	500mn	2014	Video ad.	
Microsoft	Linkedin	26.2bn	2016	Social Network	Failed hardware attempt on Nokia. Linkedin is the new bet on professional network.
	Nokia	8bn	2013	Phone	
	Skype	8.5bn	2011	Messaging	
	Mojang	2.5bn	2014	Game developer	
	Yammer	1.2bn	2012	Corp communication	
Apple	Beats Electronics	3bn	2014	Headphone	Lifestyle + Health
	HopStop.com	1bn	2013	Transit guide	
	Lattice	200mn	2017	AI	
	Gilimpse	200mn	2016	Health	
Amazon	Whole Foods Market	13.7bn	2017	Supermarket	O2O bet on Whole Foods
	Twitch	970mn	2014	Game streaming	
	Souq.com	580mn	2017	E-commerce	
	Elemental Technologies	500mn	2015	Mobile video	
	Annapurna Labs	37mn	2015	Chip maker	
Google	Motorola Mobility	12.5bn	2012	Phone	Another failed case of acquiring phone company
	Waze	1.3bn	2013	Map	
	Apigee	625mn	2016	API platform	
	DeepMind	500mn	2014	AI	

Figure 2.36: Major acquisitions: US firms.
Note: US firms invest heavily in acquiring smaller firms that do not necessarily operate in their industry.
Source: Synergy Research, Schulte Research.

Companies	Acquisition	Amount (US$)	Year	Description	Comment
Alibaba	UCWeb	4.7bn	2014	Mobile browser	Contents + data
	Youku	3.5bn	2015	Chinese Youtube	
	AGTech Holdings	2.39bn	2016	Lottery	
	SCMP	262mn	2015	Media	
	Wandoujija	200mn	2016	Android app store	
Tencent	Supercell	8.6bn	2016	Game developer	Consolidation of game industry
	China Music Corp	2.7bn	2016	Music	
	Riot Games	400mn	2011	Game developer	
	Miniclip SA	na	2015	Game developer	
	Sanook	na	2016	Thai web portal	
Baidu	91 Boyuan Wireless	1.9bn	2013	Android app store	Diverse
	Beijing Huanxiang Zongheng Chinese Literature	31.3mn	2013	Online publisher	
	TrustGo	30mn	2013	Mobile security	
Ping An	Autohome	1.6bn	2016	Auto website	Auto insurance

Figure 2.37: Major acquisitions: Chinese firms.
Note: All companies have a similar vision of the future. Chinese firms show higher degree of diversification.
Source: Synergy Research, Schulte Research.

2.12   Master Summary: Cash Flow, M&A, Capex

It is interesting to observe that given the high market capitalization and cash position, tech companies tend to invest in building their ecosystem. Banks are pursuing new technology but not doing enough in this space, partly due to the lower market capitalization, lower cash flow, lower cash reserves, high capital requirement, regulatory constraints, board composition with less tech expertise, fiduciary duties and other reasons that have been well written. Figures 2.38–2.42 show the financials.

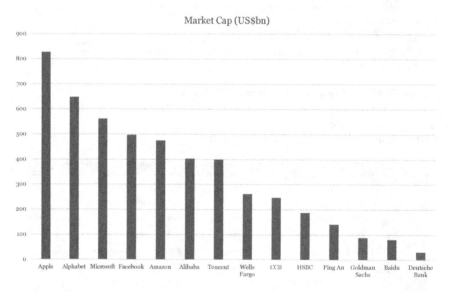

Figure 2.38:   Market cap.
Note: Tech companies are a gathering force in market cap. They dominate capital markets now. Banks can't hold a candle to tech spending.
Source: Schulte Research Estimates, Bloomberg.

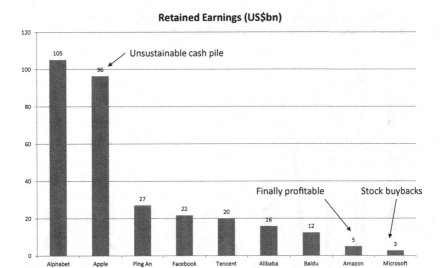

Figure 2.39: Retained earnings.
Note: Alphabet and Apple are sitting on mountains of cash. Much of it as untaxed offshore cash.
Source: Schulte Research Estimates, Bloomberg.

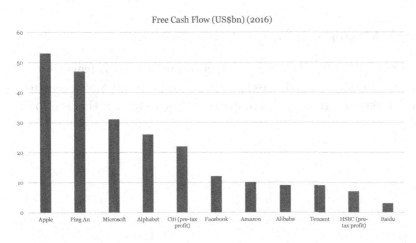

Figure 2.40: Free cash flow.
Note: Cash flows from operations — investment in operating capital (mostly fixed assets). Tech companies below away banks in cash flow ("different" numbers, but you get the point).
Source: Bloomberg, Citi Annual Report, HSBC Annual Report.

US$300 bn in "conduct costs" for banks is that much more that can't be used for investment. *Plus*, current R/D spending is woefully under invested.

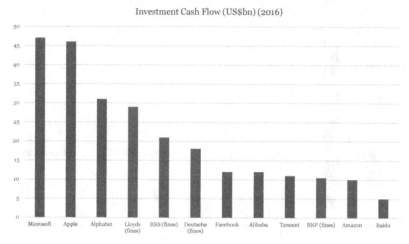

Investment Cash Flow (US$bn) (2016)

Figure 2.41: Investment cash flow (net spending in PP&E, Capex, M&A).

Source: Bloomberg.

2.13   Master Summary: Credit Check

Figures 2.42–2.49 show the financials and we shall leave it to the readers to decide for themselves. For those who are not interested in the investment into these companies, they may skip this section.

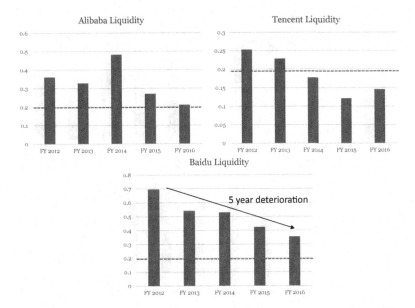

Figure 2.42: **Chinese companies liquidity (1.2 × working capital/total assets).**

Note: *Dotted line represent NASDAQ average.

Source: Bloomberg.

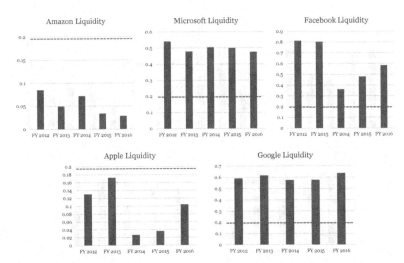

Figure 2.43: US companies liquidity (1.2 × working capital/total assets).

Note: *Dotted line represent NASDAQ average.

Source: Bloomberg.

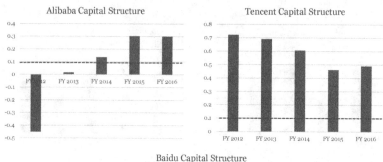

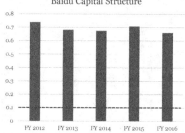

Figure 2.44: Chinese companies capital structure (1.4 × retained earnings/total assets).

Note: *Dotted line represent NASDAQ average.

Source: Bloomberg.

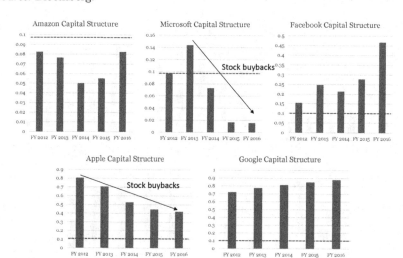

Figure 2.45: US companies capital structure (1.4 × retained earnings/total assets).

Note: *Dotted line represent NASDAQ average.

Source: Bloomberg.

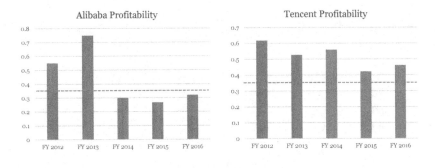

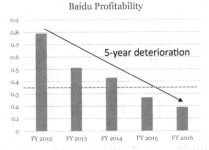

Figure 2.46: **Chinese companies profitability (3.3 × EBIT/total asset).**
Note: *Dotted line represent NASDAQ average.
Source: Bloomberg.

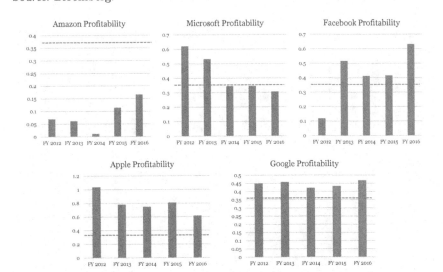

Figure 2.47: **US companies profitability (3.3 × EBIT/total asset).**
Note: *Dotted line represent NASDAQ average.
Source: Bloomberg.

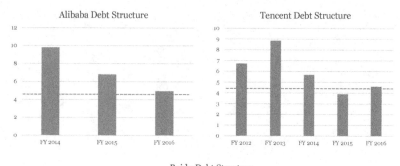

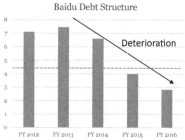

Figure 2.48: Chinese companies debt structure (0.6 × market cap/total liabilities).

Note: *Dotted line represent NASDAQ average.
Source: Bloomberg.

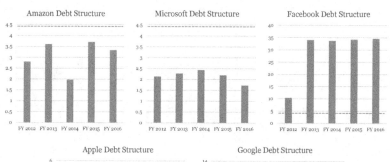

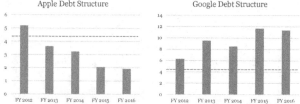

Figure 2.49: US companies debt structure (0.6 × market cap/total liabilities).

Note: *Dotted line represent NASDAQ average.
Source: Bloomberg.

PART 2

Banking and Insurance Transformation in the US, China, and Southeast Asia

Chapter 3

How the Mighty are Learning to Change: Who is Evolving and Who is Dying

In this chapter, we look at some of the transformation strategies of Chinese technology companies. In particular, we describe the advantages and challenges of China in commercializing technology into sustainable global business models. Primordial soup, or prebiotic soup, is used to describe a solution rich in organic compounds in the primitive oceans of the earth, from which life is hypothesized to have originated. According to Alexander Oparin and John Burdon Sanderson Haldane in the 1920s, this primordial soup is a fundamental aspect of the origin of life.[1] With prebiotic soup, life originated, and the first forms of life were able to use the organic molecules to survive and reproduce. We can draw a similar analogy with the FinTech creations in China just like in the 1954 movie *Creature from the Black Lagoon* (see Figure 3.1)!

---

[1] Oparin, A. "The origin of life". https://breadtagsagas.com/wp-content/upload s/2015/12/AI-Oparin-The-Origin-of-Life.pdf
Haldane, J.B.S. "The origin of life". https://www.uv.es/~orilife/textos/Haldane .pdf

Figure 3.1: A new FinTech creature from the Black Lagoon: Turning ride shares into bank shares — two kinds of financial primordial soup cooking.

China may or may not view overseas Chinese population as their own, but Chinese President Xi Jinping believes that overseas Chinese play a huge role in shaping the country's economy and politics.[2] These offshore Chinese communities or Huaqiao (华侨) may not have the affinity of their fathers or grandfathers for China as they may not speak the same language or have the same culture, especially in countries such as Singapore as the 50 years of nation-building have molded the Chinese and other ethnic groups into a nation.[3] "If, as entrepreneurs, they invest in China, it is not because of any attachment as a Huaqiao for their motherland, but as a hard business investment. The Straits-born Chinese, who did not think of themselves as Huaqiao and did not identify with China," reported by Singapore's press *Today*. Nations are fearful of the influence of China on their population.[4] However, the truth about affinity may

---

[2]https://asia.nikkei.com/Economy/Xi-s-secret-economic-weapon-Overseas-Chinese2

[3]https://www.todayonline.com/chinaindia/china/chinas-unrealistic-expectations-overseas-chinese

[4]https://www.scmp.com/comment/insight-opinion/china/article/2153771/singapore-us-overseas-chinese-are-increasingly-fearful

lie some way in between as there are more than 100 million Chinese besides Singapore's 3 million. China began to offer 5-year visas to attract foreigners of Chinese origin from February 2018,[5] as it tried to attract more Huaqiao talents to China.

According to the United Nations World Tourism Organization (UNWTO), there are more than 150 million overseas visits by Chinese tourists with an expenditure of US$2.6 billion.[6] Currently, only 7% or 99 million out of the 1.4 billion Chinese have a passport. The outbound traffic is projected to reach 400 million visits by 2030, amounting to 25% of the world's outbound traffic.

As Alipay and WeChat Pay expand beyond the country's border, the Chinese tourists are giving them a push. Tax refunds of Chinese tourists are paid through Alipay and WeChat Pay, as well as payments of taxi fare, shopping, lodging and other entrance fees outside China (see Figures 3.2–3.4).

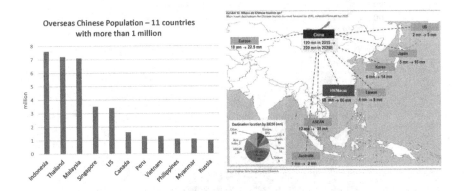

- 100% increase in PRC outbound travel.
- Includes overseas born Chinese and dependents.

Figure 3.2: PRC data diaspora: ASEAN diaspora of Chinese of 33 million.

Source: AAARI, CUNY, Schulte Research.

---

[5]https://www.straitstimes.com/asia/east-asia/china-to-issue-new-five-year-vis as-to-attract-foreigners-of-chinese-origin-from-feb-1

[6]https://www.telegraph.co.uk/travel/comment/rise-of-the-chinese-tourist/

International tourism expenditure by country
Total spending by tourists in 2016

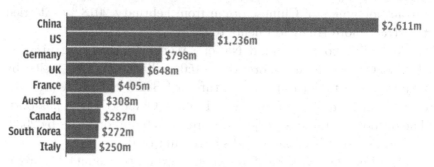

Figure 3.3:   China's international tourism expenditure.

Annual overseas visits (millions)

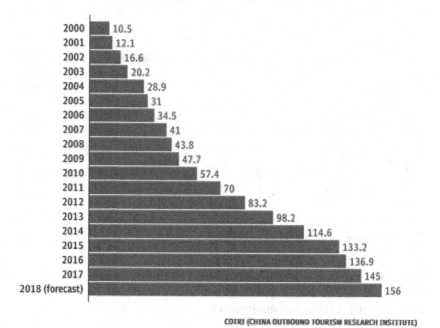

COTRI (CHINA OUTBOUND TOURISM RESEARCH INSTITUTE)

Figure 3.4:   Annual China overseas visits.

While payments are turning these payment services into banking services globally, Chinese players such as Ant Financial are taking the lead. Both SoftBank and Alibaba are dominating investment into the Indo-Pacific region for payments services, such as Ant Financial and Paytm. For retail or e-commerce investment, Alibaba is leading with investment in Tokopedia and Lazada. For rideshares, Tencent takes the lead with investment in Ola, Grab, Gojek, and Didi. Tencent is making an aggressive move in India, Indonesia, Thailand, and Singapore. Alibaba, on the other hand, is investing in areas that they are good to replicate their success in China. They are investing in e-commerce, payment and logistics sectors and spans seven key ASEAN countries. The real reward will come from data as Alibaba believes that data is raw materials and computing power is like an engine. With Alibaba Cloud expanding into the region, and the data collected via investments into payments, banking, insurance, e-commerce, Media, Food, Lifestyle, and Logistics, it is set to dominate the digital economy where data is digital assets. The old economy companies will be at a disadvantage as the essential factors of production are now AI (computing power plus data) and the ability to raise capital rather than land, labor, and capital. While land and capital are needed for production, these are secondary as without AI and robotics, companies cannot be as efficient (see Figures 3.5–3.7).

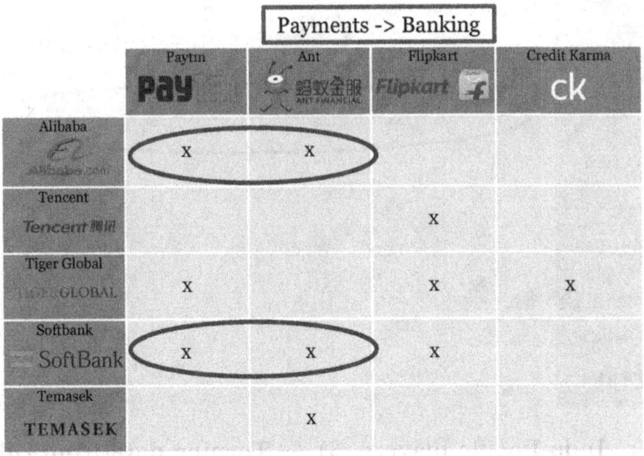

	Payments -> Banking			
	Paytm	Ant	Flipkart	Credit Karma
Alibaba	X	X		
Tencent			X	
Tiger Global	X		X	X
Softbank	X	X	X	
Temasek		X		

Figure 3.5: Indo-Pacific players (1) — Turning data from payments into banking. SoftBank and Alibaba dominate here.

Source: Schulte Research.

Retail -> Banking

	Walmart	Tokopedia	Lazada
Alibaba		X	X
Tencent	X		
Tiger Global	X		
Softbank		X	
Temasek			X

Figure 3.6: Indo-Pacific players (2) — Turning data from retail into banking. Alibaba dominates here.

Rideshare -> Banking

	Ola	Grab	GO-JEK	Didi
Alibaba		X (Didi)		X
Tencent	X	X (Didi)	X	X
Tiger Global	X	X		
Softbank	X	X		
Temasek		X	X	

Figure 3.7: Indo-Pacific players (3) — Turning data from ride shares into bank shares. Tencent dominates here.

Source: Schulte Research.

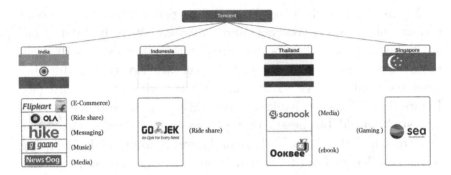

Figure 3.8: Tencent regional investments: Major bets in India.
Source: Schulte Research.

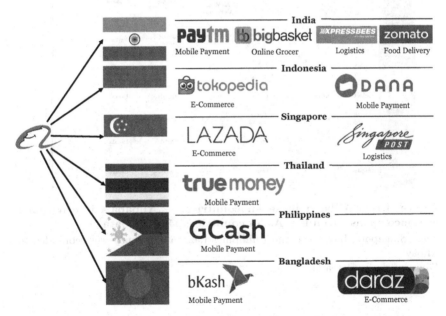

Figure 3.9: Alibaba international investments: Investment in areas they are good at (e-commerce plus payment) to replicate its success in China.

Figures 3.8 and 3.10 show the regional and international investment of both Tencent and Alibaba. With big data and sticky customers, additional low-profit margin services are drawing more customers. Tokopedia and Paytm are slowly transforming from payment services to banking services without branches (see Figure 3.11). Existing banks are not agile to provide other services and stick to old businesses that generate good profits. Given the penetration of the customer base and lower margin, banks are going to face challenges from these new creatures from the black lagoon.

	Payment	Banking	Insurance	Ecommerce	Media	Food	Lifestyle	Logistics	Cloud
China	Alipay	Ant Fortune Mybank Zhima Credit	Zhong An (12%)	Taobao Tmall 1688.com	Youku SCMP Video++ CMC UC Web Ali Music Ali Pictures	Sun Art Tmall Supermarket Hema Ele.me	Alihealth Tao Piao Piao Damia.cn Wandoujia	Cainiao	
India	Paytm (45%)	Paytm	Paytm	Paytm Mall (~30%)	DailyHunt (Planned)	Big Basket Zomato	Paytm	XpressBees	
Indonesia	DANA, Alipay	Tokopedia	Tokopedia	Lazada (>83%) Tokopedia	UCWeb	Lazada Tokopedia	Lazada Tokopedia		Alibaba Cloud
Thailand	TrueMoney	Truemoney	Truemoney	Lazada		Lazada	Lazada		
Bangladesh	bKash (20%)	bKash	bKash	Daraz (100%)		Lazada	Lazada		
Philippines	Gcash, Alipay	Gcash	Gcash	Lazada		Lazada	Lazada		
Singapore	Alipay	Singapore Post	Singapore Post	Lazada		Lazada	Lazada	Singapore Post	

Figure 3.10: Alibaba in seven countries — Alibaba's international presence spans seven key Asian countries: Massive data.
Note: Singapore, Indonesia, and Phillipines Lazada's Hellopay was rebranded to Alipay.
Source: Schulte Research.

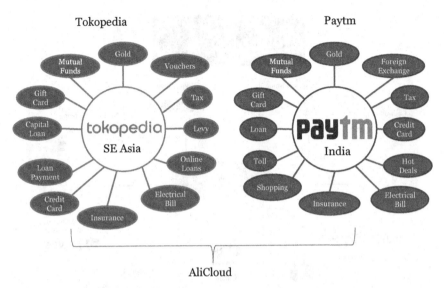

Figure 3.11: Tokopedia and Paytm digital offerings in banking. The new creature from the Black Lagoon — new reinforcing datasets.
Source: Schulte Research.

With the US$16 billion acquisition of Flipkart[7] in May 2018, Walmart is the majority owner of the Indian e-commerce company (Figure 3.12). Both Tencent and Tiger Global have board seats. Walmart also plans to use Flipkart as a "key center of learning" for the rest of its business across the world, and that includes its home market. "Not only is Flipkart innovative with the problem-solving culture that they have, but they are doing some great work both in the AI space, how they are using data across their platforms but particularly in terms of the payment platform that they have created through PhonePe. All of those things we can learn from for the future and see how we can leverage those around the international markets and potentially into the US as well," Walmart COO Judith McKenna said back in May when the deal was announced. That perhaps summarizes the real intent of the acquisition to build a Taj Mahal Mart!

---

[7]https://techcrunch.com/2018/08/20/walmart-flipkart-deal-done/

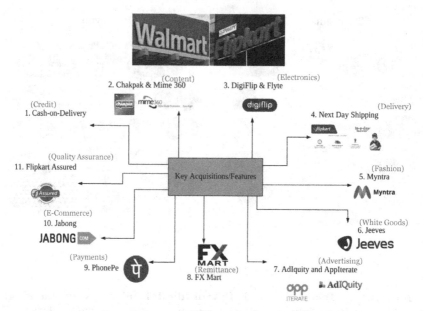

Figure 3.12: Flipkart acquisitions/features: Now this belongs to Walmart. Another new banking creature. Taj Mahal Mart.

Source: Flipkart, Schulte Research.

In 2017, Amazon Pay India was wooing local consumers with this promotional, "Select Amazon Pay balance, get 10% extra as refund amount." The Amazon Pay offers are applicable not only for transactions on Amazon's platform but also on third-party Web sites such as abhibus.com (Figure 3.13). It also has tie-ups with vendors such as Café Coffee Day, Amar Chitra Katha, Innerchef, Fasoos, and Housejoy.[8] This was in response to Amazon's biggest local rival Flipkart acquiring UPI-based payments company Phone-Pe in April 2016, and enabling payments through the wallet on the Flipkart platform in August last year to improve the digital payment experience for its customers.[9]Amazon Pay has transformed itself into a digital

---

[8]https://economictimes.indiatimes.com/small-biz/startups/amazon-opens-its-wallet-for-india/articleshow/59781514.cms

[9]http://economictimes.indiatimes.com/articleshow/59781514.cms?utm_source=contentofinterest&utm_medium=text&utm_campaign=cppst

bank offering similar services of traditional bank such as Standard Chartered India, challenging every aspect of banking services.

Figure 3.13:   Amazon India vs Standard Chartered India: The Amazon moment for banking is here. The newest bank from the primordial soup.

Amazon has made a few key acquisitions in India from online banking, finance, e-commerce, home services, publishing, insurance, to lending. Just like the biologist's Petri dish for cultivating cells, Amazon has an ecosystem of cultivated services for its banking ambition (Figure 3.14).

Figure 3.14: **Amazon's key acquisitions in India: The Petri dish for Amazon's banking ambition.**

The Chinese government has the intention of standardizing the assessment of its citizens' and corporates' economic and social reputation by 2020. The Social Credit System (社会信用体系) is a national reputation system developed by the Chinese government using social credit-scoring methods that include data from almost all aspects of individual behavior. In 2015, Ant Financial was approved by the People's Bank of China to develop Zhima Credit — a private credit-scoring platform. Zhima Credit score of between 350 and 950 is given to every user of Alipay to determine their credit standing with perks.[10] *Ant Financial did state, however, in a 2015 press release that the company plans "to help build a social integrity system". And the company has already cooperated with the Chinese government in one important way: It has integrated a blacklist of more than 6 million people who have defaulted on court fines into Zhima Credit's database. According to Xinhua, the state news agency, this union of big tech and big government has helped courts punish more than 1.21 million defaulters, who opened their Zhima Credit one day to find their scores plunging.*[11]

---

[10]https://www.wired.com/story/age-of-social-credit/
[11]https://www.wired.com/story/age-of-social-credit/

The use of big data analytics and AI has reached a level beyond any countries in monitoring behavior using behavioral data (Figure 3.15). China's blockchain patent filing accounted for 70% of all patents by 15 global names in 2017 (Figure 3.16).

Social Credit Score	Zhima Credit
Shopping behavior	Behavior trait
Participation in online consumer forums	Consumption level / preference
Identity information	Identity Information: Real name verification; info stability
Social media presence	Social Network
School and college scores	Network Influence/creditworthiness
Credit history	Credit History
Traffic or public transport violations	Loans repayment/fine history/utilities
Financial behavior	The ability to abide by the contract
Alipay/Wechat Pay	Alipay, Yu' E Bao balances; property
Political views expressed at online portals	
Medical history	
Library borrowings	
Rating on taxi app	

Figure 3.15:   Social credit score and Zhima credit score.

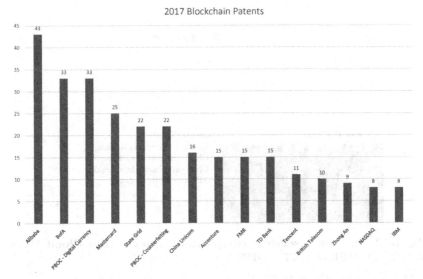

Figure 3.16:   2017 Blockchain patents by 15 global names. 70% of all blockchain patents in 2017 were from China.

Source: technode, IPR Daily, incoPat, Schulte Research.

Even more interesting is the fact that in both PRC and the US, Huawei, and ATT/Verizon are big data accumulators from technology companies. Privacy protection may be on the mind of most, but in reality, it is anyone's guess whether there will be a massive privacy invasion. With centralized control, there is always a single point of attack by hackers or single point of breach. It is bad news for privacy protection believers as there two bowls of Primordial Soup: Egg Drop Banking plus Chicken Noodle Banking. However, privacy invasion seems to be a more probable outcome unless more resources are devoted to privacy protection. When AI harms defenseless humans, it is on a massive scale! There seems to be nowhere to hide with facial recognition, digital home devices, and other monitoring equipment.

Both Cloud services are expanding rapidly, and the cloud wars are on. When it reigns, it pours. In the US, AWS, and Microsoft Azure are taking the lead, while in China, Alibaba Cloud and Huawei Cloud are leaders (Figure 3.17). We have briefly introduced both blockchain and cloud developments, and we shall refer the readers to other later chapters that will provide more details.

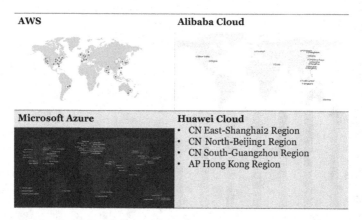

Figure 3.17: Cloud services comparison: Entering the Cloud Wars. WHEN IT REIGNS, IT POURS.

Source: Amazon, Schulte Research.

Chapter 4

New Entrants and Traditional Banks are (Finally) Trading in Lizard Brains for an Upgrade

Big banks, Insurance, and Telecom, are all under tremendous stress from digital disruption. Can they make a comeback and compete with the more nimble FinTech giants. In this chapter, we shall look at the response. DBS from Singapore, China Construction Bank in China, and Mandiri in Indonesia are trying to push back the disruptors. China Life, Ping An, SingTel, and Unicom from the insurance and Telecom sectors are also trying to be up to speed with the development in technology.

4.1 The Empire Strikes Back: Can Big Banks, Insurance and Telecom Make a Comeback in Financial Technology?

4.1.1 *Banks*

First, let us look at the advantages and disadvantages that the banks have.

Advantages to strike back are as follows:

1. massive customer base;
2. awesome data pool;
3. vast networks;
4. protected cartel, and impossible to get new licenses;

5. cheap valuations;
6. many are more than 100 years old and have established a reputation.

Disadvantages include the following:

1. Customers always complain about the poor experience.
2. Innovators dilemma: Why change when revenues are still flowing?
3. Misalignment of incentives: Banks operate for managers, not customers.
4. Old and enormous internal tech legacies which cannot be replaced.
5. Antiquated physical infrastructure which is outdated.
6. Still a paper culture — Can digital revolution invade a paper culture?
7. Protected cartel means they have no incentive to change.
8. Distribution network from the 1970s.
9. Many products are hundreds of years old.

While banks are well aware of their strengths and weaknesses, many of them are unable to act on their own either due to their legacy system, board direction, shareholders' pressure or simply tight local and international regulation. The Japanese banks are well aware of these problems and they have been struggling since the bubble burst in the 1980s. In Bank of Japan October 2017 Financial System Report, it was stated that the financial system still suffered from structural factors, such as low profitability and intense competition. In particular, Japanese financial institutions have little non-interest income and depend on net interest income as a profit source, and the number of employees and branches is excessive relative to demand.[1]

The Japanese banks and corporates were very early in Silicon Valley. The first wave was in Silicon Valley as early as the 1980s, the second wave was in the 1990s, and the third started in 2015. Companies such as Panasonic, Dai Nippon Printing, JCB, Mitsubishi Heavy Industries, and Dentsu were partners at startup incubators

---

[1] https://www.boj.or.jp/en/research/brp/fsr/fsr171023.htm/

such as Plug and Play and 500 Startups as early as 2015.[2] The "innovation tourism" brought many Japanese executives to Silicon Valley to experience the disruptive culture after Prime Minister Abe's 2015 Silicon Valley visit and Toyota's announcement of a US$1 billion investment in a local AI laboratory. Japanese were committed to open innovation way before the rest of Asia, especially after the approval of the sale of electronics giant Sharp to the Taiwanese company Foxconn. There are lessons to learn from the Japanese regarding breaking the monopolistic mindset to spur innovation.

Japanese are known for taking a long time to make decisions even after acquiring much information. They were not the preferred partners for startups that moved a lot quicker in order to scale. However, the culture changed rapidly with top management taking special interest and making frequent trips to Silicon Valley for onboarding.

While one of the authors were at Stanford, it was clear from many of the gatherings and meetings that the Japanese were learning very fast from Silicon Valley after Prime Minister Shinzo Abe spoke with former Secretary of State George Shultz and Stanford University President John Hennesy during the Silicon Valley Japan Innovation Programme. This prompted the acting head of the Freeman Spogli Institute for International Studies Professor Takeo Hoshi to say, "This Japanese administration has been focusing on changing its economy to a growth-based system built on innovation. This (Silicon Valley) is probably the best place in the world to look at that."[3]

Payments were one area that the banks were interested and on the radar of the Bank of Japan. So, it came as no surprise that the Japanese Cabinet passed a set of new bills in 2016 to recognize bitcoin as having a function similar to real money.[4] "Under the revised banking bills, bank-holding companies will be able to acquire IT-related ventures. The move is aimed at promoting innovative

---

[2]https://www.forbes.com/sites/phillipkeys/2016/12/01/third-time-could-be-the-charm-for-silicon-valley-and-japan/#2c9fc81e6731
[3]https://phys.org/news/2015-05-prime-minister-japan-silicon-valley.html
[4]https://www.japantimes.co.jp/news/2016/03/04/business/tech/japan-oks-recognizing-virtual-currencies-similar-real-money/#.XEHwLs1S82w

financial services called "Fintech" and allowing banks to embark on new settlement systems and other businesses.

The bills also include measures for helping regional banking groups consolidate their fund and system management more easily after realignment, as well as improve their business efficiency." *The Japanese Times* reported. All these were after Mt. Gox Co. abruptly shut down in 2014 after its former CEO admitted it had lost around ¥48 billion in assets.

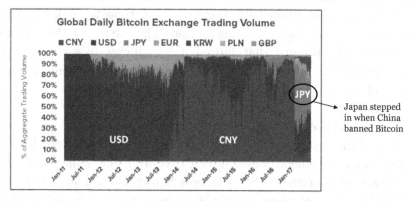

- Behind in digital payment (compared to China and Korea), so wants to leapfrog.
- In April 2016, Japan passed a law recognizing bitcoin as legal tender.
- In September 2017, FSA endorsed 11 firms for cryptocurrency exchanges. Retailers now accepting.
- Volume of bitcoin has skyrocketed. Japan replaced China.

Figure 4.1: Japan's big bet on Blockchain and Bitcoin: Macro.
Source: Coindesk.

Japan Bank Industry bet on the blockchain (Figure 4.1), and in October 2016, a 47-member consortium[5] (Mizuho, AEON, Nomura Trust, Resona) was formed to use blockchain technology from Google-backed Ripple to make payments. In March 2017, the consortium used Ripple for the cloud-based pilot — RC Cloud (real-time money transfers at low cost). Japanese were trying to reinvent the international payment system. Blockchain can reduce

---

[5]https://ripple.com/insights/forty-seven-japanese-banks-move-towards-comme rcial-phase-using-ripple/

transaction costs of remittances, corporate payments by 60% and 50%, respectively. Japanese Bankers Association (JBA), all 252 banks in the country (Figure 4.2), joined hands to develop common blockchain platform for money transfer as a possible upgrade for current payments clearing platform Zengin.[6]

Japan bank consortium
- October 2016. 47-member consortium (Mizuho, AEON, Nomura Trust, Resona).
- Blockchain technology from Google-backed Ripple to make payments.
- March 2017. Consortium used Ripple for cloud-based pilot — RC Cloud (real-time money transfers at low cost).
- Ripple: Blockchain can reduce transaction costs of remittances, corporate payments by 60% & 50%, respectively.

Japanese Bankers Association (JBA) — all 252 banks in the country
- Developing common blockchain platform for money transfer as possible upgrade for current payments clearing platform Zengin.

Figure 4.2: Japan bank industry bet on blockchain.

Japan's second-largest bank Mizuho is the most active in cryptocurrency and blockchain for financial services. It is planning to launch a Stable coin in 2019[7] (Figure 4.3).

Trade Finance
- July 2017. Mizuho completed a trade finance transaction from Japan to Australia via blockchain, covering all processes from issuing letters of credit to delivering trade documents.

Securities Transfer
- Starting from 2017, Mizuho is working on a project to utilize blockchain for securities transfer.

Digital Currency: J-Coin
- Mizuho joined a group of Japanese banks looking to launch their own digital currency called the J-Coin.
- Still in early stage, launch in time for Tokyo Olympics in 2020.
- Pegged with JPY.

Figure 4.3: Mizuho: Most active bank in blockchain.

---

[6]https://www.ccn.com/japan-banks-blockchain-trials/
[7]https://blockmanity.com/news/japanss-second-largest-bank-mizuho-to-launch-stable-coin-in-2019/

China Construction Bank is the most active in China for FinTech, as argued in Figure 4.4.

Internal Innovation
- Cloud platform to manage 10,000+ branches; improves efficiency, reduces cost.
- Slashed multi-times password to one-time password and signature.
- Big data analysis to provide SME lending.
- Reduced branch counter staff from 100,000 to 50,000
- Opened 1,300 new branches. Re-allocated 50,000 counter staffs to sales team.
- Building intelligent CCB: intelligent customer recognition, guide and processing.
- Physical robots in branches for customer services.
- Precise marketing through internet.
- Intelligent ATM with more functions, largely improving efficiency.

Credit rating based on "Technology Flow"
- CCB Guangdong launched credit rating system for tech startup loans.
- Issued 1,556 loans to tech startups which benefit from this new model.

Loan and Investment Combination
- Launched RMB 1 billion tech fund (together with Technology Financial Group).
- Equity and debt investment model to meet funding need from tech startups.

Figure 4.4:   CCB: Use, invest and lend to technology.
Source: sina.com, Red-Pulse, Schulte Research.

Realizing that the traditional way of selling insurance and banking products does not work in a digital environment, CCB started partnering with Ant Financial and IBM. Among other initiatives, direct selling was done on Ant Finance's platform. At the same time, it partners IBM to pursue new ways of selling insurance on the blockchain. The blockchain solution was developed using Hyperledger Fabric 1.0. Fabric 1.0 is the open-source production-ready blockchain software released by the Hyperledger project. The CCB project is a cross-industry consortium which sees IBM as a founding member[8] (Figure 4.5).

---

[8]https://www.ccn.com/china-construction-bank-partners-ibm-blockchain-banc assurance-solution/

Partnership with Ant Financial
- Tie-up with Ant in March 2017, leveraging big data tech, credit rating systems.
- Cooperation on credit, cross-border clearing, settlement, mobile payments.
- Cooperation in online–offline connection (channel advantage).
- Sharing the credit rating system.
- CCB's WMP directly selling on Ant Financial's platform.
- Mutual recognition of each QR code payment system.
- CCB participated in Ant Financial's $4.5 billion Series B funding in April 2016.

IBM Partnership: Blockchain-based insurance selling in banks
- HK partnered with IBM to develop blockchain platform to sell 3rd party insurance.
- Blockchain to share, record and allocate information among insurance, banks.
- Will increase efficiency and transparency with real time data and lower cost.
- Application in testing period based on IBM blockchain platform.

Figure 4.5: CCB: Partnership with Ant and IBM: Traditional way of selling insurance/banking products doesn't work!

In Singapore, DBS is among the three local banks that have started innovation laboratory to change the culture (Figure 4.6). Given the small market and financial inclusion for the ASEAN region is not on the agenda, the Singapore banks have very limited success in improving their bottom line or user base. DBS has been named the world's best digital bank by Euromoney. It remains to be seen whether the Singapore banks can challenge the FinTech giants when the Monetary Authority of Singapore relaxes the regulation for non-banks.

Fujitsu is also active in IoT and Blockchain working with Japanese banks on P2P money transfer, PT Telekom on smart cities, Tokyo Stock Exchange on trading platform, and Fukuoka Financial Group for Cloud and Mobile app fintech solutions (Figure 4.7).

Innovation Lab — DAX
- Created DBS Asia X (DAX) as innovation lab and co-working space for startups.
- Start-up engagement: Encourage FinTechsto partner with DBS staff.
- Startups encouraged to prototype ideas via 8 to 12-week programme.
- Regular API hackathons.
- Named world's best digital bank by Euromoney.

DBS Accelerator
- Collaboration with Nest, housed in "The Vault", Wan Chai.
- Digital Channel Experience
- Credit Digitization
- Customer Engagement
- Cryptocurrency
- Cybersecurity

Figure 4.6: DBS: Lab and accelerator. Leveraging Big Data, biometrics, and artificial intelligence to reinvent banking.

Source: DBS website, Schulte Research.

1. Working with Mizuho, SMFG, MUFJ on P2P money transfer with blockchain.

2. 2-year partnership with PT Telkom to deploy IoT for smart cities.

3. PalmSecure ID for Sberbank to process lunch payment in school.

4. Upgrade cash market trading platform for Tokyo Stock Exchange.

5. Fintech solution for Fukuoka Financial Group (Cloud and mobile app).

Figure 4.7: Fujitsu: Active in IoT and blockchain.

IBM is also working with various banks in partnership projects to take advantage of new technology for innovative business models (Figure 4.8).

IBM

Banks	Partnership Project
HSBC	Identify, digitize and extract key data for Trade/Receivables Finance (100 million pages of documents) with intelligent segmentation and text analytics
Mandiri	Analyze customer behavior with big data
CCB	Blockchain for selling insurance
Project Batavia (UBS, Montreal, Caixabank, Commerzbank)	Global system for trade system with blockchain to track goods and automatically release payments
BBVA, Danamon and NAB	Cross-border payments with blockchain
Crédit Mutuel Arkéa	360 degree view of customer documents across network with blockchain
Bank of Ayudhya	Streamline contract management process

Figure 4.8: IBM: Blockchain and data analysis: Doing groundbreaking work.

4.1.2 *Insurance*

Figures 4.9–4.13 show the advantages and disadvantages of insurance companies.

The advantages to strike back are summarized as follows:

1. massive customer base;
2. data advantage, especially health data;
3. the long history of established reputation;
4. hard to get new licenses;
5. cheap valuations.

The disadvantages are as follows:

1. still mostly offline through agents;
2. paper culture;

3. protected cartel means no incentive to change;
4. old products developed in the last century;
5. legacy systems: costly to change.

China Life is betting on Baidu partnership by establishing a Private Equity Fund of RMB 7 billion. There is also a strategic corporation with Baidu in building new platforms for innovative business models. Ping An, as mentioned in other chapters, is poised to become the most innovative fintech company to provide state-of-the-art banking and insurance services.

PE Fund with Baidu
- Established RMB 7 billion Internet Fund with Baidu (China life: 5.6 billion, Baidu: 1.4 billion).
- Launched in August 2017.
- Focus on mobile internet, AI, internet finance, consumer upgrading and internet+.
- Middle and late PE projects.
- Taking Advantage of Baidu's talents and resources in TMT sector.
- Culture China Life's own TMT investment team.

Cooperation with Baidu
- Strategic cooperation with Baidu on building financial platforms, robo-advisory, artificial intelligence, big data, cloud computing, and IoV.
- Robo-advisory: Analyze customer behavior and cash flow to detect user's financial need and risk preference, providing customized investment products.
- Credit rating: Credit rating based on big data analysis of user's search data and data from partners.

Figure 4.9: China Life (1): Bet on Baidu partnership?
Source: Xinhua, Red-Pulse, Schulte Research.

Company	Amount (USD million)	Description
Didi Chuxing	600	Largest ride-hailing firm in China
China Unicom	3,280	10% share in Unicom's SOE reform
BOCOM	NA	First bank with SOE reform
Guangfa Bank	3,480	43.7% shares bought from Citi; largest shareholder; complete financial ecosystem (insurance + bank + funds); channel advantage
Starwood Capital	2,000*	Enhance its oversea investment capacity

Figure 4.10:   China Life (2): Active investments.

Note: *Lead investor for a round total $2 billion.

Future Plan

Step	Target
1. Online	"internet+", taking advantage of internet
2. Intelligent	AI to achieve intelligent operation and services
3. Blockchain	100% digitalized with blockchain, big data, AI, IoT and Cloud

3 Intelligent Life Services

Sector	Project
1. Car	automatic car claims
2. Lifestyle	customer service robots, intelligent voice navigation
3. Finance	intelligent finance with Guangfa bank

Figure 4.11:   China Life (3): Future plan.

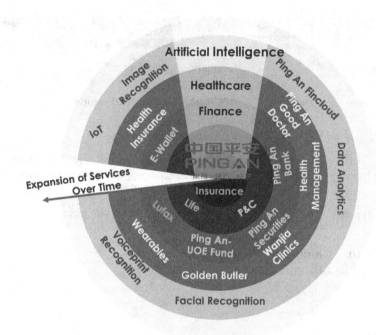

Figure 4.12: Ping An has made greater strides into becoming an autonomous financial ecosystem than any other company.

Figure 4.13: Ping An healthcare ecosystem: Connecting doctor, patient, pharmacy, clinic, hospital, bank.

Source: Ping An, Schulte Research Estimate.

4.1.3   Telecom

Figures 4.14–4.19 show the advantages and disadvantages of telecom companies to strike back. The advantages are as follows:

1. physical devices at home (routers) can be an entrance point for IoT;
2. brand awareness for the call and mobile internet data;
3. content provider for TV to gather lifestyle data on the massive user base;
4. huge cash piles.

The disadvantages are as follows:

1. margin pressure;
2. too big to change?
3. elephantine corporate culture;
4. fixed line business dead.

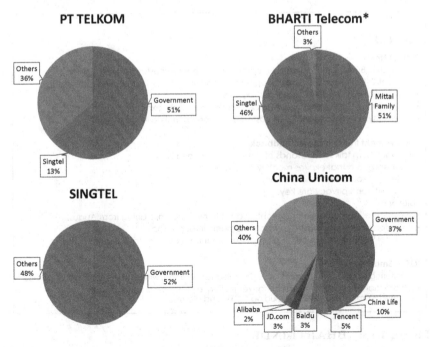

Figure 4.14:   Ownership: Singtel active in India and Indonesia.

Partnership	Description
NTT	Established ATII to focus on virtual infrastructure technologies. Three initial projects: 1) High Value-added Network Services 2) Server Platform Virtualization 3) Flexible Access Network Virtualization
Content	Partnership with Dreamworks, HBO, Viu, Singapore, Hooq, CatchPlay
Mediahub	Support content and advertising industry
Netflix	In talks with Netflix to bring Netflix to Indonesia
Fujitsu	Apply IoT to smart cities, healthcare, manufacturing, logistics
Indigo	Incubator to provide 30 startups with seed capital of $19,530 each, up to $150,000

Figure 4.15:　PT Telkom.

---

bharti

Project Next

In July 2017, Bharti Airtel launched Project Next to improve customer experience.
- Spent 2400 hours observing/mapping traffic flows in stores, redesigning Airtel stores.
- Concepts are: Share, Create and Experience.
- Created a "Social Wall" where all Airtel users can share experience with others.
- New MyAirtel App to make all customer service on mobile in real time.

Improvement from customer feedback

Feedback from talk to thousands of customers in innovation lab.
- Unused data that expires monthly is a waste.
- Need of social network function.
- Security and privacy are key.

Solutions
- Customer can carry unused monthly data to next month, starting from August.
- Customer can share data to family members through app.
- "Airtel Secure" Insurance: to protect customer from accidental/liquid damage.

IOT — Smart Home
- Entering smart home market as its broadband/fixed line/wireless are in millions of homes.
- Potential to connect with heating, ventilation, air conditioning, lighting, entertainment and safety including CCTV and alarms.

Figure 4.16:　BHARTI-INDI.

Source: ET Telecom.

Internal - Innovation	
Singtel Innov8	— Singtel's own venture capital fund, with USD 250 million AUM — Presence in Singapore, Silicon Valley, Tel Aviv and China — Network capability, next-gen devices, digital content, customer experience
IoT	Launched nationwide IoT network for enterprise customers across Singapore.

External - Partnership	
Alibaba	Cloud interconnection with Alibaba Cloud
Go Ignite	Alliances with Deutsche Telekom, Orange and Telefonica, launching global start-up programs. First winners are as Follows. — Blueliv for cyber security — Contiamo and Kentik under big data analytics — Platform.sh for content delivery — Streamdata.io for customer experience enhancement
SingCash	Partnership with hub providers in Myanmar to provide money transfer for Myanmar citizens in Singapore, with Singtel Dash app.

Figure 4.17: Singtel: Innovation through investments and partnerships with Deutsche Tel., Orange, Telefonica, China, India, Indonesia.

Source: Singtel website, Schulte Research.

SOE Reform — trial for the country
- First SOE for mixed-ownership reform.
- $11.7 billion share sale to bring in private strategic investors.
- Employee Incentive shares to increase efficiency.

Redundancy Cut
- Cut headquarters staff from 1,787 to 891 people, reducing 50% of personnel.
- Total departments were cut from 23 to 20.
- Provincial-level managerial branches were cut from 697 to 516, down 26%.

New Initiatives
- Set up Big Data subsidiary in September 2017 to analyze , to leverage 470 million users.
- China Unicom processes 600 billion internet records everyday.
- Data backed: credit investigation, client development and precise marketing.
- Cooperation with Alibaba for opening up Cloud computing resources.

Figure 4.18: China Unicom (1): Frontline of SOE reform — 600 billion data/day!

Source: China Daily, Red-Pulse, Schulte Research.

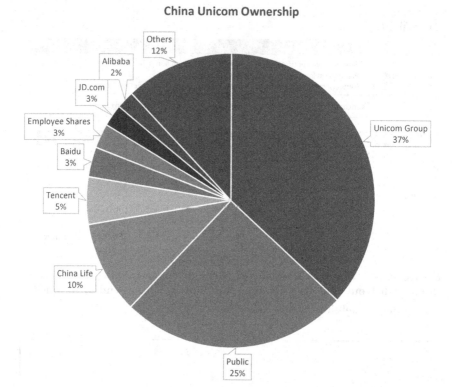

Figure 4.19: China Unicom (2): Ownership after SOE reform. Partnership with strategic investors for innovation.

Source: China Daily, Red-Pulse, Schulte Research.

For the telecom companies, it is interesting to observe that unlike the Chinese fintech companies, the Chinese telecom companies have been slow to react. This may partly be due to the slow reform. Singtel, on the other hand, has been initiating much innovation via partnerships and acquisitions. PT Telkom has been working on new projects with partners.

4.2 Conclusion

In conclusion, the battles are still on with banks, insurance, and telecom companies striking back. The top Chinese banks have been fairly successful given the huge and high net worth client base that they have. However, the profit margin of banks is fast eroding because of the excesses in lending to real estate and other less efficient corporates. A lot of the earnings from the innovative activities are not able to patch up the losses that are occurring of these excesses. This is a warning to many of the banks in the region that have huge exposure, directly or indirectly, to the real estate sector while enjoying good profit margin. The desire to lend to infrastructure projects in the Belt Road Initiative for long-term stable income stream may also be subject to positive cash flow from these projects that may not necessarily be a given fact. Of course, given the large numbers of the employees that these institutions are employing, it is right in some sense that they are too big to fail. However, this is only to the extent that the authority can afford the losses by printing or maintaining the monopolistic structure. The verdict is still out there, but the faith of most of the banks will in some sense be tied to the top of the pyramid customers that they have. On the other hand, the regional and smaller banks may have a better chance to grow if emerging technology is available to them, especially open source blockchain software that can assist them to scale not only locally but internationally aided by Big Data and AI. Given the large population of unbanked, the future looks brighter for those banks that can ride on emerging technology.

The top insurance companies, on the other hand, seems to be innovating along well with long-term capital to experiment. Many projects are incubated and have been listed on the exchanges to have a liquid exit, or having a reasonable valuation. As for the telecom companies, the capital expenditure and depreciation may be a drag as emerging technology continues to squeeze existing margin. Branching out to financial activities may be an option, but so far, it is proven to be challenging.

Collaboration, acquisition and joint venture may seem the easier road to innovation. Legacy issues regarding regulation, processes, previous investment, equipment and less technical workforce are hindrance to innovation. Mindset is another hurdle, but there are larger obstacles such as lack of expertise. Fiduciary duties will prevent a board without technical expertise to avoid innovation perceived to be "risky". Having stakeholders that chase short-term interest is another hindrance that these institutions faced. Financial strength and existing profitability are double-edged swords. They may sometimes have the opposite effects of becoming a hindrance when it comes to innovation. On the contrary, desperation may instill a sense of present danger and trigger the survival instinct.

The verdict is still out there, but the odds are at stake against the incumbents that are unable to change.

Chapter 5

Banking on the Transformation of Blockchain

5.1 Why Banks Lost and How to Fix It

No technology will thrive along and the next phase of accelerated disruption will come from the convergence of technology especially in the areas of AI, Blockchain, Cyber Security, Cloud, Data Technology, Drone, Devices (IoT), and Environment-friendly technology. In short, the convergence of ABCDE technologies. In any industrial revolution, there are winners and losers. Now, it seems that most of the incumbents, especially banks, are receiving the brunt of the disruptive forces. Banks and financial institutions, being blessed with good earnings and legacy regulation, have not been able to capitalize on their past successes. They will be facing severe tests going forward. However, being comfortable with growing profits from the Quantitative Easing and real estate loans have hindered the changes that are much needed. With difficulties in a mindset change to invest billions in technologies that will possibly not see any significant profits will be the death nail in the coffin. However, all is not lost as some changes in banks have shown some aspirations to become new FinTech and TechFin companies to provide banking and insurance services in a customer-focussed way.

Any institutions that can capitalize on convergence in one exponential technology will realize that taking advantage of the law of accelerating convergence will enable them to realize that they can augment the existing strategy to include other exponential technologies with much less effort. Disruptive forces are at their strongest at the intersections of emerging and scalable technologies. Once the mindset of an institution changes, it will be easy to imagine and take advantage of combinatory scenarios, amplify innovation and its disruptive power. These big opportunities emerge when new and scalable technologies are combined.

Web 3.0[1] is the convergence of AI, IoT, and Blockchain. AI cannot function alone without privacy and security protection by Blockchain, neither can it function alone without the dynamic and collection of real-time data via IoT. Similarly, Blockchain without a real-world application will not be able to define its maximum impact on the social and economic system. IoT that collects data will not be able to give intelligence monitoring and comfort from malicious elements without blockchain and AI.

This chapter will give an idea of how some of the banks are taking advantage of blockchain technology in anticipation of the fourth industrial revolution.

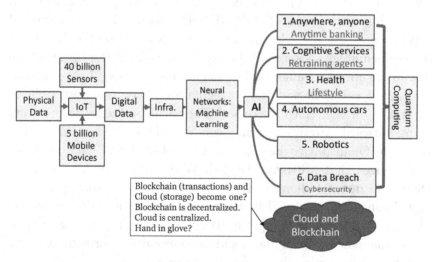

Figure 5.1:   Artificial intelligence described on a single chart, showing how blockchain is the foundation in BANKING and INSURANCE.

Source: Schulte Research Estimates.

---

[1]https://medium.com/@lbildoy/web-3-0-convergence-of-artificial-intelligence-iot-and-blockchains-ac3e72c3a1

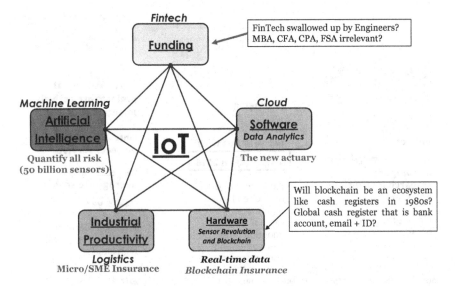

Figure 5.2: The whole ecosystem: CFAs, MBAs, CPAs and FSAS may become irrelevant: The engineers will take over.

Source: Schulte Research Estimates.

Four technologies of fourth Industrial Revolution will reshape banking:
1. Artificial Intelligence
2. Robotics process automation   4. Blockchain
3. Internet of Things

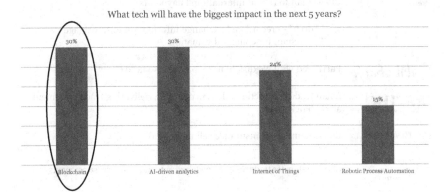

Figure 5.3: Fourth Industrial Revolution: IBM's Big Idea. The ultimate end game is blockchain.

Source: Money Live, IBM, Schulte Research.

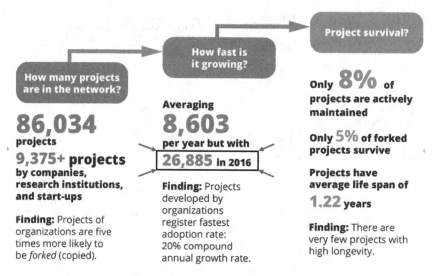

Source: Deloitte analysis of GH Torrent data and GitHub API data, as of October 12, 2017.

Deloitte Insights | deloitte.com/insights

Figure 5.4: Blockchain projects in Github: Astounding growth in technology but small "bucket shops" bound to fail (90% failure rate).
Source: Deloitte, Schulte Research Estimates.

Bank	Blockchain Pilot
Santander	Blockchain to record international payments
ABN·AMRO	1. "Torch" — record and exchange information for mortgage 2. Blockchain for collateral valuation
Rabobank	Partnered with Nexuslab, micropayments via IoT
MIZUHO	Asian trade transactions (letters of credit, delivering trade documents all on blockchain)
China Construction Bank	Blockchain-based insurance selling in banks
Project Batavia (UBS, Montreal, Caixabank, Commerzbank)	Global system for trade to track goods and automatically release payments

Figure 5.5: Case Study 1: What banks are doing.
Source: Money Live, IBM, Schulte Research.

Description	Goal
All personal info in one ID	"Once only" entry
Decentralized platform to record/view data	Individual controls access to private data
Being residents without visiting	Moe population; get best virtual talent
Foreigners logged in to use services	Bring in firms without physical presence
Paramedics access patient records	Full info of patients; pre-register tests
Voting online	Easy — vote anywhere
Cybersecurity training	Making economy more secure
Blockchain behind entire system	Protect & track all data changes
Connected smart device network	Cross-verification to improve security

Figure 5.6: Case Study 2: What countries are doing. What if you could build a blockchain city?

Source: The New Yorker, Schulte Research.

Function	Description	Goal
e-Estonia ID	All personal info in one ID	"Once only" entry
X-Road	Decentralized platform to record/view data	Individual controls access to private data
Individual E-residency	Being residents without visiting	Moe population; get best virtual talent
Firm Digital Residency	Foreigners logged in to use services	Bring in firms without physical presence
e-ambulance	Paramedics access patient records	Full info of patients; pre-register tests
i-voting	Voting online	Easy — vote anywhere
Locked Shields	Cyber security training	Making economy more secure
K.S.I. A Blockchain	Blockchain behind entire system	Protect and track all data changes
Mesh Network	Connected smart device network	Cross-verification to improve security

Figure 5.7: Estonia has done this already!

Source: The New Yorker, Schulte Research.

A. Customers have an absurd experience
1) HALF of UK customers: stop onboarding process during application
2) HALF of UK customers: I'd buy more services if paper wasn't needed

A. Banks have an absurd experience
1) HALF of banks: NO onboard digital channels
2) 20% of banks' onboarding: 24 hours++; Alibaba loan in 3 minutes.

Only 10% of banks can achieve full omni-channel integration!! Why?

Figure 5.8:   Why is bank so hard to change? Digital experience: Banks are falling behind by standing still.

Source: Money Live, Appway, Schulte Research.

1. **Precontemplation**: "I'm Fine."
Get information; point out need for change from many sources

2. **Contemplation**: "What if." think about change; what if; get an ideal and agree on it.

3. **Preparation**: "How." what kind, how, when, timeline, resources, support, regimen. Agree and move.

4. **Action**: "Let's go." Proposals, budgets, partnerships, affiliations, external help. Agree and move.

5. **Maintenance**: 'Milestones.', Disciplined quarterly assessment. Discussion of progress. Self assessment.

6. **Relapse**: "Failure". Must be without anger or bitterness. Move on. Gently point out need for change. Agree to move on.

1.Precontemplation
2. Contemplation
3. Preparation
4. Action
5. Maintenance
6. Relapse
Go back to 1.

Figure 5.9:   Six steps to change: How to change a person in denial is same as changing an organization in denial.

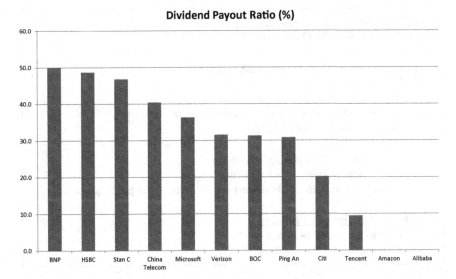

Figure 5.10: Dividend payout ratio: Banks and telecom have higher payout ratio.

Can banks survive:

1. If search engines undermine myth of cross-selling high margin products?

2. If information is easier to get; can customers get the best price and product?

3. If "AI assistants" accelerate this trend by using algorithms to comparison shop?

4. If Blockchain allows two-party funding of working capital?

All of these are happening now at an accelerated pace.

Figure 5.11: Banks are falling further behind by these four trends.
Source: Money Live, Infosys, Schulte Research.

---

1. One-stop money management
Make better decisions based on all financial information

2. Better advice based on behavior analysis
- "safe to spend" alerts
- AI-powered analysis of spending habits
- Real-time account data to avoid overdraft or charges
- Plan to achieve personal goals

3. Help clients get the best deal every day
- From simple banking to retail loyalty program
- Protect clients instead of penalizing their mistakes for profits (overdraft)

4. Make the customer's life easy
- Anticipate customer needs
- Partnership with lifestyle firms

Figure 5.12: The GOAL: New Customer Relationship that's transparent, fair, easy: Treat each person as if he is different! (They are).

Source: Money Live, Smart Communications, Schulte Research.

Fight with tech giants is unwise — they have this down to a fine science.
Can banks survive through:
- Launch of niche subsidiaries? (GS Marcus for HNW)
- Specialization in certain segments? (JPM Finn for millennials)
- Casting a wide net by experimenting with several startups?

Advantage of a niche subsidiary	Example
1. Cheaper than replacing outdated legacy systems in multiple jurisdictions	HSBC
2. Avoid resistance inside the existing organization	Stan C
3. Building new brand is easier than rescuing old brands with negative news	Deutsche
4. Dual branding strategies to serve different target customers	JPM
5. Rebuild a team with customer first company value	DBS
6. Less harm to parent company if pilot fails	Google Wallet

Figure 5.13: The solution is in front of our eyes: Digital only subsidiaries. How to achieve the goal: Ubiquitous, easy, open platforms with rewards!

Source: Money Live, Infosys, Schulte Research.

	Company	Niche subsidiary	Specialization
US	J.P.Morgan	finn by CHASE ○	Millennials
	Goldman Sachs	Marcus: BY GOLDMAN SACHS ™	Personal loans
China	Alibaba Group	网商银行 MYbank	Personal loans/SME
	Tencent 腾讯	WeBank 微众银行	Personal loans
	中国平安 PING AN 保险·银行·投资	陆金所 Lufax.com 中国平安集团成员	Wealth Management/P2P

Figure 5.14: Niche subsidiaries (1) — China and US.
Source: Money Live, Infosys, Schulte Research.

	Incumbent	Niche subsidiary	Specialization
Europe	UniCredit Group	buddybank	Mobile users
	Santander	openbank	Digital bank
	BNP PARIBAS	Hello bank!	Millennials
	ING	YOLT	App
Aus	nab	UBank backed by nab	Online/home loans

Figure 5.15: Niche subsidiaries (2) — Europe and Australia.

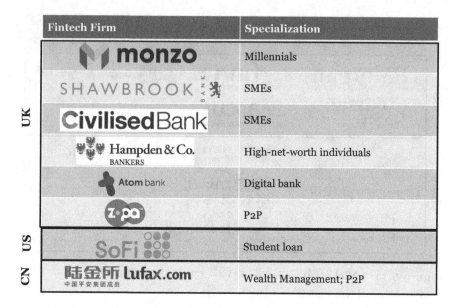

	Fintech Firm	Specialization
UK	monzo	Millennials
UK	SHAWBROOK BANK	SMEs
UK	CivilisedBank	SMEs
UK	Hampden & Co. BANKERS	High-net-worth individuals
UK	Atom bank	Digital bank
UK	zopa	P2P
US	SoFi	Student loan
CN	陆金所 Lufax.com 中国平安集团成员	Wealth Management; P2P

Figure 5.16: Specialization is also important for Fintech startups.
Source: Money Live, Infosys, Schulte Research.

Future Plan

Step	Target	Method
1. Online	"internet+", taking advantage of internet	Internal
2. Intelligent	AI to achieve intelligent operation and services	Internet Fund
3. Blockchain	100% digitalized with blockchain, big data, AI, IoT and Cloud	Baidu

Figure 5.17: Case Study 1: China Life: Three ways: 1. Big partnership; 2. Internal; 3. PE fund.

The choices of financial services for consumers are growing exponentially, and consumer expectations are higher with new fintech products:

Company	Description
◎ SIMPLE	Onboard users in 3 mins, "no fee" banking with smart budgeting
ᘔ STARLING BANK	1. Analytics of customer's spending habits. 2. Rewards for customer feedback 3. Evolves with more data
Atom bank	Reevoo: feedback platform to understand customers' concerns
monzo	Updates service daily based on behavior & feedback

Figure 5.18: Case Study 2: More choices and higher expectations.

Source: Money Live, Infosys, Schulte Research.

Many fintech firms originally aimed to compete with incumbent banks. Instead, they built partnerships due to the lack of scalability.

Fintech Firm	Collaboration	Bank Partnership
ᘔ TransferWise	Retail fx	Starling Bank
monzo	"Mobile first"	TD Bank
ezbob	SME lending: Esme Loans	RBS
XENOMORPH	Data analytics.	HSBC

Issues
1. Will tech giants (BABA, Tencent, G.A.F.A.M.) with vast customer bases, solid reputation, better technology cooperate with incumbents? That train left. NO. 2. After PSD2, will incumbents be relegated to "low-margin infrastructure provider", while tech giants take over high-margin distribution businesses? Likely.

Figure 5.19: Case Study 3: From competition to collaboration.

Source: Money Live, Infosys, Schulte Research.

> *"The unstoppable force of blockchain is barreling down on the entrenched, regulated and ossified infrastructure of modern finance. Their collision will shape the landscape of finance for decades to come".*
> — Don Tapscott, Blockchain Revolution
>
> Read Chapters 3 and 4 — they will blow you away!

Figure 5.20: Blockchain is the next challenge: Get ready.

Blockchain use cases in banking
- Instant and cost-effective payments
- Digital identity
- Trade finance
- Data access around large organizations
- KYC
- IoT-activated micropayments
- Compliance
- Fraud detection
- Insurance product selling

The most revolutionary use of blockchain is removing working capital from the bank and creating a bilateral trading relationship!

Figure 5.21: Blockchain — this is where the fourth Industrial Revolution is heading!

Source: Money Live, IBM, Schulte Research.

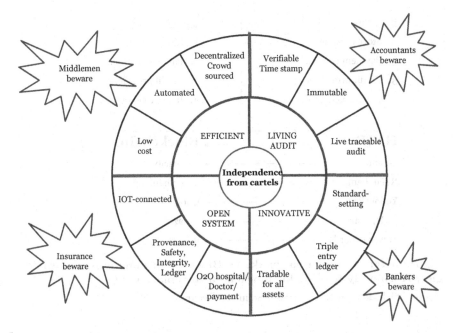

Figure 5.22: **How blockchain cuts through entrenched cartels: Turn your physical world into a digital reality ASAP.**

Source: "Blockchain Revolution", Schulte Research.

Function	Product	Industry at risk
Move, store, value stuff	bonds, deposits, goods, services	Retail Banking
Bet on future values	insurance, hedging, arbitrage	Derivatives Trading
Funding Assets	capital, dividends, rents	I Banking, Asset Management
Insurance	gauge risk of home, life, assets	Insurance
Verify ID and credit	secure system	Credit ratings, ID (government?)

Figure 5.23: **Blockchain can replace ALL financial players just as Amazon replaced ALL forms of retail. You can join this or fight it.**

Source: "Blockchain Revolution", Schulte Research.

Banks are:	Blockchain can be:
1. Store of value	1. Store of value
2. Payment company	2. Payment company
3. Credit providers	3. Credit providers
4. Income generators	4. Income generators

HTTP ——————————→ **Blockchain**

Steam ——————————→ Electricity

Retail banking ——————————→ Cyber banking

Yellow Pages ——————————→ Google

B&N, Sears ——————————→ Amazon

Tower Records ——————————→ itunes

Figure 5.24: Commercial banks vs Blockchain: Functions and analogies. Absolutely feasible for blockchain to replace banks.

Source: "Blockchain Revolution", Schulte Research.

Millennials	Homo Digitalis – 1980-2000	Product
1. Financial	1 less work more meaning	Mental health, holistic life
	2 little wage power	Higher Volume over lowerprice
	3 student loan burden	Debt relief insurance
2. Professional	1 pure mobility.No office, pc	Temporary plans
	2 institutional distrust	Digital subsidiaries are perfect
	3 job hopping "Gigs"	Partnership, collaboration
3. Personal	1 non-owners	Any time anywhere products
	2 quality of life	Lifestyle IS the product.
	3 pure digital in everything	ZERO paper
4. Worldview	1 Questions boundaries, structures	Seeking solutions in diverse apps
	2 No black/white – gray	LGBT, womens, Internat'l products
	3 Grew up w/ Terror: wiser, stronger	Tougher, grittier advertising

Figure 5.25: Boomers vs Millennials in banking: Technology is doing what it does its way because millennials are doing what they do.

Chapter 6

Blockchain Disrupts Http and SWIFT

6.1 Introduction

In 2016, Bangladesh Bank attackers hacked SWIFT software, stole US$81 million[1] and vanished without a trace. Belgium-based cooperative SWIFT maintains a messaging platform for banks remittances and over 3,000 organizations use the service. The "custom malware" Dridex has no impact on SWIFT's network or core messaging services, but it is a dent on a financial system that is far too centralized. It could have been worse if all of the 35 SWIFT requests were met and US$1 billion would have been out of the bank's account.

Nevertheless, officials say US$30 million of money weighing roughly 3,300 pounds disappeared from the Philippines' casino after the transfer of part of the funds to Philippines.[2] The Bangladesh Bank robbery is now known as the Bangladesh Bank cyberheist. The US has charged a North Korean computer programmer with hacking the Bangladesh Bank, and also in connection with two other global cyber attacks, the WannaCry 2.0 virus, the 2014 Sony Pictures attack.[3]

---

[1]https://www.bankinfosecurity.com/report-swift-hacked-by-bangladesh-bank-a ttackers-a-9061
[2]https://www.bankinfosecurity.com/report-swift-hacked-by-bangladesh-bank-a ttackers-a-9061
[3]https://news.abs-cbn.com/business/09/07/18/us-charges-north-korean-in-ban gladesh-central-bank-sony-hacks

Two years later in 2018, the cyber attacks are causing alarm with (i) stolen information on 500 million Starwood guest records of Marriott; (ii) a massive data breach at Google+ that exposes 50,000 users; (iii) 50 million Facebook accounts were compromised; (iv) 2 million customer personal data stolen in T-Mobile Hack; (v) 14 million records of Verizon subscribers were exposed; (vi) a quarter of the population (1.5 million) of Singapore data were stolen and many other incidents.[4]

The Economist magazine and others have also been raising alarms about digital dictatorship. There are digital experiments with a new form of social control by governments such as North Korea, China, Vietnam, and Russia. At the same time, the Facebook–Cambridge Analytica data scandal has raised the fear that Cambridge Analytica has used millions of people's Facebook profiles data for political purposes.

All these brought to mind two major issues. The first, is a digital dictator unavoidable? Second, is privacy dead? Will the fear of digital dictatorship, cybersecurity and privacy protection lead to a more decentralized financial system and Internet? Will decentralized system and blockchain offer more secured financial system and privacy protection?

6.2   Blockchain and Distributed Ledger

Let us try to understand blockchain in a non-technical way. Blockchain can be explained in many different ways according to the architectural design. Blockchain is cryptographically secured, but the data are not necessarily encrypted. Each block is "write-once and append-only" in computing language, and by this, we mean that the blockchain is written and added on to the previous block of data forming a chain of blocks. Blockchains are distributed to every node and completely or partially replicated. Blockchains are therefore distributed ledgers that consist of a chain of blocks of cryptographically signed transactions. Each block is formed by

---

[4]https://www.campussafetymagazine.com/technology/21-devastating-cyber-at tacks-2018/slideshow/11/

consensus among the nodes or participants and linked or chained to the previous block using cryptography techniques that efficiently summarized those transactions from the previous blocks in a single one-way asymmetric hash.

In plain language, all the nodes or participants are to verify the block of transactions through a consensus coded according to a set of predetermined rules of how the consensus is being formed. Once the block is formed using a variety of cryptographic techniques, it contains historical information of all the transactions in a hash. A hash function is any function that can be used to map data of arbitrary size to data of a fixed size. The values returned by a hash function are called hash values, hash codes, digests, or simply hashes. A cryptographic hash function is a special class of hash function that has certain properties which make it suitable for use in cryptography. It allows one to easily verify that some input data maps to a given hash value, but if the input data is unknown, it is deliberately difficult to reconstruct it (or any equivalent alternatives) by knowing the stored hash value.[5]

In its purest form, blockchain is a decentralized architecture that allows users to transact and creates secured transaction records that are costly to change or simply unchangeable.

A Merkle root is the "top hash" of a Merkle tree or "hash of all the hashes" of all the transactions that are part of a block in a blockchain network. A hash tree or Merkle tree is a tree in which every leaf node (e.g., L1, L2, L3, and L4 in the diagram) is labeled with the hash of a data block (L1, L2, L3, and L4), and every non-leaf node (Hash 0-0, 0-1, 0-2, and 0-4, as well as Hash 0 and Hash 1) is labelled with the cryptographic hash of the labels of its child nodes. "These hash trees allow efficient and secure verification of the contents of large data structures. A Merkle tree is recursively defined as a binary tree of hash lists where the parent node is the hash of its children, and the leaf nodes are hashes of the original data blocks"[6] (see Figures 6.1 and 6.2).

---

[5]https://en.wikipedia.org/wiki/Hash_function
[6]https://en.wikipedia.org/wiki/Merkle_tree

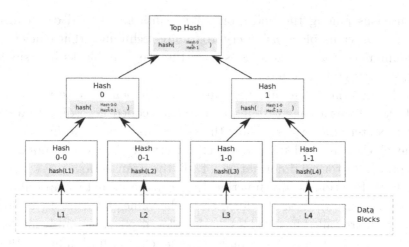

Figure 6.1: Hash trees.
Source: Wikipedia.

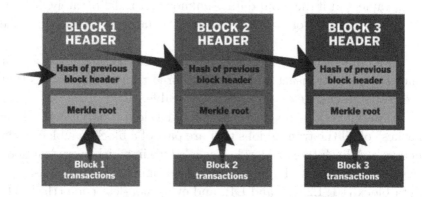

Figure 6.2: Simplified bitcoin blockchain.
Note: **With blockchain technology**, each page in a ledger of transactions forms a block. That block has an impact on the next block or page through cryptographic hashing. In other words, when a block is completed, it creates a unique secure code, which ties into the next page or block, creating a chain of blocks, or blockchain.
Source: Computer World and Network World.[7]

---

[7]https://www.computerworld.com/article/3191077/security/what-is-blockchain
-the-most-disruptive-tech-in-decades.html

It is important to note the difference between distributed database, distributed ledgers, and blockchains. A distributed database is a database that is replicated across multiple nodes or all nodes depending on the design and needs in order to have a consistent view of the database at any one state. The decision to update is never made by a centralized authority or master database unilaterally. Distributed ledgers, a subset of distributed databases, are designed to withstand malicious nodes. For example, with Byzantine fault-tolerant systems, certain percentages of malicious nodes are tolerated so that the database can still be synchronized. Blockchains, while sharing the same adversarial threat design, bundle transactions into blocks, broadcast data to all participants and cryptographically chained the secure blocks. Blockchains are a subset of distributed ledgers. Not all distributed database are distributed ledgers, and not all distributed ledgers are blockchains even though these terms are used interchangeably causing unnecessary confusion.

There are other confusions such as distributed and decentralized. Some define a distributed system as one in which processing is shared across multiple nodes, but the decisions may still be centralized while decentralization means that there is no single point of decision-making or attack or failure.[8] However, given that blockchain is a design architectural, these definitions will only cause more confusion. Under these definitions, a centralized system is a subset of the distributed system.

Others will view a distributed system as one that is the most decentralized in the sense that every node has independent decision capabilities with an almost equal endowment, such as computing power. A decentralized system will mean that there is no one single point of attack that will cause failure, but there may be multiple centers. In this sense, a distributed system is a more stable decentralized system than just a multi-center centralized system. Here, a distributed system is a subset of the decentralized system.

---

[8]https://medium.com/distributed-economy/what-is-the-difference-between-dec entralized-and-distributed-systems-f4190a5c6462

We may also make the differentiation between online and offline. Having a decentralized computing system is not inconsistent with offline centralized governance that may decide how the codes are written and how regulation can influence online behavior. Having a centralized system, on the other hand, does not mean that it is inconsistent with offline decentralization. In reality, there is no perfect distributed system. It is just a matter of design and the degree of decentralization.

There is much literature on blockchain, its basics and applications,[9],[10] but this chapter will focus on the blockchain implementation and whether the days of centralized systems such as SWIFT and HTTP are numbered. SWIFT[11] is a global member-owned cooperative and the world's leading provider of secure financial messaging services. More than 11,000 financial institutions are using SWIFT's messaging services in more than 200 countries and territories around the world. SWIFT enables secure and automated financial communication between users, and it plays the role of standardization. HTTP means Hyper Text Transfer Protocol and is the underlying protocol used by the World Wide Web. This protocol defines how messages are formatted and transmitted, and what actions Web servers and browsers should take in response to various commands.[12]

The value chain in Figure 6.3 is going to be substantially different in the future with decentralization, living audit, all assets tradeable, automation, and a secure network. Whether it is governments, financial service providers, or social enterprises, new technology will transform the way of doing businesses. As more data will be

---

[9]Lee, D. K. C.; R. Deng. (2017). *Handbook of Blockchain, Digital Finance, and Inclusion*, Vol. 1 and 2. San Diego, United States: Elsevier, Academic Press.

[10]Lee, D. K. C. (2015). *Handbook of Digital Currency*. San Diego, United States: Elsevier, Academic Press.

[11]https://www.swift.com/about-us

[12]https://www.webopedia.com/TERM/H/HTTP.html

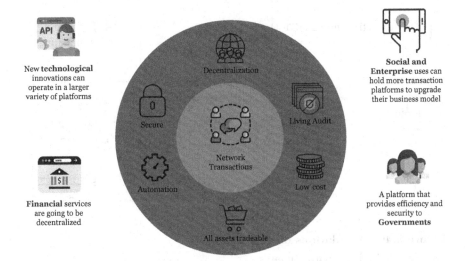

Figure 6.3: **Blockchain is improving every part of the value chain.**

collected via IoT and digital devices, blockchain will become an important base layer infrastructure to store the data securely with privacy protection. Beyond the base layer, we can store the app using decentralized technology, and we can form organizations that have no centralized control. Decentralized Application (Dapp) and Decentralized Autonomous Organization (DAO) are enablers for peer-to-peer value exchange without a middleman or centralized control. These apps and organizations will be very costly for anyone to hack, centralize the governance, and storage. As blockchain tokens are rewards to incentivize collaboration of parties that may distrust or are unknown to each other, peer-to-peer exchange of values via smart contract without third party may improve cost efficiency with new business models.

In this chapter, we will use the approach of the foundations of blockchain in eight categories, namely, Protocols, Layers (First, Second and Multi), Tokens, Computation, Finance, Enterprise, Social, and Government. We will focus on the uses of needs of each industry (see Figures 6.4 and 6.5).

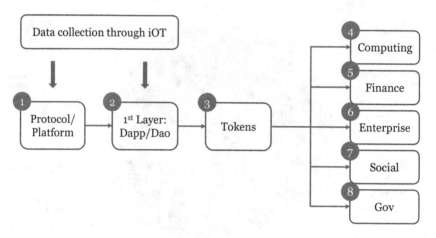

Figure 6.4:   Foundations of blockchain.

Source: Schulte Research Estimates.

Industry/ Technology	Uses and needs
1. Protocol	Block time, proof of stake, protocol flexibility
2. Layers	Transaction speed, API flexiblity
3. Tokens	Transaction speed, smart contract capability
4. Computation	Transaction speed, storage capacity
5. Finance	Transaction speed, governance cost reduction, asset ownership verification
6. Enterprise	Transaction speed, supply chain data transparency, admin cost reduction
7. Social	Application diversity, privacy/cybersecurity
8. Government	Cybersecurity, data management

Figure 6.5:   Applications and uses for each industry.

Source: Schulte Research.

6.3   Blockchain Protocols

Among the major players, Ethereum is the current leader with a large number of developers working and a great number of people watching the progress (Figure 6.6).

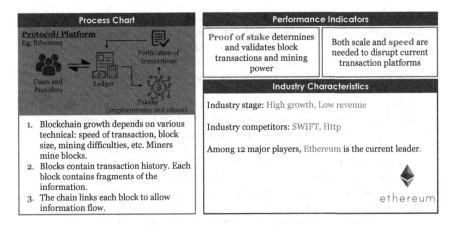

Figure 6.6: Blockchain protocol: The Ethereum Blockchain.

Source: Ethereum, Blockgeeks, Statista.

While there are many blockchain protocols with various degrees of decentralization, the table below shows a comparison among four protocols: EOS, Ripple, Ethereum and Bitcoin. The four chosen protocols have some underlying implications, which is the more decentralized and socially scalable, the slower the speed (Figure 6.7). Bitcoin, which is the most scalable socially, has the slowest speed among the four. Nick Szabo (2017) defined what is social scalability, and his view on Bitcoin's success is interesting[13]:

> *The secret to Bitcoin's success is certainly not its computational efficiency or its scalability in the consumption of resources. Specialized Bitcoin hardware is designed by highly paid experts to perform only one particular function — to repetitively solve a very specific and intentionally very expensive kind of computational puzzle. That puzzle is called a proof-of-work, because the sole output of the computation is just a proof that the computer did a costly computation. Bitcoin'spuzzle-solving hardware probably consumes in total over 500 megawatts of electricity. And that is not the only feature of Bitcoin that strikes an engineer or businessman who is focused on minimizing resource costs as highly quixotic. Rather than reduce its protocol messages to be as few as possible, each Bitcoin-running computer sprays the Internet with a redundantly large number of "inventory vector" packets to make very*

---

[13]http://unenumerated.blogspot.com/2017/02/money-blockchains-and-social-sc alability.html

sure that all messages get accurately through to as many other Bitcoin computers as possible. As a result, the Bitcoin blockchain cannot process as many transactions per second as a traditional payment network such as PayPal or Visa. Bitcoin offends the sensibilities of resource-conscious and performance-measure-maximizing engineers and businessmen alike.

Instead, the secret to Bitcoin's success is that its prolific resource consumption and poor computational scalability is buying something even more valuable: social scalability. Social scalability is the ability of an institution — a relationship or shared endeavor, in which multiple people repeatedly participate, and featuring customs, rules, or other features which constrain or motivate participants' behaviors — to overcome short-comings in human minds and in the motivating or constraining aspects of said institution that limit who or how many can successfully participate. Social scalability is about the ways and extents to which participants can think about and respond to institutions and fellow participants as the variety and numbers of participants in those institutions or relationships grow. It's about human limitations, not about technological limitations or physical resource constraints. There are separate engineering disciplines, such as computer science, for assessing the physical limitations of a technology itself, including the resource capacities needed for a technology to handle a greater number of users or a greater rate of use. Those engineering scalability disciplines are not, except by way of contrast with social scalability, the subject of this essay.

Protocol	Crypto Volume	Block Time	Transaction Speed Rank
Block.one (EOS)	828,948,913	500 ms	1
Ripple	39,244,312,603	3.5 seconds	2
Ethereum	99942221	15.3 seconds	3
Bitcoin	17081188	10 minutes	4

Figure 6.7:   Protocol comparison.
Source: CoinMarketCap.

Ethereum is an open blockchain platform that lets anyone build and use decentralized applications that run on blockchain technology.[14] The Ethereum Virtual Machine (EVM) is the runtime environment for smart contracts in Ethereum (Figure 6.8). A runtime environment is the execution environment provided to an application or software by the operating system. Smart contracts are programs that execute exactly as they are set up to by their creators or programmers. Computer scientist and cryptographer, Nick Szabo,[15] originally described the idea of a smart contract as a kind of digital vending machine. In his famous example, he described how users could input data or value, and receive a finite item from a machine, in this case, a real-world snack or a soft drink. Smart contracts, not necessarily a legal contract, allow the performance (execute and enforce online) of credible transactions without third parties. These transactions are trackable and irreversible because they are governed and enforced by computer codes.

Unlike simple apps, decentralized applications (dApps) are applications that run on a P2P network of computers rather than a single computer. Some have explained dApps as a "blockchain enabled" website,[16] whereas the smart contract is what allows it to connect to the blockchain. Instead of an API[17] connecting to a Database, a smart contract is connecting to a blockchain. Others have defined dApps as an application run by many users on a decentralized network with trustless protocols. Designed to avoid any single point of failure, they typically have tokens to reward users for providing computing power.[18]

Similar to Ethereum, EOS is a blockchain platform for the development of decentralized applications (dApps). Instead of verifying the state of the network at any given time, machines or nodes verify the series of events (transactions or simply messages) that have

---

[14]http://ethdocs.org/en/latest/introduction/what-is-ethereum.html
[15]http://www.fon.hum.uva.nl/rob/Courses/InformationInSpeech/CDROM/Literature/LOTwinterschool2006/szabo.best.vwh.net/smart.contracts.html
[16]https://blockchainhub.net/decentralized-applications-dapps/
[17] An application programming interface (API) is a set of routines, protocols, and tools for building software applications. An API is a software intermediary that allows two applications to talk to each other.
[18]https://en.wikipedia.org/wiki/Decentralized_application

occurred so far to keep track of network state. Simply put, the network can scale to one million transactions per second out of the gate on a single machine through "consensus over events", with theoretically infinite scaling possible in parallel between multiple machines.[19]

Ripple, on the other hand, does not have a blockchain. Ripple is open sourced and has own patented technology: the Ripple protocol consensus algorithm (RPCA).[20] It is highly centralized and its token XRP is premined with Ripple Labs owning 61%. Many have raised the issue that there is no utility value in XRP and required for other products that Ripple Labs has released. According to critics, Ripple runs contrary to the spirit of cryptocommunity which is to disintermediate, decentralize, democratize and also to ensure that transactions cannot be frozen!

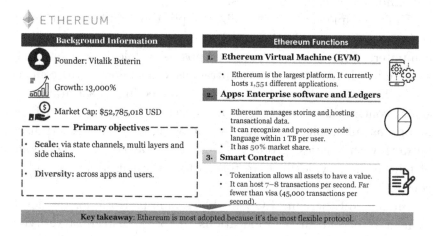

Figure 6.8: **The protocol snapshot 1: Ethereum (1).**

Source: Ethereum, Blockgeeks, Coindesk.

A Decentralized Autonomous Organization (DAO) is one of the most complex forms of a smart contract, where the rules of the decentralized organization are embedded into the code of the smart

---

[19]https://coincentral.com/what-is-eos/

[20]https://cointelegraph.com/ripple-101/what-is-ripple#who-are-ripple-labs-investors

contract.[21] There are solutions to scaling in Ethereum, but they are risks associated with each approach.[22]

"Orphan blocks are those blocks which are not accepted into the blockchain network due to a time lag in the acceptance of the block in question into the blockchain as compared to the other qualifying block. Orphan blocks are valid and verified blocks, but are rejected ones (Figure 6.9). They are also called detached blocks as they exist in isolation from the blockchain."[23]

"A hash tree or Merkle tree is a tree in which every leaf node is labelled with the hash of a data block and every non-leaf node is labelled with the cryptographic hash of the labels of its child nodes. Hash trees allow efficient and secure verification of the contents of large data structures."[24]

We will refer readers to other references for the discussion of Merkle Tree and Hash.[25]

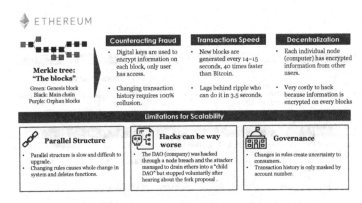

Figure 6.9: The protocol snapshot 1: Ethereum (2).

Source: Quora, Medium.

[21] https://blockchainhub.net/dao-decentralized-autonomous-organization/
[22] https://medium.com/coinmonks/scaling-solutions-on-ethereum-explained-d97 0b66e28e5
[23] https://www.investopedia.com/terms/o/orphan-block-cryptocurrency.asp
[24] https://en.wikipedia.org/wiki/Merkle_tree
[25] https://hackernoon.com/merkle-trees-181cb4bc30b4
   https://blockonomi.com/merkle-tree/

EOS (Figure 6.10) attempts to solve scalability with delegated proof-of-stake (DPoS). DPOS is a type of governance model for validating block units in a blockchain system to prevent network attacks and faulty transaction recording. Twenty-one parties are pre-selected by EOS as delegated nodes and users are asked to vote on "witnesses" to confirm blocks that are produced by these super delegated nodes. Supernodes are assigned mining power based on the amount of cryptocurrency they hold in DPOS, unlike PoW that forces network participants to expend computing resources by mining randomly generated cryptographic hashes to verify blocks.[26]

PoS and DPoS have been widely criticized for making the rich richer, enabling subjectivity and centralization of network decision-making. "Staking model skeptics caution that PoS and DPoS super-majorities can band together to unduly influence collective interests, making validation slower and less neutral than in PoW blockchains."[27,28]

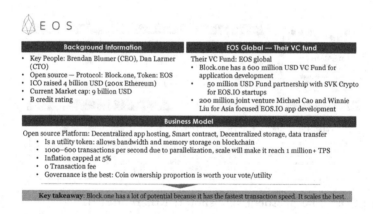

Figure 6.10: The protocol snapshot 2: Block.one (EOS).
Source: EOS.IO, Steemit, Quora.

---

[26]https://www.coindesk.com/early-execs-leave-block-one-the-peter-thiel-backed-crypto-startup-behind-eos

[27]https://medium.com/@matteoleibowitz/eos-dont-believe-the-hype-c472b821e4bf

[28]https://www.coindesk.com/early-execs-leave-block-one-the-peter-thiel-backed-crypto-startup-behind-eos

Figures 6.11–6.58 are self-explanatory. Figures 6.59 and 6.60 are included in the appendix.

6.4 Blockchain Layer

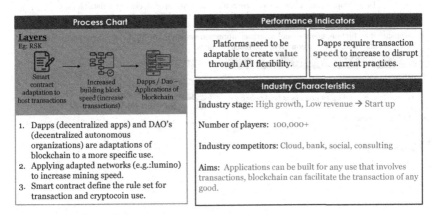

Figure 6.11: Blockchain's 1st layer: (DAPP and DAO) — Improve speed.

Source: Blockchainhub.

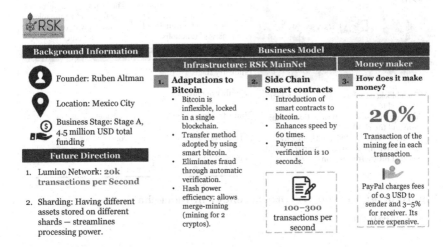

Figure 6.12: The layer snapshot: Rootstock (RSK).

Source: RSK.

Has a patent for the "data block", its to allow data on the blockchain to be accessed through one block	Has partnered with: 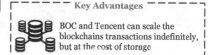

Business Model

Aims to speed up transaction time
- Before, a new block stores all transactions of the previous.
  - Now: Data compressing system packs transactions from multiple blocks into a data block.
- Data blocks form per transaction request and will be hosted on a different storage system.
  - Data is run through a hash function with a hash value.
- To ensure it is still tamper-proof and secure: compression system in the blockchain records the relationship between data block, compression event and blocks involved.

Business Model	**Key Advantages**
Benefits: • Less data stored on each block • Transaction is sped up as process is held only on one block • Mining time increases	BOC and Tencent can scale the blockchains transactions indefinitely, but at the cost of storage

Figure 6.13: Data storage — BOC.

Source: Coindesk.

6.5 Blockchain in Cryptocurrency

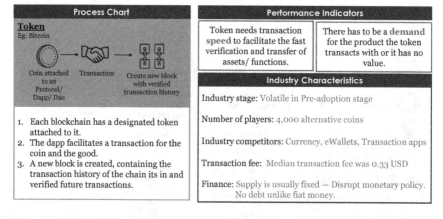

Figure 6.14: Token (cryptocurrency) in blockchain.

Source: Medium.

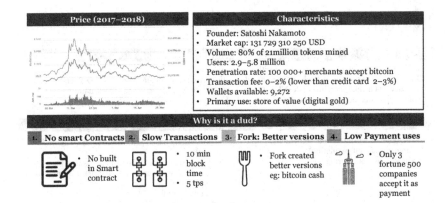

Figure 6.15: Snapshot: Bitcoin.

Source: Coinmarketcap.

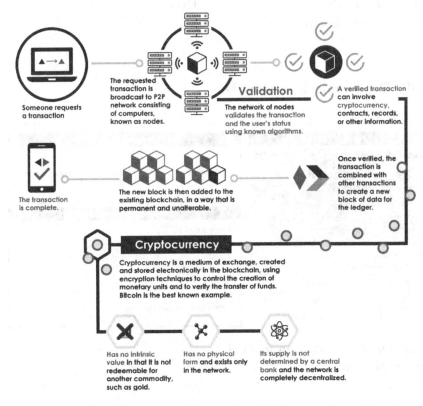

Figure 6.16: A cryptocurrency blockchain: The way it works.

Source: Solartribune.

Background: Utility settlement coin project

- Blockchain based digital cash partnered with Clearmatics
- USC will have variations → Each attached to a domestic currency → Collateral with cash → Spending currency will be like spending cash → Used to trade traditional financial assets via blockchain
- Not a cryptocurrency, it's the first digital cash backed by an financial asset

Benefits of trading on Blockchain	Aims
• Instantaneous within minute settlements • Tamper-proof (self-regulating) • No more need for clearing house	• Blockchain based platform to issue by the end of 2018 • Bonds — Smart bond (eliminating need for governance, lower operational risk, instantaneous settlement) • Equities trading

Key takeaway: USC eliminates the need for slow traditional banking processes while ensuring its own self-regulation (cutting costs) and secure transactions.

Figure 6.17:   Utility settlement coin: UBS.

Source: Coinmarketcap.

6.6   Blockchain in Computing

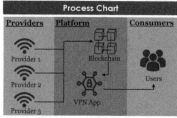

Process Chart	Performance Indicators
Providers Platform Consumers	Industry relies on speed, price and reliability to take on larger providers.

How it works?

1. People who have extra bandwidth (WIFI network) become providers.
2. Blockchain platform connects the users.
3. App facilitates the transaction of coin — network bandwidth, this is paid according to the amount of network used.
4. User receives vpn service cheaper and faster.

Industry Characteristics

Industry stage: Introductory

Characteristic: Revenue making (tapping into a new source of bandwidth network) but can be cost reduction and used as a bandwidth distribution tool.

Industry competitors: VPN providers

How will it increase capacity: More providers will increase the network available. Enhancing layer increases speed.

Figure 6.18:   Blockchain in VPN.

Source: Mysterium.

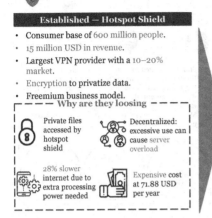

Established — Hotspot Shield

- Consumer base of 600 million people.
- 15 million USD in revenue.
- Largest VPN provider with a 10–20% market.
- Encryption to privatize data.
- Freemium business model.

Why are they loosing

Private files accessed by hotspot shield

Decentralized: excessive use can cause server overload

28% slower internet due to extra processing power needed

Expensive cost at 71.88 USD per year

Disruptor — Mysterium Network

Founder: Robertas Visinkis

Location: Switzerland

Seed A: 14 million USD

Creating a VPN Sharing Economy

- Open source, decentralized VPN software allowing users to be both the providers and the users, hosted by Ethereum.
- Each provider allows users to use their extra unused network, connected, recorded and payed through the blockchain.
- Token called MYST.

How do they make money

1. 1–3% transaction fee for MYST currency (variable to amount) is taken from when a provider shares his internet and is connected through their blockchain platform.
2. 0–2% smart contract fee for withdrawing money

Figure 6.19: Snapshot: VPN.

Source: Mysterium, Hotspot Shield.

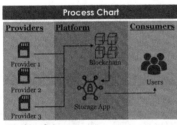

Process Chart

Providers | Platform | Consumers

Provider 1

Provider 2

Provider 3

Blockchain

Storage App

Users

How it works?

1. People who have extra storage space on their computer (memory) become providers.
2. Blockchain platform connects the users.
3. App facilitates the transaction of coin — memory is paid according to the amount used.
4. User receives storage service cheaper, faster, encrypted and can access it anytime.

Performance Indicators

Industry relies on speed, security and reliability to take on corporate cloud providers. Its cheaper for users because users pay per kb stored, there is no excess charge and wasted space.

Industry Characteristics

Industry stage: Introductory

Characteristic: Revenue making (tapping into a new source of memory storage) but can be cost reduction if used for memory distribution or maximizing memory usage capacity utilization.

Industry competitors: Cloud providers

How will it increase capacity: More providers will increase the storage available. It needs to expand blocks to increase transaction speed to hold more users.

Figure 6.20: Blockchain in file storage.

Source: Stori.

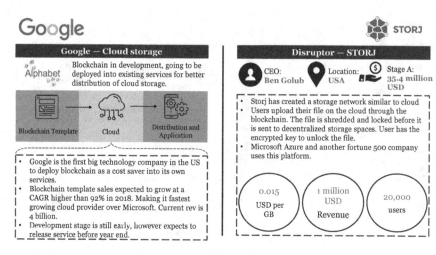

Figure 6.21: Snapshot: File storage.

Source: Google, Stori.

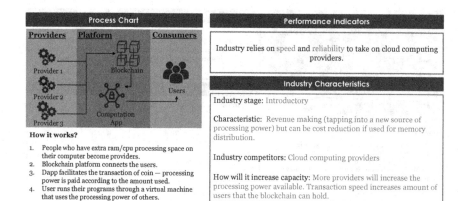

Figure 6.22: Blockchain in computation.

Source: Elastic, Medium.

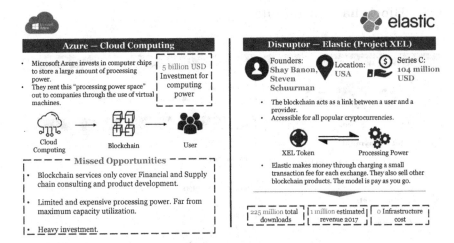

Figure 6.23: Snapshot: Computation.

Source: Microsoft, Elastic, Project EXL.

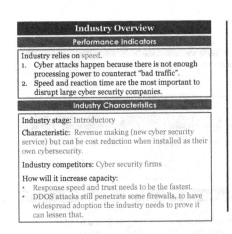

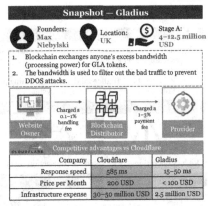

Figure 6.24: DDOS cyber security.

Source: Gladius.

6.7   Blockchain in Finance

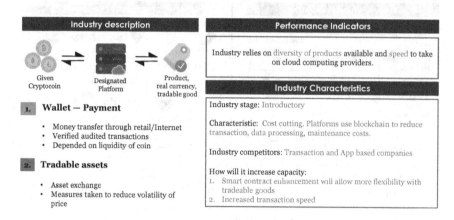

Figure 6.25:   Blockchain in platform technology.

Source: Quoine, Tech Crunch.

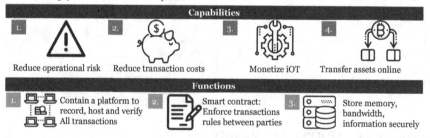

Figure 6.26:   Distributed ledger technology: Foundation of blockchain financial services.

Source: DHL, Schulte Research Estimates.

- Lack of standards and secure medium for information exchange.

- Trade data and assets are concealed, making valuable data hard to get access to and verify. Risk of fraud and need for excessive compliance and auditing.

- Manual, costly, time-heavy, error-prone and redundant processes.

- Aging technology makes client integration slow, complex and costly. Solutions are inflexible and difficult to improve.

- Large pool of trade assets not accessible.

- Tedious and long process to originate trade assets.

Figure 6.27: DLT solves these problems in finance.

Source: DHL, TradeEIX.

6.8 Applications for Financial Services

coinbase

Benefits of trading on Blockchain	Typically, crypto trading is expensive
• Founder: Brian Armstrong and Asiff Hirji • Cryptocurrency tradable: 150+ billion • 20 million customers • 32 countries supported • Revenue: 1 billion USD • Funding (total 217 million USD): 2015 75 million USD, 2014 series A 5 million, 2013 Venture capital firms 25 million	Bitcoin, Ethereum, Litecoin, Bitcoin Cash, Ripple Median Transaction Fee historical chart

Business Model

- Problem: Customer was not able to withdraw money right away (ACH bank transfer system that takes 3–5 working days).
- Products: Broker (bitcoin, bitcoin cash, Ethereum, Litecoin), Opensource development, Cryptocurrency wallet.
- Software: "Segwit" lowers transaction fees by 20% and speeds up network by increasing blocksize.
- Transaction fee: Low (0.25%, market average is 1.5%).
- Nine patents in bitcoin exchange, hot wallets for bitcoin, tips buttons.
- Market share: 17%, behind bitfinex which has 38%.

Key takeaway: Decreases costs but does not speed up transactions.

Figure 6.28: Brokerage — Coinbase.

Source: Coinbase, bitcoins.

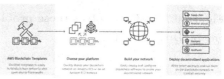

AWS Blockchain financial services: **ECS** and **EC2 instance**
Current Service Description

- Price range: 0.005–5.5 USD per hour. Bills are charged per second.
- Product: Renting virtual cloud space for blockchain projects and running systems.
  - Runs (by cutting intermediary): Clearing, Settlement, Cross-border transactions and other financial transactions.
- Protocol use: Ethereum (completely open-source) and Hyperledger Fabric (more centralized control).
- Customers: T Mobile, PWC, Intel, Delloite and Guidewire.
- Partners: Kaleido, Sawtooth, Corda R3, PokitDok, Quorom, Blockapps, ShoCard, Luxoft.
- Future aim: 5 years projection to have 70 million banking customers (by Bain & co).

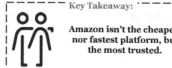

Key Takeaway:
Amazon isn't the cheapest nor fastest platform, but the most trusted.

Figure 6.29: Financial service — Amazon.

Source: AWS, CNN.

ripple

What is a Distributed Ledger Technology (DLT)

- Founder: Arthur Britto, David Schwartz, Ryan Fugger
- Product: Money transfer, currency exchange, remittance
- Total funding: 93.6 million USD
- Stage: Beta testing (post seed A round)
- Customers: 75 financial institutions

What is a Distributed Ledger Technology (DLT)

- Transaction speed: 3–5 seconds (bank transfer is 3–5 working days)
- Handle 1500 transactions per second (Fastest blockchain)
- Charges 1/1000 ripple (around 0.001 cents) per transaction

Current Obstacles

- Currently ripple has a limited supply available, stages in selling will cause volatile prices. Banks may also get currency for free.
- No established dedicated function and use yet.

Key takeaway: Decreases costs and speeds up transactions. Best for this. But comes with most risks.

Figure 6.30: Money transfer — Ripple.

Source: Ripple, Coindesk, Android authority, Cryptovest.

BANK OF CHINA

BOC Hong Kong using blockchain for 85% of its real estate appraisal

Description

Has 150 dedicated staff working on the application

Blockchain applications
- Property valuations can be done in seconds, previously took days
  - Appraisers issue certificates on the blockchain and can be viewed transparently
- Conduct real estate deals on the blockchain
  - Cheaper process, involves less governance

Future expansions
- Trade Finance, ID management, cross-border payments

Key takeaway: Real estate processing is much faster for clients and less need for governance.

Figure 6.31: Real estate — BOC.

Source: BOC, Cryptovest.

About		Products	
Private P2P payment on blockchain platform: • 29 countries in NA and EU, expanding to Asia • Acquired Poloniex exchange for 400 million USD, which has millions of customers • Handles 4 billion USD worth of crypto exchange per month • 400 employees • Circle trade earned 60 million USD rev in 2017 • Total rev: 250 million USD • Location: USA, Boston • Founders: Jeremy Allaire, Sean Neville • Funding 246 million USD		1. Circle invest	• Brokerage of 7 cryptocurrencies • Created a cryptoindex
		2. Circle Pay	• P2p money transfer system without transaction fee • Instant settlement • FX rate for: USD, Euro, GBP
		3. Circle Trade	• Trade volume: 2 billion USD • Cryptotrading fund, minimum investment of 250k USD
┌─ Key Takeaway: ─┐ Circle offers a range of financial services that are cheaper than traditional ones. They have more flexibility and have faster transaction speed.		4. Poloniex	• API for open source trading algos • Cold storage for deposits • Access to new tokens and blockchain tech

Figure 6.32: P2P payment — Circle.

Source: Circle, Schulte Research.

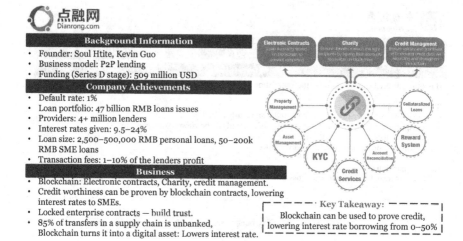

Background Information

- Founder: Soul Htite, Kevin Guo
- Business model: P2P lending
- Funding (Series D stage): 509 million USD

Company Achievements

- Default rate: 1%
- Loan portfolio: 47 billion RMB loans issues
- Providers: 4+ million lenders
- Interest rates given: 9.5–24%
- Loan size: 2,500–500,000 RMB personal loans, 50–200k RMB SME loans
- Transaction fees: 1–10% of the lenders profit

Business

- Blockchain: Electronic contracts, Charity, credit management.
- Credit worthiness can be proven by blockchain contracts, lowering interest rates to SMEs.
- Locked enterprise contracts — build trust.
- 85% of transfers in a supply chain is unbanked, Blockchain turns it into a digital asset: Lowers interest rate.

Key Takeaway:
Blockchain can be used to prove credit, lowering interest rate borrowing from 0–50%

Figure 6.33:   Microfinancing (P2P lending) — Dianrong.

Source: Reuters, Coindesk, Dianrone, Blockchain innovation conference.

Patent application for credit score generation on the blockchain

Application Description

- Credit ratings can be built through historical browsing history and other predictive data.
  - Using more than just credit ratings, it can include social media, browsing history, etc.
- Create an immutable record.
- Transparent credit calculator ratings.
- Allow users to improve their score quicker.
- Transactions will be carried out through smart contract automatically with no human interaction.

Key takeaway: Blockchain can process credit ratings faster, cheaper, and better.

Figure 6.34:   Credit — RBC.

Source: Coindesk, RBC.

DTCC

Company Financial Performance	Functions and Platform
Revenue: 960 million USD	Blockchain version that can host derivatives trading worth 11 Trillion for credit • Platform for credit default swaps reporting • Upgrading current platform (TIW) • Digital settlements
AUM 3.5 billion USD	• Partnered with: IBM, r3, Axoni • Protocol from Ethereum

Aims

- Streamlining governance and risk management
- Clearing derivatives in seconds (before it could take weeks)
- Become a network operator instead of a data warehouse
- Apply this to all asset classes

Key takeaway: DTCC is using blockchain to make governance cheaper and faster than existing platforms.

Figure 6.35: Derivatives platform and trading governance — DTCC.

Source: Reuters.

ICBC 中国工商银行

Patent filed for blockchain system to improve efficiency of certificate of issuance

Process
1. Certificate issuers matches with a users credentials digitally.
2. Data will be encrypted and moved into the blockchain where all nodes will be updated.
3. Entities that require the certificate approval will be activated.
4. Selective decryption with user credentials occurs to see necessary documents to streamline flow.

Benefits
- Eliminates counterfeit issues
- Makes process automated and efficient
- Automatically stores data

Key takeaway: Blockchain improves verification speed and document security.

Figure 6.36: Commercial banking — ICBC.

Source: Bitrazzi, Coinbase.

QUOIN≡

Background Information	Business Model	
Founders: Mario Gomez Lozada, Mike Kayamori Location: Singapore Business Stage: Stage A, 123 million USD total funding	**Aim**	1. Aims to provide **liquidity** to all cryptocurrencies by being a **regulated** exchange. 2. Gives **individual investors** the access to trade any financial product.
	Product 1: World Book	1. **Smart order routing**: Utilizes blockchain to provide low latency and real-time feeds. 2. **Matching engine**: A enhanced DAO on blockchain capable of hosting 1 million + transactions per second.
Performance Indicators 1. 12 billion USD transactions per year 2. One of few blockchain companies audited by an accountant firm(Deloitte) 3. 30+ available cryptocurrencies 4. Market cap: 284 million USD 5. Revenue: 3 million USD	**Product 2: Prime Brokerage**	1. **Credit facilities**: Liquid blockchain platform is used to lend currency, using crypto or fiat currency as collateral. 2. **Fiat management**: Leveraging an extensive relationship with multiple banks to offer liquidity to platform users.
	How they generate revenue	1. **QASH**: The cryptotoken that will be used for all financial services. 2. **Fees**: Trading transaction fee + Currency conversion fee.

Figure 6.37: Trading: Quoine (liquid).

Source: Liquid, Medium, Quoine.

IBM ❀ UBS

Project details
• Global Finance Platform based on Blockchain build on IBM Blockchain Platform • Bank participants: UBS, BMO, CaixaBank, Commerzbank, Erste Group • Stage: Pre-launch stage

What is it?
- Transaction platform for companies to do global trade on any assets.
- Automatically records transactions.
- Smart contracts ensure payment from buyer (eliminate need for FOB terms).

Aims
- Security: Open account transactions lead to uncertain payments from buyers, this is eliminated.
- Support multiparty and cross-border trading networks.
- Speed: letter of credit from bank is no longer needed.
    - 7-day transaction is turned into a 1 hour.

Key Takeaway:
Assets can be settled and exchanged at a faster rate, without credit from a bank needed.

Figure 6.38: Trade finance: Project Batavia — IBM and UBS.

Source: IBM, Supply Chain Dive.

AIA: Bancassurance	Ping An: R3
Platform Hyperledger Fabric • Sharing policy data → speed up transactions • Encryption → only documents needed for process can be viewed • Transparency → Users can view who sees their documents	• Ping an is Chinese first member of R3 consortium (research into blockchain applications in finance) • Collaboration of blockchain research with 200+ other firms

AXA	Zhong An
Platform Ethereum • Smart contract connected to global air traffic control database → 2-hour delay means automatic reimbursing • Cuts insurance paperwork → no wasted time • Fast transaction → Notified transaction minutes after landing (before 1–3 days)	Zhong An: Shanghai Blockchain Enterprise development alliance Anlink platform • Facilitate all insurance transactions • More secure than others → own Anlink platform • Smart contracts → Faster transactions • 20+ partners

Key takeaway: Zhong An has the most developed (fast) and integrated blockchain platform.

Figure 6.39: Insurance industry has a lot of activity: Zhong An winner.

Source: AIA, Blockchain, Ping An, Cointelegraph, The Digital Insurer.

PWC is the first big 4 firm to have develop an blockchain based accounting system. It is cheaper, more accurate and faster than traditional auditing practices.

Product	Uses

• Transaction platform and control/ risk mitigation
  • Payment process
  • Verify transactions
	• Online identification
	• Supply Chain auditing and verification
	• Transactions auditing
	  • Logs all transactions on blockchain
	  • Developed testing controls and measures for companies to monitor their own transactions

Figure 6.40: Accounting — PWC.

Source: PWC.

About	Vaultchain Platform
• Founders: Fraser Buchan, Matthew Trudeau, Mike Haughton • Location: USA, New York • Funding: 9 million USD • Stage: Venture round • Partners: Royal Canadian Mint will provide storage platform for physical gold	Tradewind has developed a platform called Vaultchain that facilitates gold transactions • Speeds up transaction • Reduces transaction cost • Increase liquidity of gold

Business Model
• Reduce transaction fee: Gold bars have 3–5%, GLD 1.5%, Vaultchain 0.15%. • Ownership: Tradewind offers the physical gold bar option to own the good directly and have it delivered.

Key takeaway: Investors can own the goal physically, have cheaper transaction fees and faster transaction processes.

Figure 6.41:   Gold transaction — Tradewind.

Source: Bloomberg, Tradewind, Crunchbase.

6.9   Blockchain for Enterprise

Problem:	High cost, slow supply chain transactions, unsure payment timings.
Blockchain Solution:	Reduction of cost, faster processes due to less governance.

Description

Supply management for Samsung's global supply network
First company to integrate it into daily operations. This is particularly for shipping.

Use for:
• Processing time cut.
• Cargo tracking, condition of goods can be tracked on the blockchain.
• Eliminate document fees for shipping (costs on average double compared to other transports).
• Capable of managing 488,000 tonnes of air cargo and more than a million containers .
• Launch to shipping time reduced by more than half.
   • Easy to respond to shifting demands.

KPI

Reduce shipping cost by 20%

Figure 6.42:   Supply chain.

Source: Supply Chain Dive, IBM, Azure, Asia Times.

Microsoft
Azure

Problem:	Unreliable suppliers with no trace of accountability.
Blockchain Solution:	iOT sensors to monitor real-time data about transport condition.

Description

Azure has a blockchain consulting/development service for supply chains.

Current industry problems: 69% don't have visible supply chains, 65% experience supply chain disruption, 41% use excel to track supply chain movement.

Microsoft blockchain covers whole supply chain process
* Substandard goods: 1 in 10 medicines transported are substandard, blockchain stores all production data so its visible to consumers, can reduce 1 million deaths per year.
* Transport monitoring: Microsoft has iOT sensors to monitor goods and store real-time data of transport conditions. Aims to reduce 200 million tones of food from being spoilt.
* Smart contract and iOT to increase accountability of, reducing 85% late deliveries. 10% of freights contain inaccurate, smart contract makes supplier liable.

Figure 6.43: Food/medical distribution.

Source: Supply Chain Dive, IBM, Azure, Asia Times.

WWF

Problem:	500 million workforce fishing industry, 30% of worlds fish come from over fishing illegally. Economy has lost 23 billion USD and is supported by slave labor.
Blockchain Solution:	Tracking when, where and how the fish was caught for retailers and consumers to see.

Description

WFF has partnered with Consensys and SeaQuest Fiji to create a blockchain application to track the buying and selling of fish across the whole supply chain.
* Distributors and retailers can see where the fish comes from as each order has to be logged into the blockchain.
  * Information is tamper-proof: it can view what vessel caught it, what was the fishing method.
  * Aims to push out slave labor by making each transaction in the whole supply chain transparent.
* Shoppers can view where each fish comes from

| Supplier | Transport | Retailer | Consumer |

Figure 6.44: Food security and legitimacy.

Source: Invest In Blockchain.

Problem:	Music industry is dominated by record labels who take 15–20% of the earnings.
Blockchain Solution:	No middleman, any artist can publish their own songs on the blockchain and have a record of how many times it was played and include a streaming payment system.

About	Description

• Stage: Beta testing
• Team: 13 people
• Business model: Publishing fee of 10$ USD + Small transaction fee of <1%
• Customers: 1,400 users
• Target customers: Middle-class artist who want to make 50–80k USD a year
 	
┌ ─ ─ ─ ─ Key Takeaway: ─ ─ ─ ─ ┐
⎪ Artists get more credit over record label ⎪
⎪ cartels. ⎪
└ ─ ─ ─ ─ ─ ─ ─ ─ ─ ─ ─ ─ ─ ┘
	• UJO Music: Cutting a 300-day process of getting discovered
	• Immogen Heap: Supply chain management in music to usher payments.
	• RAC: Individual artists can license Copyright music and publish music on the Ethereum protocol, adapted to be an app for viewers to listen to music on.
	  • Automates payments per views
	  • Decentralized database of rights
	• Giraffage: ERC20 tokens to facilitate payments and interactions between fans and artists .

Figure 6.45:   Copy-right: Music.

Source: UJO, Music Alley.

Problem:	Charity expenditure is sometimes misspent or corrupt. 70% growth rate YoY.
Blockchain Solution:	Transparency, a ledger that tracks the inflow and outflow of money where each transaction is confirmed and verified by user, charity and donator.

Description

Background information:
• Founders: Alex Shashou, Dmitry Koltunov, Julie Ulrich, Justin Effron
• Location: USA, New York
• Funding: 34 million USD total funding
• Stage: Early ventures, series B
• Other applications: Can have other applications for VC, bond, PE investors

Business process:
• Monitor charities spending on the blockchain, donations will be frozen until they prove the they spent money effectively.
• Donors can see how each transaction is spent.
• How do they make money: Undisclosed transaction from donors who donate to charities.

Figure 6.46:   Charity.

Source: Alice, fundraising UK, Medium.

civic

Problem:	Cyber attacks have affected 54% of all companies in the world, costing an average of 5 million USD per attack. 1.5 billion people in the developing world lack legal identity.
Blockchain Solution:	Identity can be managed and stored on the decentralised network; its harder to hack and you can authorize what who can see what.

About Civic	Description	
• App: Secure Identity Platform • Founders: Jonathon Smith and Vinny Lingham • Location: California, San Francisco • Funding: 35.8 million USD • Stage: Seed rounds	• Identity platform is an identity wallet that encrypts users information on the Blockchain. • When asked to give person information, QR code can be scanned and only selected info will be transferred. • Identity can be verified and accessed through biometric scanning. • Utilizes utility token called CVC and Rootstock layer. • Significantly lower fees than privacy management companies, only requires a small transaction fee.	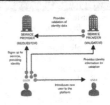

Figure 6.47: Identity management.

Source: Crunchbase, Civic, Invest In Blockchain.

ChronoBank.io

Problem:	Verification of employee data is time-consuming, labour-heavy, and costly.
Blockchain Solution:	Allow safe storage of employee data, making HR processes such as setting up medical benefits easy, fast, and cheap.

About	Description
• Application: Recruitment platform • Founder: Sergei Technology • Location: Australia, Sydney • Funding: 9.4 million USD • Stage: Post ICO	• Labor-hour tokens tokenize hourly wages which can be converted to national currency or used on multiple blockchains. • Employers post job postings and freelancer headhunters process the interaction. They will be given a rating which will build up their reputation and their pay. • Platform charges a commission fee of 1% per transaction. • Transactions capable of occurring within 1 second.

Figure 6.48: Recruitment.

Source: Crunchbase, Chronobank.io.

About	Functions
• Stage: Active • Founder: E.G Galano, Joseph Lubin • Funding: 6 million USD • 600+ Employees • Enterprise Ethereum Alliance member	Production studio which consults and builds decentralized applications using Ethereum. 1. Consensys Labs (main business): Incubator for Ethereum startups. 2. Solutions (main business): Building, development and implementing consultants team of private blockchain solutions.
Industries addressed — Web 3.0	3. Consensys Academy: Education institution for Ethereum knowledge.
• Music • Prediction markets • Custody • Supply chain • Expert networks • Education • Fractional real estate ownership	4. Consensys Ventures: (VC): Worth 50 million USD. Invests in blockchain companies. Most recent round, invested in six companies. 5. Ethereal: Summit that gathers blockchain humanitarians as a networking place.
Web 3.0 — the decentralized WWW — will be anchored in persistent portable identity	**Key takeaway:** Consensys challenges the consulting industry; offering high diversity and technical blockchain solutions.

Figure 6.49:   Blockchain solutions — Consensys (1).

Source: Consensys, Schulte Research.

Blockchain Solutions

Project	Customer	Description
Blockapps Strato	Microsoft Azure	Blockchain verification database
Project Consensys	Delloite	Digital Bank
Blockstack Labs	Microsoft	Data protection app
Zug ID	Swiss government	Identification documentation
Adchain	MetaX	Advertisment feedback database

Incubator Startups

Consensys Startup	Partner	Description
Kaleido	AWS	Blockchain Cloud Business
Viant	GSK	Pharmaceutical supplychain tracking
Grid+1	NA	Electricity provider with market registration
Gitcoin	NA	Software developer platform for freelancers to do projects for companies who need technology help.
Civil	NA	Decentralized journalism platform with a new revenue model. Earns money by connecting citizens who want fact-based news to fund journalism through Ethereum.

Figure 6.50:   Blockchain solutions — Consensys (2).

Source: Consensys, Schulte Research.

Problem:	Advertisements are hard to track reach and success rate, control audience viewership and prevent advertising fraud.
Blockchain Solution:	Real-time data about advertisements platform use, audience reach, click through rate and legal verification will be displayed on a platform without fraud.

About	Business model due to blockchain
• Stage: Active, seed rounds • Funding: 15.7 million USD • Location: USA, New York • Founders: Carolina Abenante Esq, Mark Grinbaum • Partnership with NASAQ • Platform is on Cloud and Blockchain • Transaction fee: From 50% to 5%	• Transparent and trusted advertising market to buy, sell and trade advertising inventory (ad space). • Benefits to publishers: Reducing transaction and handling fees, increase sell-through, retain higher cost per thousand views. • Benefits to advertisers: Platform discovery is more versatile, less price volatility, lock in premium advertising inventory in advance.

Key takeaway: NYIAX allows advertising space to be traded transparently for better for publishers to advertisers, providing larger exposure to more platforms.

Figure 6.51: Advertising — New York interactive advertising exchange.

Source: NYIAX, Ethnews, Crunchbase, NASDAQ.

6.10 Blockchain for Social Media

About	Description — Blockchain
• Founder: Nikolai Durov, Pavel Durov • Location: UK, London • Funding: 1.7 billion USD ICO • Stage: Active, post ICO • Partners: Royal Canadian Mint will provide storage platform for physical gold	• 200 million active users, 70 billion messages daily • Plans to spend 400 million USD over next 3 years • 10 million downloads of dapps on Google Play • Largest ICO, filecoin is second and only raised 257 million USD
Functions	Ton: Telegram Open Network (Crypto Wallet) • Cryptocurrency name: Gram • Usage: Dapps and Wallet
• Primary function: Messaging platform • Secondary function: Payments, File storage, Censorship-proof browsing, Host dapps on their protocol called MTProto, API	• Can handle millions of transactions per second (larger than Paypal), using hypercube routing. • Uses multi-blockchain architecture for infinite sharding (splitting when size becomes bigger to increase capacity). • No hashing, just smart contract and proof of stake to maximize efficiency.

Key takeaway: Telegram has the processing power and functions far more diverse and powerful compared to Facebook due to revolutionary blockchain technology, it is primed to take over.

Figure 6.52: Messaging/social network — Telegram.

Source: Crunchbase, Telegram, Recode, Techcrunch, Tokendata.io.

Function
• Blockchain platform using AI and iOT
• Partner with Hangzhou Realty Admin Bureau

How does it work?
- Rent a flat using Facial recognition and instant payment when you leave and lock the door.
- Possible due to smart contracts that use iOT sensors to precisely store information of entering and leaving times on the blockchain.
- Hosts the transaction based on the time to eliminate time wastage when you leave.

Key takeaway: Home-sharing becomes more convenient, efficient, transaction secured and cheaper. This can apply to ride sharing as well.

Issues Addressed

1. Fraud

Facial recognition and immutable records prevent fraud transactions.

2. Data control

Extensive data collection for travel, movement, location.

3. Transparency

Home owners and consumers understand agreement and liability.

Figure 6.53:   Home rental — Ant Financial (Alibaba).

Source: SCMP, AntFinancial, Alibaba.

About
• Founder: Alan Vey, Annika Monari
• Location: UK, London
• Funding: 20 million USD
• Stage: Active, post ICO

Blockchain Functions

1. Core Protocol
- Smart contract enforces rules set by provider
- Flexible adaptations depending on need for economic model

2. Service Layer
- API that allows enterprises to create their own smart contracts to sell tickets

3. Application Layer
- Flexibility that can use the protocol or a wrap (application) for interface

Business Model

1. Voting → Consumer Transparency

- Legitimize all events on the protocol.
- Verification of all members on the ticket supply chain.
- Ensure transactions are seen by all members.

2. Token → Fraud Protection

- Right holders and consumers have a source where they can verify the ticket application.
- Authorizing off-chain fees, terms and conditions.

3. Secondary market → Enforcing Rights

- Proof of stake (verification and voting) mechanisms that prevent secondary market entry.
- Consumer protection from additional fees.

Figure 6.54:   Ticketing — Aventus Systems.

Source: Crunchbase, Aventus.

IBM

Problem:	Fragmented medical data management, 23.5% of records didn't match symptoms.
Blockchain Solution:	Enables single EHR, and accurate timely diagnostic recording improvement by 24% and increase net profit margins from average 45% to 61%.

What is addressed	Description
EHR (Electronic Health Record)	• 16% of healthcare providers expected to use blockchain this year • Medical record immutable and within one system • Billing record is flawlessly audited, no admin errors
Counterfeit drug prevention	• Manufacturing iOT sensors can pick up ingredient tampering • 10% of total medicines transported are counterfeit, this will reduce this and ensure a transparent manufacturing process record
Clinical trial test results	• Provides accountability and transparency to clinical trial patient results • Uses electronic data capture (EDC) • Data from researches save 22% more time with data transfer
iOT with Medical equipment	• iOT can collect Patient health data and store it on the blockchain for a record • Can also track and opioids use

Figure 6.55: Healthcare data management — IBM.

Source: IBM, Medium, Healthitanalytics.

6.11 Blockchain for Government

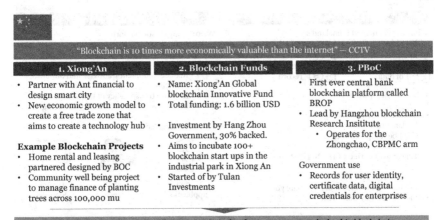

"Blockchain is 10 times more economically valuable than the internet" — CCTV

1. Xiong'An	2. Blockchain Funds	3. PBoC
• Partner with Ant financial to design smart city • New economic growth model to create a free trade zone that aims to create a technology hub **Example Blockchain Projects** • Home rental and leasing partnered designed by BOC • Community well being project to manage finance of planting trees across 100,000 mu	• Name: Xiong'An Global blockchain Innovative Fund • Total funding: 1.6 billion USD • Investment by Hang Zhou Government, 30% backed. • Aims to incubate 100+ blockchain start ups in the industrial park in Xiong An • Started of by Tulan Investments	• First ever central bank blockchain platform called BROP • Lead by Hangzhou blockchain Research Insititute • Operates for the Zhongchao, CBPMC arm Government use • Records for user identity, certificate data, digital credentials for enterprises

Key takeaway: China is heavily investing in its local governments to take lead in blockchain implementation into all city services.

Figure 6.56: Smart city — Xiong An, China.

Source: Coindesk, CCTV, SCMP, Cointelegraph, Technode.

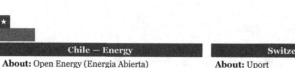

Chile — Energy	Switzerland — Voting
About: Open Energy (Energia Abierta)	**About:** Uport
• Ethereum ledger to record energy sector statistics	• Government issued digital identity
• Expected to use 100k+ different nodes	• Ethereum ledger system to issue digital
• To increase	verification to citizens
• Cyber security	• Creating a digital passport
• Integrity	
• Traceability	Extensions
• Public information transparency	• Spring 2018 — eVoting system
• Uses iOT to collect data	• Official Zug citizen ID
• Average market prices	• E-signatures
• Electricity generating capacity	• Payment fees (such as parking)
• Hydrocarbon prices	
• Marginal costs (mark ups)	
• Compliance rules	

Key Takeaway: Chile solves their main concern of someone hacking their energy system.

Key Takeaway: Switzerland aims to streamline e-business transactions by using blockchain to govern and host them.

Figure 6.57: Government data management.

Source: Coindesk, Medium, Ethnews.

USA — Security	HKMA trade platform
Who: Department of homeland security	**Who:** HKMA trade platform
• Uses blockchain for cyber security	• Proof of concept
	• Digitize paper-intensive processes by smart contracts

 Security storage of data   Immutable record

1. US border patrol

• Storage of security footage
• Updated real time
• Can't be tampered with

2. Trade and office of trade relations

• Proof of concepts
• Facilitating movement of legitimate goods
• Increase supply chain visibility

• Digitize paper-intensive processes by smart contracts
  • Reduce fraudulent trade risk and duplicate financing
  • Improve transparency of transaction
  • Legal cross-border regulatory enforcing

Industries targeted:

Mortgage applications   Trade finance   Digital Identity

Figure 6.58: Government — Secure data/transaction storage.

Source: Medium, DHS, Opengovasia.

Appendix

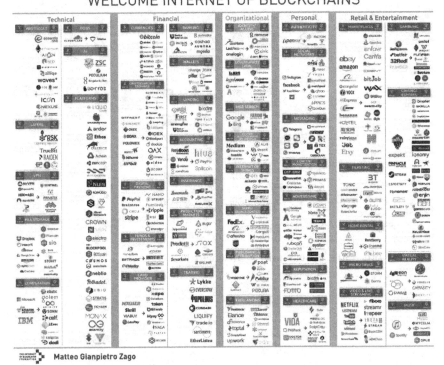

Figure 6.59: WEB 2.0 → WEB 3.0 comparison landscape.

Figure 6.60: 50+ blockchain real-world use cases.

PART 3

Global Leadership in the Transformation with AI and Banking/Insurance

Chapter 7

Alibaba vs Tencent: The Great Race

7.1 The Battle of Chinese Tech Giants

(a)

(b)

Figure 7.1: Alibaba vs Tencent.

In this chapter, we will be looking closely at the comparison between Alibaba and Tencent. Figures 7.1–7.2 show the theme of this chapter that depicts the competition between Alibaba and Tencent. Analysts seem to suggest that Alibaba beats Tencent on products, diversity, geography, and valuation! As of August 2018, US-listed Alibaba and Hong Kong-listed Tencent are both registering negative returns for the year of −5.8% and −22.1%. Amazon is up an amazing 66.5% with Apple's share price up 30.8%. Chinese tech companies are still relatively small if we compare them with the American tech companies. In particular, the largest company Apple is 2.8 times and 2.5 times larger than Tencent and Alibaba. Regarding Price to Book (P/B) value, Alphabet and Facebook are 5.1 and 5.9 times, respectively. Overall, tech companies are not cheap and Amazon is relatively above average at 27.2 times for P/B and Price to Earnings (P/E) ratios. As for Price to Sales (P/S) ratio, Alibaba and Tencent are both above average at 10.2 and 9.3 times.

Ticker	Company	Mkt Cap ($bn)	P/B	P/E	P/S	Asset/ Equity	YTD Return
BABA US Equity	Alibaba	421	7.3	26.2	10.2	1.6	−5.8
700 HK Equity	Tencent	385	8.9	30.0	9.3	2.0	−22.1
MSFT US Equity	Microsoft	830	10	25.3	7.5	3.1	26.5
AMZN US Equity	Amazon	952	27.2	63.3	4.5	4.7	66.9
FB US Equity	Facebook	471	5.9	18.7	9.8	1.1	−7.6
GOOGL US Equity	Alphabet	814	5.1	21.4	6.6	1.3	11.3
AAPL US Equity	Apple	1069	9.3	16.8	4.4	2.8	30.8
Average		706	10.5	28.8	7.5	2.4	14.3

Figure 7.2:   Valuation.
Source: Bloomberg, CNN, Schulte Research.

According to UBS and Schulte Research,

1. Alibaba is winning the Southeast Asia expansion race: leads in four of five largest e-commerce markets.
2. Alibaba has proved the company can innovate through the success of Ant Financial and move this technology to other countries.

3. Alibaba's adtech is reputed to be the best in the world, giving them the highest revenue per user per minute.
4. Alibaba has an unparalleled reputation in Asia-Pacific markets, with triple-digit cloud growth and easy establishment of partnerships.
5. Alibaba's core business in China is still healthy and growing — it has a runaway Number 1 market share.
6. Alibaba has excellent financial parameters and is not bogged down in the anti-gaming crackdown that Tencent is enduring.

Alibaba International is focusing on expanding its e-commerce and mobile payments internationally. With a strong grip in the PRC e-commerce market, Alibaba is expanding into the Asia-Pacific payments and cloud computing markets throughout Asia. While multinationals fight to control India, Alibaba has taken their expansion a step further by establishing a commanding lead in other Asian digital markets. Alibaba is going a step further by diversifying its subsidiaries into fully diversified financial conglomerates — Paytm, Lazada, and Tokopedia. In mobile payments, Alibaba is ranked number 1 in India, Thailand, Bangladesh and number 2 in the Philippines. In e-commerce, Alibaba is number 1 in Indonesia, Thailand, Bangladesh, Philippines, number 2 in Singapore, and number 3 in India (Figure 7.3).

Countries of Interest	Mobile Payments Industry Ranking	E-Commerce Industry Ranking
India	#1	#3
Indonesia	-------	#1
Thailand	#1	#1
Bangladesh	#1	#1
Philippines	#2	#1
Singapore	-------	#2

Figure 7.3: Alibaba International: E-commerce and mobile payments focused expansion.

Source: Times of India, Schulte Research.

In India, Alibaba international investments are across mobile payment (Paytm), online grocery (Bigbasket), logistics (XpressBess), and food delivery (Zomato). In the region, Alibaba's investment in e-commerce includes Tokopedia, Lazada, and Daraz. In mobile payments, Alibaba has invested in Dana, Truemoney, GCash, and bKash. It is interesting that in Singapore, it has been increasing its stake in Singapore Post over the years. However, its investments are beyond and into banking, insurance, media, lifestyle, and cloud in seven key Asian countries (Figures 7.4–7.7).

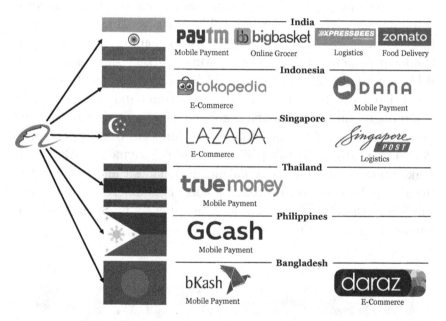

Figure 7.4:   Alibaba international investments.

	Payment	Banking	Insurance	Ecommerce	Media	Food	Lifestyle	Logistics	Cloud
China	Alipay	Ant Fortune Mybank Zhima Credit	Zhong An (12%)	Taobao Tmall 1688.com	Youku SCMP Video++ CMC UC Web Ali Music Ali Pictures	Sun Art Tmall Supermarket Hema Ele.me	Alihealth Tao Piao Piao Damia.cn Wandoujia	Cainiao	
India	Paytm (45%)	Paytm	Paytm (For Dengue)	Paytm Mall (~30%)	DailyHunt (Planned)	Big Basket Zomato	Paytm	XpressBees	
Indonesia	DANA, Alipay	Tokopedia (Planned)	Tokopedia (Planned)	Lazada (>83%) Tokopedia	UCWeb	Lazada Tokopedia	Lazada Tokopedia		Alibaba Cloud
Thailand	TrueMoney	Truemoney (Planned)	Truemoney (Planned)	Lazada		Lazada	Lazada		
Bangladesh	bKash (20%)	bKash	bKash (applied)	Daraz (100%)		Lazada	Lazada		
Philippines	Gcash, Alipay	Gcash (Micro-Lending)	Gcash (Planned)	Lazada		Lazada	Lazada		
Singapore	Alipay	Singapore Post (Standard Chartered)	Singapore Post (AXA)	Lazada		Lazada	Lazada	Lazada	Singapore Post

Figure 7.5: **Alibaba's international presence spans seven key Asian countries.**

Note: Singapore, Indonesia, and Philippines Lazada's Hellopay was rebranded to Alipay.

Figure 7.6: **Map of GDP growth shows Asia-Pacific leadership.**
Source: IMF, Schulte Research.

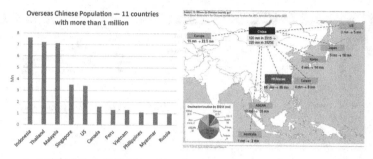

- 100% increase in PRC outbound travel

Figure 7.7:   PRC diaspora: ASEAN diaspora of Chinese of 33 million.
Source: AAARI, CUNY, Schulte Research.

Alibaba' expansion is into the largest and fastest-growing Asian economy (Figure 7.8). The Asia-Pacific region is the most economically important segment of the global economy for three key reasons:

(1) **Largest Economy:** 35% of the global GDP;
(2) **Fastest Growth:** 5.7% real GDP growth rate;
(3) **Highest Population:** 57% of the world.

The Asia-Pacific region will remain the leader in these three categories through 2023. Figure 7.9 shows the explosive growth of e-commerce.

	GDP: 2017 (USD bn)	GDP: 2023E (USD bn)	CAGR: 2018–2023E
India	$2,600	$4,700	7.9%
Indonesia	$1,000	$1,500	5.5%
Thailand	$460	$650	3.7%
Singapore	$320	$440	2.7%
Philippines	$310	$520	6.9%
Bangladesh	$260	$440	7.0%
China	$12,000	$22,000	6.1%

Figure 7.8:   Sustained GDP growth in countries of interest.
Source: IMF, Schulte Research.

	Retail E-Commerce Sales: 2017 (USD mn)	Retail E-Commerce Sales: 2021E (USD mn)	CAGR: 2017–2021E
India	$21,000	$50,000	25%
Indonesia	$8,200	$18,000	22%
Thailand	$3,500	$6,800	18%
Singapore	$2,400	$3,600	10%
Philippines	$420	$820	18%
China	$1,100,000	$3,000,000	28%

Figure 7.9: Explosive e-commerce growth in these countries gives a positive outlook for Alibaba's expansion.

Source: Times of India, Schulte Research.

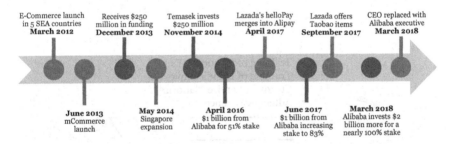

Figure 7.10: Lazada history.

Source: Lazada, Schulte Research.

Lazada's value comes from its experience, data and market leadership (Figures 7.10–7.12).

(1) It has a shopping behavior database on 200 million SEA visitors monthly, a two-time primary competitor.
(2) Its experience in markets with immature infrastructure helps Alibaba expand Rural Taobao program.

(3) Its experience in markets with varied cultures provides lessons for expansion out of homogenous China.
(4) Its SEA e-commerce is rapidly growing.
(5) Lazada leads SEA e-commerce.

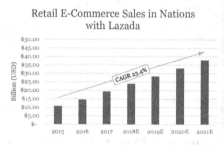

Countries With Lazada	Lazada's Market Position
Indonesia	#1
Philippines	#1
Malaysia	#1
Vietnam	#1
Thailand	#1
Singapore	#2

Figure 7.11: Lazada's value comes from its experience, data, and market leadership.

Source: Lazada, Schulte Research.

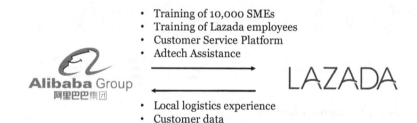

- Training of 10,000 SMEs
- Training of Lazada employees
- Customer Service Platform
- Adtech Assistance

- Local logistics experience
- Customer data
- Local cultural understanding

Figure 7.12: Mutual exchange between Alibaba and Lazada.

Source: Forrester, Schulte Research.

Figure 7.13: Alibaba has four key assets in India.

Alibaba is a major player in India with 3 years of experience (Figures 7.13 and 7.14). Since early 2015, Alibaba has aggressively invested in India spending roughly US$2 billion to date with plans to invest an additional US$8 billion over the next 4 years. Even now, Alibaba has a strong position within the Indian digital marketplace.

Figure 7.14: Alibaba is a major player in India with 3 years of experience.
Source: Times of India, Statcounter, Schulte Research.

Mobile payments are growing fast in India driven by policy changes and smartphone penetration. Two factors have increased use of mobile payments in India (Figures 7.15–7.19):

1. **Policy:** India's 2016 demonetization move and creation of a national digital payment architecture has incentivized the use of digital payments.
2. **Smartphone Penetration:** 25% of Indians currently use smartphones, up from 15% in 2015.

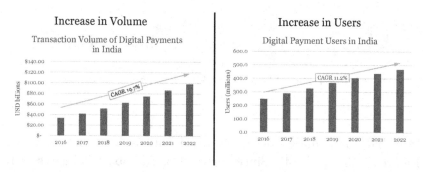

Figure 7.15: Mobile payments are growing fast in India driven by policy changes and smartphone penetration.

Source: Statista, Schulte Research.

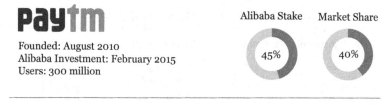

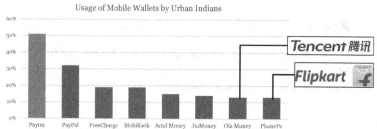

Figure 7.16: Alibaba dominates Indian mobile payments via Paytm investment.

Source: eMarketer, Paytm, India NPCI, Schulte Research.

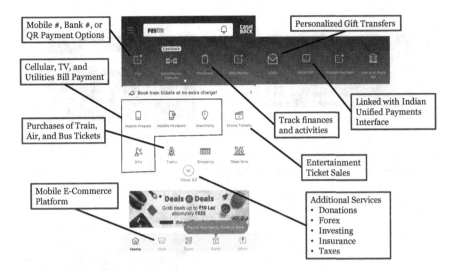

Figure 7.17:   Paytm homepage.

Source: Paytm, Schulte Research.

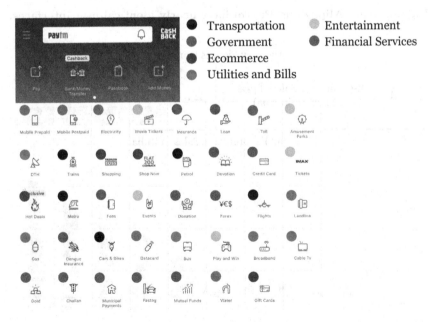

Figure 7.18:   Paytm app covers a comprehensive array of needs for the Indian consumer.

Source: Paytm, Schulte Research.

Figure 7.19: **Alipay and Paytm have nearly identical user interfaces.**

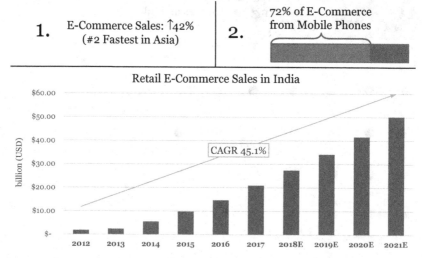

Figure 7.20: **India e-commerce is growing rapidly, driven by smart-phone penetration.**

Source: eMarketer, Schulte Research.

Alibaba is behind the curve with Indian e-commerce, but is making up ground fast (Figures 7.20–7.26). Alibaba has invested primarily in two Indian e-commerce ventures, Big Basket and Paytm Mall, which had a combined net sales of US$240 million. Despite Amazon and Walmart's leading position, Alibaba's e-commerce sales more than doubled in 2017.

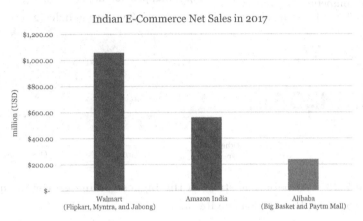

Figure 7.21: Alibaba is behind the curve with Indian e-commerce, but is making up ground fast.

Source: Indian Times, Statista, Schulte Research.

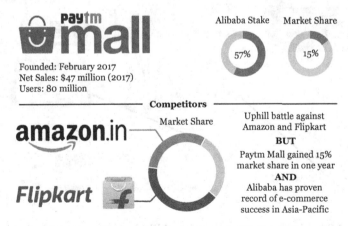

Figure 7.22: Paytm Mall is bringing to India the same successful strategies used by Taobao.

Source: Bloomberg Quint, Indian Times, Forrester, Schulte Research.

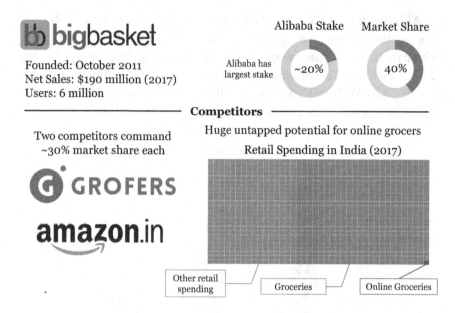

Figure 7.23: Alibaba has stake in the biggest online grocer in India.
Source: Statista, Reuters, Schulte Research.

Figure 7.24: Integration will drive Alibaba's success in India.
Source: Paytm, Schulte Research.

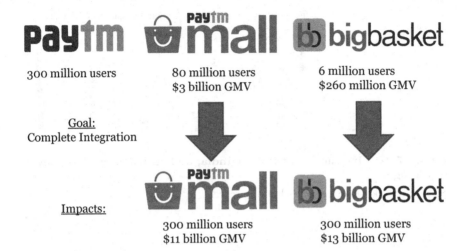

300 million users   80 million users   6 million users
$3 billion GMV $260 million GMV

Goal:
Complete Integration

Impacts:

300 million users   300 million users
$11 billion GMV $13 billion GMV

Figure 7.25: Integration solves user disparity between Alibaba's Indian ventures, enables market leadership.

Source: Statista, Reuters, Schulte Research.

Figure 7.26: Alibaba in Indonesia.

Indonesia is the next India, and Alibaba has first mover advantage (Figures 7.27–7.29). Alibaba is thinking a step ahead of its competitors by investing early in Indonesia because of the following reasons:

1. Indonesia's importance in SEA.
2. It is the largest e-commerce market.
3. It is the fastest growing e-commerce market.
4. It is the most populous nation.

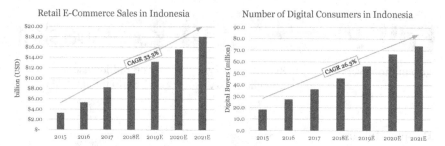

Figure 7.27: **Indonesia is the next India, and Alibaba has first mover advantage.**

Source: eMarketer, Statista, Schulte Research.

Top 10 E-Commerce Websites in Indonesia (May 2018)

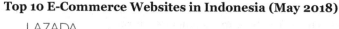

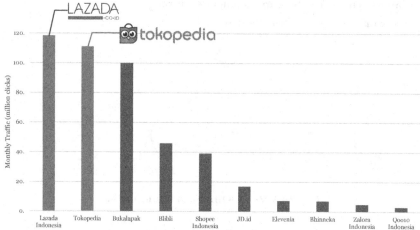

Figure 7.28: **Alibaba has half of Indonesia's e-commerce with stakes in two largest sites.**

Source: SimilarWeb, Schulte Research.

Alibaba Stake   Market Share

~100%   26%

<u>Founded</u>   <u>1st Alibaba Investment</u>
March 2012   April 2016

Alibaba Stake   Market Share

<50%   24%

<u>Founded</u>   <u>1st Alibaba Investment</u>   Undisclosed
February 2009   August 2017   Minority Stake

Figure 7.29: Tokopedia and Lazada are Alibaba's investments in Indonesian e-commerce.

Note: Market share determined by portion of monthly e-commerce visitors in May 2018.

Source: ecommerceIO, eMarketer, Schulte Research.

Tokopedia resembles a fusion of Paytm, Etsy, and Amazon with two services (Figures 7.30–7.38):

(1) A commission-free platform for businesses to sell their products online and provide logistics support and other business services.

(2) A digital payments ecosystem that provides payment options for a variety of online services excluding P2P and POS payment.

Revenues are earned through memberships, transaction fees, and paid advertising. Alibaba acquired a minority stake in Tokopedia with their $1.1 billion investment in August 2017.

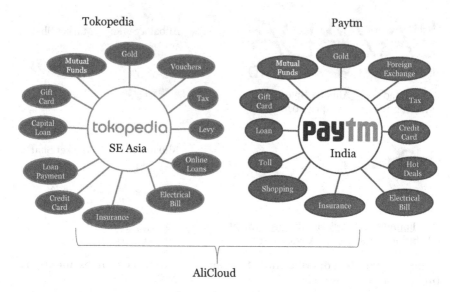

Figure 7.30:   Tokopedia and Paytm digital offerings in banking.
Source: Company Web, Schulte Research.

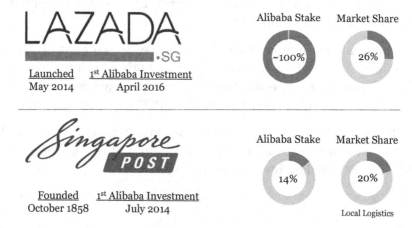

Figure 7.31:   Alibaba is using Singapore investments to boost logistical abilities in SEA.

Note: Market share determined by portion of monthly e-commerce visitors in May 2018.

Source: ecommerceIO, eMarketer, Schulte Research.

Alibaba Stake

20%

<u>Founded</u>   <u>1st Alibaba Investment</u>
2003   November 2016

Users: >1 million

- Digital payments leader in Thailand

- Owns 9,000 7-Eleven branches in Thailand

Truemoney has utilized marketing technology to cut costs while growing users

60%
decrease in
cost per install

62%
lower cost per
acquisition than
industry benchmarks

Figure 7.32:   True money

Source: Statista, Reuters, Schulte Research.

Alibaba Stake   Market Share

~100%   54%

<u>Founded</u>   <u>1st Alibaba Investment</u>
March 2012   April 2016

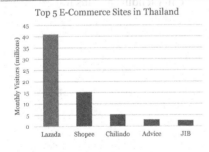

Top 5 E-Commerce Sites in Thailand

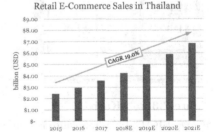

Retail E-Commerce Sales in Thailand

Figure 7.33:   Lazada Thailand.

Note: Market share determined by portion of monthly e-commerce visitors in May 2018.

Source: ecoomerceIO, eMarketer, Schulte Research.

Figure 7.34: Alibaba faces no competition in Bangladesh mobile payment and e-commerce.

Launched	1ˢᵗ Alibaba Investment
July 2011	April 2018

Alibaba Stake   Market Share

20%   80%

30 million users

Applications have been filed to
enter lending and insurance

Figure 7.35:   bKash.

Source: BRAC, bKash, Schulte Research.

Launched | Alibaba Acquisition
2012 | May 2018

Alibaba Stake 100%   **Market Share** 88%

Daraz leads Pakistani and
Bangladeshi e-commerce markets

Acquired primary
competitor, Kaymu,
in June 2016

Figure 7.36: Daraz.

Source: Statista, Schulte Research.

Founded | 1st Alibaba Investment
March 2012 | April 2016

Alibaba Stake ~100%   **Market Share** 55%

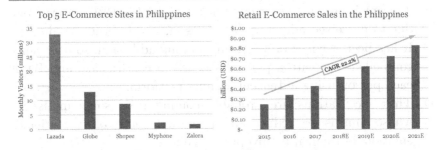

Figure 7.37: Lazada Philippines.

Note: Market share determined by portion of monthly e-commerce visitors in May 2018.

Source: ecommerceIO, eMarketer, Schulte Research.

GCash

Alibaba Stake

45%

Founded	1st Alibaba Investment
October 2004	February 2017

Users
5 million

- Strategic partnership with PayPal for international money transfers

- GCredit lending functions like a digital credit card

Figure 7.38: GCash.
Source: Company Web, Schulte Research.

The Tencent Executive Summary (Figures 7.39–7.40) is given as follows:

A. Tencent's international footprint is much weaker than Alibaba. It relies on morphing Wechat in each country as well as smaller investments.

B. Tencent is enduring a severe crackdown by the PRC government due to perceived addiction by adolescents to internet gaming.

C. It has been accused of fomenting gaming addiction, so it is being compelled to create a police database to confirm age and ID of each person using a Tencent gaming platform.

D. These accusations are probably coming from some ministries whose aim is to reduce the perceived sense of educational, emotional or cultural "myopia" of children.

E. Despite the drop at a 1 year low, valuations are not compelling.

F. Alibaba is a much more diverse geographical and product suite of offerings.

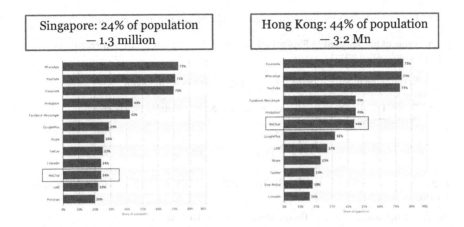

- WeChat has 1 billion AMU
  - 100 million AMU International — 20 million in Malaysia (2/3 of pop.)

Figure 7.39: **South-east Asia WeChat Users — 24% of Singapore and 44% of Hong Kong.**

Source: Statista, Schulte Research.

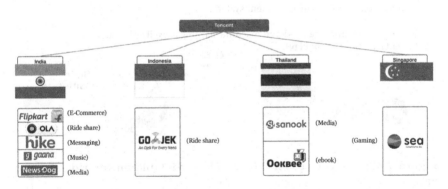

Figure 7.40: **B. Tencent regional investments.**

Source: Schulte Research.

We will leave the readers to explore Figures 7.41–7.56 based on the above charts, which are self-explanatory. Tencent investments span across India, Indonesia, Thailand, and Singapore.

- Flipkart is India's leading e-commerce marketplace (40% market share).

- Tencent participated in Flipkart's $1.4 billion funding round (with Microsoft and eBay in April 2017).

- In the same month, Alibaba invested $45 million more in Paytm.

- Tencent continues to be an investor in Flipkart (6% shareholder post Walmart deal).

Figure 7.41: Flipkart (April 10, 2017) — $1.4 billion for Amazon of India.

Source: Flipkart, Reuters, Techcrunch, Schulte Research.

- Ola is India's leading ride-hailing service, operating in 110 cities.
  - Uber operates in 30 cities.

- Ola FY17 Total Revenue — US$110 million.

- Tencent lead Ola's $1.1 billion funding round (with Softbank in October 2017).

- Ola money, is Ola's own payment system.

- Ola is also investing in AI and machine learning as well as in-car entertainment (Ola Play).

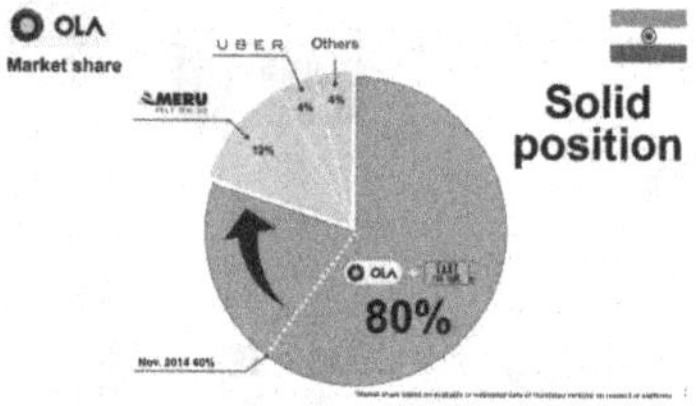

Figure 7.42: Ola (October 11, 2017) — $1.1 billion for ride-sharing app.

Source: Ola, Reuters, Medianama, Schulte Research.

- Hike is a messaging app with/over 100 million users.
- Tencent led Hike's $175 million funding round in August 2016.

- In 2017, Hike launched Hike Wallet — mobile payments service
  - Hike Wallet is now compatible with Ola Cabs.

- However, Paytm (Alibaba) and PhonePe (Flipkart) crowd the mobile payments market — 300 and 100 million, respectively.
- WhatsApp, Facebook, and Gmail dominate messaging.

Figure 7.43: Hike (August 16, 2016) — $175 million for mobile payments.
Source: Techcrunch, CNN, Schulte Research.

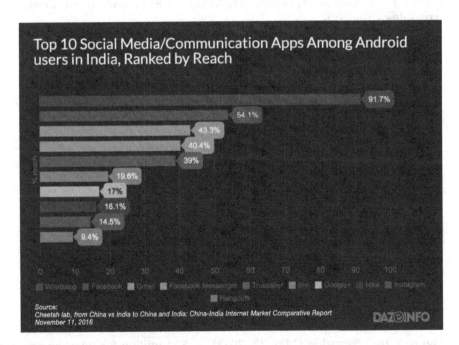

Figure 7.44: Hike cont.
Source: DazeInfo, Schulte Research.

- Music streaming service launched by Times Media.

- Tencent lead Gaana's $115 million funding round in February 2018.

- Crowded Market — Saavn (Tiger Global); Hungama (Xiaomi); Apple Music; Spotify (Tencent).

- Total revenue Indian digital music — US$126 million.

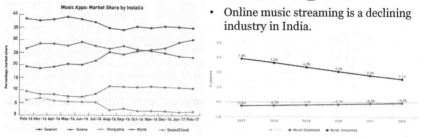

- Online music streaming is a declining industry in India.

Figure 7.45: Gaana (February 27, 2018) — $115 million for music streaming (Spotify).

Source: Techcrunch, Statista, Medium, Schulte Research.

- NewsDog is an India-focused vernacular news aggregator with 50 million readers.

- Tencent lead NewsDog's $50 million funding round in May 2018
  - Tencent's first investment in the Indian news industry.

- Tencent missed the rise of news aggregators in China (Bytedance and its news aggregator app Toutiao).
  - Bytedance backs one of NewsDog's competitors—Dailyhunt.
  - UC News, another competitor, is owned by Alibaba.

Figure 7.46: Newsdog (May 22, 2018) — $50 million for news aggregator (Daily Mail of India).

Source: NewsDog, Techcrunch, Schulte Research.

- GO-JEK specializes in ride-hailing, logistics and digital payments and operates in 50 cities in Indonesia.
  - Indonesia's first unicorn

- Tencent lead GO-JEK's US$1.2 billion funding round — 2017.

- Participated in another round of funding US$1.5 billion in February 2018 with Google, Temasek and BlackRock.

- Valuation — US$6 billion.

Figure 7.47: GO-JEK (May 2017) — $1.2 billion for ride-hailing, logistics, and digital payments.

Source: GO-JEK, Techcrunch, Kr-Asia, ASEAN Post, Schulte Research.

Map of cities GO-JEK operates in

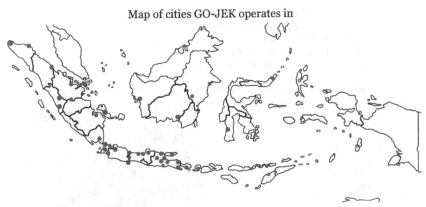

Notable cities: Jakarta, Surabaya, Bandung, Medan, Semarang, Makassar, Lampung, Padang **(By Population)**

Figure 7.48: GO-JEK (May 2017) (1) — Java, Sumatra, Sulawesi (it's everywhere).

Source: GO-JEK, Schulte Research.

- GO-JEK's payment service and e-wallet, Go-Pay, was launched in April 2016

- GO-JEK AMU: 15 million
- Go-Pay AMU: 8 million

Mobile Internet Users in Indonesia Who Use Digital Cash*, by Service, Dec 2016
% of respondents

Services used		
Mandiri e-money	43.8%	
Flazz	39.1%	(BCA)
T-Cash	29.1%	(PT Telkomsel)
Go-Pay (from Go-Jek)	27.1%	
Rekening Ponsel	15.6%	(Bank CIMB Niaga)
Line Pay	15.6%	

No 36.5%

Yes 63.5%

Use digital cash^

Figure 7.49:   GO-JEK (May 2017) (2).

Source: GO-JEK, The Jakarta Post, eMarketer, Schulte Research.

- Sanook is a multi-faceted entity with services that include:
  - Web Portal — Sanook.com (36 Mn AMU)
  - News Portal — NoozUp
  - Digital Music — JOOX (22 Mn total users)
  - E-Commerce — Saybuy/Dealfish (250,000 users/day)

- Tencent acquired 49% of Sanook for US$10 million USD in August 2010.
- In December 2016, Sanook was renamed Tencent Thailand.

Figure 7.50:   Sanook (August 2010) — Web Portal 36 million AMU.

Source: Technode, The Nation, Schulte Research.

- In June 2017, Tencent announced its US$19 million investment in Thai eBook startup, Ookbee, as joint venture.
  - Launch of digital content platform Ookbee U
  - Ookbee U: comics, music and blogs
- Ookbee has 8 million users across Thailand, Indonesia, Malaysia, the Philippines and Vietnam.
  - 4 million AMU
  - 90% Thai eBook market share

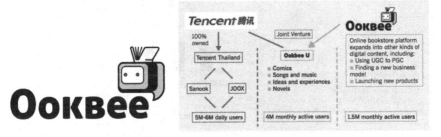

Figure 7.51: Ookbee (June 2017) — ebook company 90% market share.

Source: TechinAsia, The Nation, SCMP, Schulte Research.

- Founded in 2009 as Garena, Sea Ltd. is an internet platform provider — primarily shopping and online games.

  - Garena rebranded itself, Sea Ltd., following a US$550 million funding round in 2017.
  - Garena, the gaming branch, accounts for 70% of total revenue.
  - Garena offers free games and makes money via sale of in-game virtual items.

- Sea ltd. branched into e-commerce and digital financial services through respective services, Shopee and Airpay.

- Sea ltd. went public in October 2017 selling 49 million ADS at USD $16.25 USD/ADS.

- Tencent controls 30% of Sea ltd. voting power.

Figure 7.52: Garena/Sea Ltd. — Gaming company (League of Legends, Fifa, Point Blank).

Source: Bloomberg, The Motley Fool, Schulte Research.

- Sea Ltd. has been operating at a loss:

    - Costs tripled to US$263 million for the quarter ended in December (analysts cost expectations — US$201 million).
    - Total revenue only grew 41% to US$124 million.

- Market Cap. US$5.4 billion

Figure 7.53:   Garena/Sea Ltd. (1).

Source: SCMP, Bloomberg, Schulte Research.

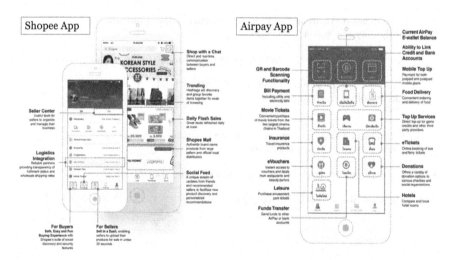

Figure 7.54:   Garena/Sea Ltd. (2).

Source: Sea ltd, Schulte Research.

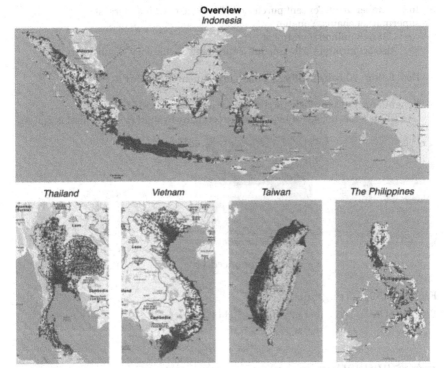

Heatmap of our Users Across our Three Core Platforms in their Respective Markets as of June 30, 2017

Figure 7.55: Garena/Sea Ltd. (3).

Source: Sea ltd, Schulte Research.

- In December 2017-Tencent purchased 5% of one of China's largest supermarket chains, Yonghui
  - US$638 million investment
  - Also acquiring 15% in Yonghui Yunchuang Technology — subsidiary

- Deal follows Alibaba's US$2.9 billion investment in Sun Art Retail Group

- Yonghui operates through 580 brick and mortar stores in China
  - China's fourth biggest hypermarket company by market share –first is Sun Art

Figure 7.56: Yonghui.

Source: SCMP, Bloomberg, Yonghui, Schulte Research.

7.2   Summary

Many are unaware that both Alibaba and Tencent have extended their wings to many countries in Asia either via investment, acquisition, joint venture, and collaboration. It presents an opportunity for the Chinese tech giants to learn very early in the game to penetrate different markets in the region. Each of these markets has a different culture, languages, regulation, practices, politics, and processes. While many may have encountered Alipay and Wechat, the two giants are very much in the stealth mode bringing along with them the technical expertise and experience in the social media and e-commerce business. It does not matter if we prefer Alibaba or Tencent, they are here to stay and their presence will be increasingly felt.

Chapter 8

Ping An: Blazing New Trails and Connecting the Dots

PINGAN

8.1 Introduction

Ping An is the most progressive insurance firm globally in creating an O2O platform. While Alibaba and Tencent have acquired many companies in Asia, Ping An is behind in searching for collaboration partners and acquisitions. Given Ping An's resources, the market has high expectations of Ping An Bank to be better structured and perhaps restructured and recapitalized. Being based in Shenzhen has its advantages, but far from Beijing's policymakers. Following is the summary of this chapter.

The positives about Ping An are as follows:

1. expansion into fast-growing industries O2O (Online to Offline);
2. increase in online users and rapid integration of bancassurance model;
3. undervalued share price from overreaction to banking struggles;
4. awareness of and efforts to ameliorate credit risk.

Some key features of Ping An are as follows:

1. The first insurance company to create a complete horizontally integrated O2O offerings in the growing Healthcare and Fintech industries to sustain revenue growth.

2. Rapidly integrating the full hospital/clinic experience to include the doctor, pharmacy, bank and insurance company as a one-stop shop.
3. However, it is hampered by legacy issues in old bank/insurance models, cost problems, guzzling capital bank, and excess labour capacity.
4. Ping An was behind the curve compared to Baidu, Alibaba, or Tencent in online data acquisition; these three process 10 times more data daily than Ping An has acquired in total.

Ping An is catching up fast by focusing on the following:

1. Offline health data and expanding into (a) Healthcare and (b) Lending.
2. **Healthcare:** Good Doctor (with their health cloud) to gain additional value from their actuarial expertise.
3. **Lending:** Puhui is using access to longer credit histories to outperform competitors in P2P lending and financial cloud services.

We hope that the 39 slides from Figures 8.1–Figure 8.39 will give an idea of the potential of the company. This chapter is entirely written by Paul Schulte, assisted by Gavin Liu.

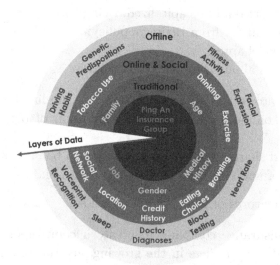

Figure 8.1:   Ping An has unparalleled offline data access and is among the first to actively integrate it and see it as one pool.

	Ping An	Competition
User Profile	Name authenticated with over 3,300 data fields	Credit Karma
Record Access	Credit Ratings, Government Databases, Medical Records, Partner Banks	Alibaba
Data Types	Online and Offline: Medical, Biometric, Location, Financial, Social, Genetic, Behavior	Unmatched
Data Size	8.5 Petabytes (1.7x Data in Library of Congress)	Alibaba, Baidu, Tencent all superior
Chat/Blog	Microblog discussion forum on health news/topics	Tencent

Figure 8.2: Ping An's data universe vs competitors: Unmatched access to data biometric, genetic, location, financial, social, and medical.

Source: Ping An, Wolfram Alpha, Schulte Research.

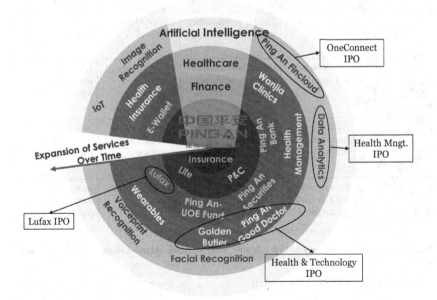

Figure 8.3: Ping An has made greater strides into becoming an autonomous financial ecosystem than any other company.

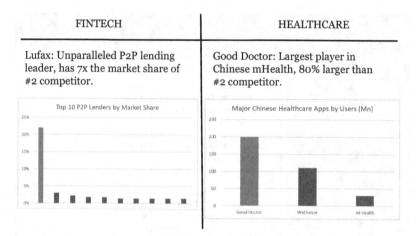

Figure 8.4:   Two subsidiaries prove Ping An can win in Fintech and Healthcare.

Source: Wang Dai Zhi Jia, Ping An Healthcare Technology, Schulte Research.

How O2O Improves...	Health & Life Insurance	P & C Insurance	Fintech	Healthcare
1. Customer Experience	↑67% repeat customers	Leader in customer satisfaction	12-hour loan approval	97% satisfaction rate
2. Efficiency	↑7.1% Sales per agent	Onsite assessment under 10 minutes for 95% of crashes	↓30% Interbank trading time; ↓ 20% cost	Imaging analysis time ↓99%
3. Cost Reduction	99% of claims handled online	↓85% in claims leakage	Stopped RMB 300 billion in fraud for partner banks	>8x consultation capacity
4. Accuracy	70% of claims paid in under 30 minutes	↓60% fraud	Credit loss ratio ↓60%	90% accuracy forecasting flu patterns

Figure 8.5:   How exactly does O2O and the Data Grab benefit Ping An?

Source: Ping An, Schulte Research.

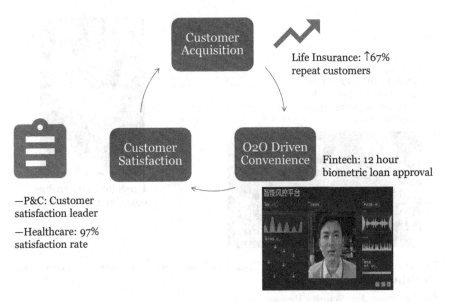

Figure 8.6: O2O has substantially increased customer experience at Ping An.

Source: Ping An, Schulte Research.

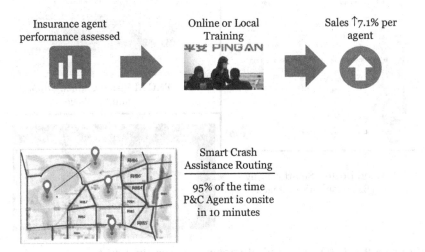

Figure 8.7: O2O has revolutionized insurance processes by increasing efficiency.

Source: Ping An, Schulte Research.

Medical imaging
analysis time down from
20 mins to 10 sec, with
accuracy increased from
60% to 98%.

↓ 30% Interbank trading
time; ↓20% cost

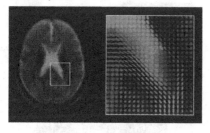

Figure 8.8: O2O efficiency has also provided major speedups in Healthtech and Fintech.

Source: Ping An, Schulte Research.

Life and Health Insurance: 99% online

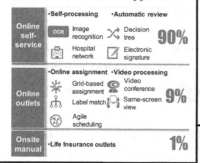

P&C AI Vision Claims: ↓85%
claims leakage

Good Doctor: Smart routing
↑8x consultation capacity

Figure 8.9: O2O data and efficiency has reduced costs.

Source: Ping An, Schulte Research.

P&C
Auto Fraud ↓60%

Life and Health
70% of claims paid in
30 minutes

Fintech
- Credit loss
  ratio ↓60%
- Fraud reduced
  RMB 300
  billion

Healthcare
90% accuracy in
disease forecasting

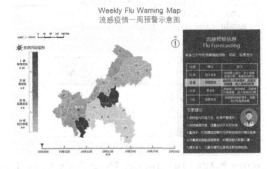

Figure 8.10:   O2O has enabled Ping An to improve accuracy even at a faster pace.

Source: Ping An, Schulte Research.

Segment	How O2O/Big Data Enabled Segment	Result
Lufax	Superior risk profiling of borrowers enabled industry leading rates without increased credit risk	Est. RMB 10 billion profit in 2017 for undisputed P2P lending leader Lufax
Good Doctor (1833.HK)	Online smart routing using offline medical data has boosted doctor efficiency and patient satisfaction	Est. US $7.5 billion EV for Good Doctor and a leading market share in industry with 39% CAGR
Property and Casualty	Mobile app allows P&C customers to upload images of auto damage and receive more accurate claim without an agent	RMB 6 billion reduction in claims expenses

Figure 8.11:   The financial statement impact of O2O and Big Data.

Source: Frost & Sullivan, JP Morgan, DBS Vickers, Schulte Research.

75% of revenue, 53% of income, and 40% of balance sheet come from insurance. Income has grown across both insurance segments for at least 4 years.

Despite not leading in written premiums for either segment, Ping An's consistency across segments (+ strong margins) have made it the largest insurer globally by market cap.

Ping An's insurance activities:
1. Life and Health Insurance
2. Property and Casualty Insurance

Percentage of Ping An's Net Income

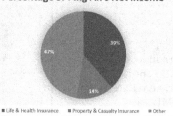

Ranking by Written Premiums	Life and Health	Property and Casualty
Ping An	#2	#2
China Life	#1	#4
PICC	#6	#1

Figure 8.12:   Insurance: Ping An's Core Business.

Source: Ping An 2017 Annual Report, Schulte Research.

Life and Health Insurance: 54% of revenue and 38% of net income.

O2O Applications:

- Product offerings and sales pitches catered to client based on data history
  - Repeat buyers ↑67%
  - 12 month lead conversion rate over 20%

- Data driven agent performance assessment increases productivity
  - ↑7% sales per agent

- Biometric ID for policy purchasing and customer service
  - Fully automated in 90% of cases
  - Online handling for 9% of cases
  - In-person visits only required 1% of the time

Figure 8.13:   O2O increases life and health insurance customer retention and cuts costs.

Source: Ping An, Schulte Research.

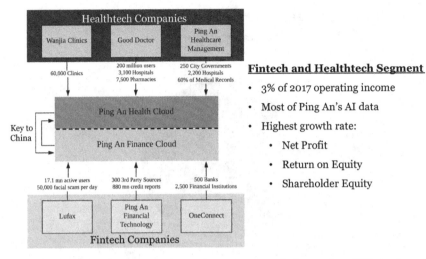

Fintech and Healthtech Segment

- 3% of 2017 operating income
- Most of Ping An's AI data
- Highest growth rate:
  - Net Profit
  - Return on Equity
  - Shareholder Equity

Figure 8.14: **Data acquisition structure reveals key role of Fintech and Healthtech segment.**

Source: Ping An (Annual Report, Shareholder Presentations), Schulte Research.

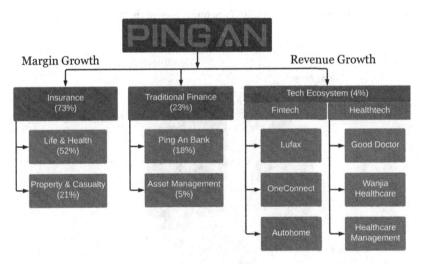

Figure 8.15: **Ping An's structure and distribution of revenue (%).**

Notes: Bank is 18% and tech is 4%. This should reverse in coming years. Distribution of revenue, excluding other business and eliminations.

Source: Ping An Annual Report, Schulte Research.

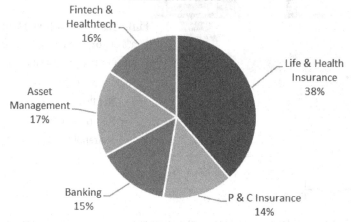

Figure 8.16:   Ping An's net income distribution by business segment.

Notes: Banking hogs the balance sheet but is small contribution to income. Distribution of net income attributable to shareholders, excluding other business and eliminations.

Source: Ping An Annual Report 2017, Schulte Research.

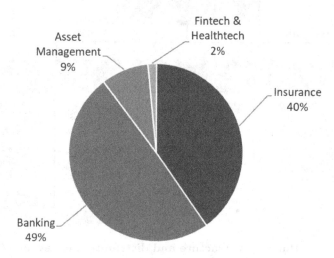

Figure 8.17:   Ping An's asset distribution by segment (%).

Notes: Bank is 50% of balance sheet but 15% of income. The bank needs to go! Distribution of assets, excluding other business and eliminations.

Source: Ping An Annual Report 2017, Schulte Research.

Two revenue drivers for Ping An's Fintech segment:
1. Lufax — China's leading P2P/Wealth Management lending service.
2. OneConnect — World's largest financial cloud with unbeatable O2O data.

P2P Lending (Lufax)		Cloud Solutions (OneConnect)	
Product	**Solution**	**Product**	**Solution**
Mass Market Lending	Competitive rates on loans	Smart Banking Cloud	Platform and tech solutions for small and medium banks
Middle-Upper Class Wealth Management	Risk analysis on P2P lending	Smart Investment Cloud	Corporate risk profiling for asset managers
Government Financial Solutions	Market for public institutions to trade assets and liabilities	Smart Insurance Cloud	Technology solutions for small and medium insurers

Figure 8.18: Creating new solutions to age old problems.

Note: Two new models: DATA to create financial products and storage for that data.

Source: Ping An, Schulte Research.

MARKET SHARE OF CHINESE P2P LENDERS

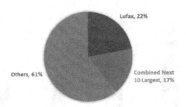

1) Lufax is **undisputed market leader** in P2P loans with 22% market share (second largest P2P lender's share is 3%).

2) Uses Ping An's cloud data analytics to offer better rates to customers without increased credit risk.

3) In Q1 of 2018, Lufax's loans under management grew by 5.3%.

Figure 8.19: Lufax's leads the P2P marketplace by a mile.

Source: Wang Dai Zhi Jia, DBS, Schulte Research.

1. Lufax to grow modestly over remainder of 2018 despite a P2P lending stall.

2. Force consolidation of lenders from 1,800 in June to 800 by 2018 and 200 by 2021.

3. This consolidation allows Lufax's ability to gain market share.

4. Higher proportion of secured loans than competitors, insulating it from regulations that primarily target unsecured loans.

5. In 2019, the regulatory environment is expected to become more supportive.

Figure 8.20:   Massive consolidation in P2P means good prospects for Lufax.

Source: Schulte Research.

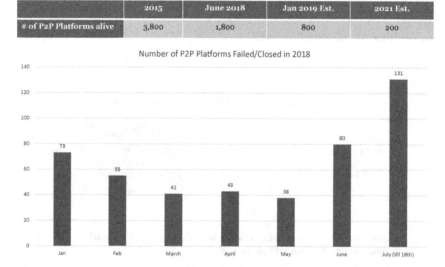

	2015	June 2018	Jan 2019 Est.	2021 Est.
# of P2P Platforms alive	3,800	1,800	800	200

Number of P2P Platforms Failed/Closed in 2018

Figure 8.21:   131 P2P platforms defaulted or closed in the first two weeks only of July 2018.

Source: Financial Times, Wang Dai Zhi Jia, Bloomberg, SCMP, China International Capital Corporation, Schulte Research.

<u>Market</u>

- 10.2% average yield encouraged investor growth.
- 50 million registered users.
- 1.3 trillion RMB in outstanding loans.

<u>Cause of Collapse</u>

China's deleveraging drains liquidity in high risk area like shadow banking and P2P. Tougher regulation for P2P is on the way.

DISTRIBUTION OF P2P LENDING

<u>Aftermath</u>

Investors typically recover only 20% of their investment after years of investigation.

Small businesses and individuals have now increasingly struggled to acquire financing.

Figure 8.22:   The deleveraging is real. Liquidity is tight.

Source: Financial Times, Wang Dai Zhi Jia, Bloomberg, SCMP, China International Capital Corporation, Schulte Research.

OneConnect is the world's largest financial cloud platform offering Fintech solutions to small and medium-sized financial institutions and charging performance fees. As a secondary business, OneConnect performs corporate risk profiling for asset managers.

	Smart Banking Cloud	**Smart Insurance Cloud**	**Smart Investment Cloud**
Consumer Centered	O2O platform connecting FI's to consumers	Smart verification service enables smaller insurers to use biometric identification	--------
Business Centered	Risk assessment of SME financing	--------	--------
Financial Institution Centered	Online, blockchain based interbank trading platform	O2O damage verification and claim approval	Data+AI driven corporate risk profiling for asset managers

Figure 8.23:   OneConnect is exporting Fintech to small and medium financial institutions.

Source: Schulte Research.

OneConnect's Smart Banking Cloud helps small to medium-sized banks and financial institutions use Fintech to improve customer experience and run their business more efficiently. The Smart Banking cloud accounts for almost all of OneConnect's business.

Advantages	FI to Consumer	FI to SME	FI to FI
Customer	Provides a platform for consumers to compare and choose FI's	Provides convenient, remote contract signing and same day approval	Trading platform for interbank transfers with a 30% faster trading time at 20% lower cost and 100% use of blockchain
Financial Institution	Efficient lending via biometrics and risk prevention via cloud	Complete and dynamic risk assessments throughout process	

Figure 8.24: OneConnect: Smart banking cloud.

Source: Schulte Research.

OneConnect's Smart Insurance platform:
- 23 insurer partners
- Two services use O2O to boost customer experience and company ops
  - Smart Claim: AI image based loss assessment for car crashes (5 million)
    - Rapid investigator arrival and claim issuing
    - Recognition of vehicle type (90% of models)
    - Damage recognition and part pricing in few seconds
      - 20 million part prices, 5 million work hour processes, 60k vehicle types
      - Adjusts costs based on region
    - AI Risk prevention
      - Prevents fraud and claims mistakes
    - Benefits: 3 billion reduced risk leakage, 40% increase claims efficiency
  - Smart Verification: using biometrics to speed up claims and policy sales
    - New policy surrender rate down to 1.4% from 4% industry avg.
    - 3 days -> 30 minutes for claim receiving
    - 90% of customer service dealt with via smart verification

Figure 8.25: Smart insurance cloud.

Source: Ping An Press Release, Schulte Research.

Despite OneConnect's 62% increase in partnerships during FY 17, Q1-18 saw slowdown to only 4% increase in partnerships.

Credit inquiries also dropped from 287 million in Q4-17 to 180 million in Q1-18.

OneConnect had a US$650 million Series A round in early 2018 (valuation of US$7.4 billion.). However, a string of failed IPOs pushed back timeline of going public.

Expect sluggish growth: OneConnect provides a valuable service for small to medium-sized financial institutions, but until the global markets settle, technological integration is a secondary concern for most of these financial institutions.

Figure 8.26:   OneConnect: Future prospects.

Source: Schulte Research.

Facial Recognition: 99.8% Accuracy, Best in the World
- ID confirmation reduced fake identities to 0%
- Emotional state detected in 1 second: 98.1% accuracy

Voiceprint Recognition
- ID detection with 99.7% accuracy
- Emotion recognition

500 data scientists employed
20,000 R&D staff members

End sale surrender rate reduced from 4% to 1.4%.

Figure 8.27:   Biometric data.

Source: Ping An Tech Innovation Presentation, Schulte Research.

Expanding into PRC healthcare industry to secure market share in a rapidly growing and fragmented segment of the economy.

Turning healthcare into a core business for years. The industry is finally poised for a small number of players to enter, reform, and control Chinese healthcare.

Ping An's experience and data from insurance (combined with their extensive financial prowess and technology) can allow them to be a healthcare market leader.

The outcome of this expansion attempt will have a greater impact on the long-term future of Ping An than any other segment of their business.

Figure 8.28: Ping An wants Chinese healthcare, Outcome will define company future.

Source: Frost & Sullivan, Schulte Research.

Three Facts about Chinese Healthcare

1. Healthcare expenditure will continue to increase rapidly.

2. Healthcare providers cannot meet demand without efficiency improvements.

3. Market concentration will increase across healthcare segments.

Figure 8.29: Chinese healthcare: Industry overview.

Source: WHO, BMI, Schulte Research.

A diabetes epidemic, an aging population, and long-term health effects from poor air quality guarantee that healthcare expenditure in China will rise for the foreseeable future.

Healthcare spending is estimated to grow to over US$1 trillion in 2020 with a CAGR of nearly 12% over the period from 2017–2022.

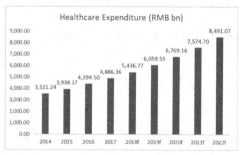

Figure 8.30: Healthcare expenditure is guaranteed to grow, and fast.

Source: BMI, IBIS, Schulte Research.

The healthcare system will be unable to meet increased demand for healthcare without a fundamental change in the system.

The average duration of a doctors appointment in China is 8 minutes. The average wait time at a Chinese doctor's office is 3 hours.

Patient routing is problematic: 8% of hospitals are responsible for nearly half of the total visits.

Figure 8.31: Growing inability to meet market demand.
Source: Frost & Sullivan, IBIS, Schulte Research.

The government is beginning to realize how incredibly expensive sick people are.

Without cost reduction in healthcare services, the government's social insurance program would begin running a deficit as early as 2020.

The fragmentation throughout the healthcare industry causes a high level of inefficiency that the government fully intends to stop.

Figure 8.32: Government regulations forcing industry consolidation.
Source: Frost & Sullivan, IBIS, Schulte Research.

The three market drivers: internet healthcare will have a bigger share of a larger market with fewer players.

Penetration rate of online consultations was only 1.8% in 2016, but will grow at 45.2% CAGR 2017–2021E.

Ping An is positioned to dominate the rapidly growing and consolidating healthcare industry:
- Government push for fewer, larger players in marketplace.
- Regulations evolving to allow online diagnoses, prescriptions, and pharmaceutical sales.

Ping An's Strategy:
- Use AI and massive online–offline dataset to improve treatment.
- Stay involved in every step of consumer healthcare value chain.

Figure 8.33: Ping An is perfectly suited to lead China's healthcare reform.
Source: Frost & Sullivan, IBIS, Schulte Research.

Ping An Good Doctor is an integrated one stop portal for
1) online consultations information,
2) online sale of standardized health services,
3) medicine and health e-commerce,
4) health management services.

Figure 8.34: Ping An Good Doctor.

Note: This ecosystem epitomizes O2O efficacy as the world's largest health platform centered around an app.

Source: Good Doctor, Pre-IPO Presentation, Schulte Research.

Four Business Activities	Revenue mn RMB (% of Total)	Gross Profit mn RMB (% of Total)	Gross Margin
Consumer Healthcare	655.4 (35%)	304.2 (49.7%)	46.4%
Family Doctor Services	242.2 (13%)	142.5 (23.3%)	11.7%
Health Mall	896.1 (48%)	104.6 (17.1%)	58.8%
Health & Wellness Management	74.3 (4%)	60.8 (9.9%)	81.8%
Total	1,868	612.1	32.8%

Figure 8.35: Ping An Good Doctor revenue streams.

Source: Frost & Sullivan, IBIS, Schulte Research.

One of China's few underdeveloped industries: ePharmacies have only 1.5% of market share compared to 31.8% in the United States.

Regulatory consolidation push on pharmaceutical sales supply chain offers tremendous growth potential in an industry where the five largest companies only control 6% of market share.

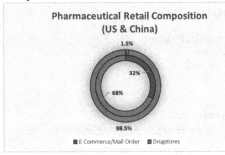

Chinese pharmaceutical retail has grown at 13.0% annualized rate with revenue expected to be US $142 billion in 2018.

Ping An deliberately holding back on ePharmacy offerings until they're confident in sourcing.

By 2020, ePharmacy will drive Good Doctor's revenue.

Figure 8.36:   Good Doctor: Health Mall.

Source: SCMP, IBIS, Schulte Research.

	Pharmaceutical Industry
Current Industry State	Fragmented
Government Concerns	Inefficiencies, poor drug quality
Government Response	• Quality control on manufacturers • Strict requirements for retailers • Two-invoices policy: medication can only pass hands twice from factory to hospital
Impact on Industry	Consolidation (est. 20% market share up for grabs)

Figure 8.37:   Regulatory environment changes to pharmaceuticals.

Source: Ping An Vitality, Schulte Research.

Vitality Program: Points Based Reward Program for Healthy Living
1. Earn Points through Healthy Living
   - Online Assessments
   - Non-Smoker Declaration
   - Go for a Physical
   - Connect Wearables
2. Increase Status Based on Points Earned
   - Five Tiers:
     - Blue, Bronze, Silver, Gold, Platinum
3. Earn Rewards for Healthy Choices
   - Leisure and Lifestyle Cash Back
   - Health Related Rewards
     - Wearables
     - Gene Tests
     - Appointment Discounts

Figure 8.38: Ping An's vitality program.

Source: Ping An Technology Conference, Schulte Research.

Wanjia Clinics: Think OneConnect for healthcare providers
- Provides cloud services for clinics throughout China
  - Electronic medical record management
  - AI enhanced diagnostic support
- One in three clinics in China use Wanjia (60,000 partner clinics)

Ping An Healthcare Management: City OneConnect's Healthcare Segment
- Provides cloud services for the PRC's public insurance system
  - Involved in coverage of 800 million citizens and 250 cities
- Disease forecasting in Chongqing and Shenzhen
  - Predicts flu patterns and provides city with information on where to position health resources one week in advance

Figure 8.39: Good Doctor is the lynchpin for healthcare entry, but Ping An's involvement includes government and care provider support.

Source: Ping An Technology Conference, Schulte Research.

Chapter 9

Huawei and the Global Landscape in 5G and Cloud

9.1 Huawei: The R&D Nerve Center of China

Huawei Technologies Co. Ltd. is headquartered in Shenzhen, Guangdong since 1987. It is officially a private employee-owned company specializing in telecommunications equipment, networking equipment, and consumer electronics. Founded by a former engineer in the People's Liberation Army Ren Zhenfei, it has over 180,000 employees in 161 countries with 80,000 in Research and Development as of December 2017 with 36 joint innovation centers in no fewer than 14 countries.[1] In each country Huawei operates, 70% of staff is hired locally.[2] Its 2017 revenue and net cash flow from operating activities were US$92.55 billion and US$14.77 billion respectively with a cumulative average growth rate of 26% and 44% since 2013 (Figure 9.1).

---

[1] https://www.huawei.com/en/about-huawei/sustainability/win-win-development/develop_love

[2] https://www.linkedin.com/pulse/were-huawei-your-partner-choice-hong-eng-koh-%E9%AB%98%E5%AE%8F%E8%8D%A3-/

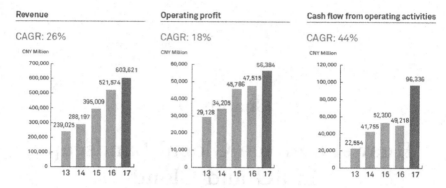

Figure 9.1:   Financials of Huawei.

Source: Huawei.

In the wireless network domain, Huawei has launched an end-to-end 5G solution, including commercial products of wireless networks, transport networks, core networks, and customer premise equipment (CPE). It joined forces with carriers and mainstream device chipset vendors to complete the interoperability development testing (IODT).[3] In the cloud domain, it has launched a range of cloud services such as enterprise communications and IoT platforms, as well as storage as a service. Huawei innovates in the AI, cloud computing, and Big Data domains together with the customers and partners. Huawei also uses innovative chip technologies and architectures to promote enterprise digitization and intelligent transformation.

Huawei is expanding its AI strategy with its new AI Developer Enablement Programme that includes an investment of US$140 million for the enhancement of the skills of AI developers.[4] Huawei is planning to support a million AI developers and partners for the next 3 years to take advantage of the US$380 billion AI industry by

---

[3]https://www.huawei.com/en/about-huawei/corporate-information/research-development

[4]https://sbr.com.sg/information-technology/news/huawei-invest-us140m-in-global-ai-developer-programme

2025.[5] Huawei is known for its new products in the technology world that includes their own developed AI chips in its Ascend series,[6] the Atlas Intelligent Computing Platforms,[7] and mobile data centers MDC 600.[8]

In AI and Big Data, through Enterprise Intelligence (EI) platform and distributed services based on Big Data, database, and AI technologies, Huawei helps enterprises boost productivity through data innovation in two main areas:

(1) In public safety, Huawei created a "traffic brain" for the city of Shenzhen to manage traffic lights, improve traffic flow, and increase automated detection efficiency of traffic violations.
(2) In finance, Huawei has launched an integrated data warehouse platform to reduce the cost for capacity expansion of the data warehouse system and has bolstered the performance of the core analytics function.

Huawei Technologies Company made headlines after the surprise Canadian detention of its CFO Ms. Meng Wanzhou in the last quarter of 2018. Analysts opined that more countries would blacklist its switches, routers, and phones out of growing concern that they could be spied on by foreign agencies. On the other side of the fence, technologists and businesses in China are focusing on opportunities as China pushes to create an industry that is less dependent on cutting-edge US semiconductors and software.[9]

---

[5]https://sbr.com.sg/information-technology/news/huawei-eyes-backing-1-million-ai-developers-in-3-years
[6]https://www.huawei.com/en/press-events/news/2018/11/huawei-ascend-310-world-internet-conference-2018-award
[7]https://www.huawei.com/en/press-events/news/2018/10/atlas-intelligent-computing-platform
[8]https://e.huawei.com/hk/news/it/201810151059
[9]https://www.businesstimes.com.sg/technology/inside-huaweis-secret-headquarters-china-is-shaping-the-future

Some argue that a "Silicon Curtain" has emerged that both the US and the Chinese will focus on using their own produced chips and not provide chips to one another. Huawei is only second to the US San Jose based Cisco System Inc in telecom and networking equipment. Huawei is also second to the Korean's Samsung Electronics Co in the smartphone market according to IDC Statistics in September 2018 with 15% world's shipment. Perhaps China is just seeking for self-sustaining and uninterrupted supply in semiconductors, which is the core technology of the future.

Whatever the speculation, Huawei is the icon of the rise of China in the fourth industrial revolution. It is leading in the semiconductor efforts in customizing chips for cloud and AI applications in the areas of image, voice recognition and autonomous vehicles. Huawei spent US$13 billion on research and development in 2017, an increase of more than 17% from the previous year. The sales of its chip design unit HiSilicon reached US$7.6 billion in 2018. However, China is not all smooth in its growth in the cloud business and AI applications. Its AI technology is still lagging behind the US despite the production of several chips including the Ascend, a specialized chip with high computing performance. The AT&T deal to distribute Huawei smartphones through the carrier was canceled,[10] and six top US intelligence chiefs told the Senate Intelligence Committee that they would not advise Americans to use products or services from Huawei.[11]

The largest customer is the Chinese government, and Huawei can become the "nerve center"[12] for smart cities with powerful chips set that can process video footage for as many as 16 cameras, a 400%

---

[10]https://www.scmp.com/news/china/diplomacy-defence/article/2127528/colla pse-huawei-deal-att-will-threaten-china-us-trade

[11]https://www.cnbc.com/2018/02/13/chinas-hauwei-top-us-intelligence-chiefs-caution-americans-away.html

[12]https://e.huawei.com/sg/publications/global/ict_insights/201806041630/focu s/201808170835

https://www.huawei.com/en/press-events/news/2017/11/Huawei-Smart-City-Nervous-System-SCEWC2017 jump in processing power. Huawei will continue to dominate the growth in 5G, customized chipset, and AI technology. The domestic demand will continue to drive Huawei's research, and development with expansion into South America, Africa, and Asia rather than the USA, Europe, Australia, and New Zealand.

Figure 9.2 shows the speculation of how Huawei and ATT/ Verizon are linked to the Government. IQT's partners include the Central Intelligence Agency, National Security Agency, and Department of Defense, and others in the intelligence and defense communities.[13] However, there is no evidence to suggest that there is any truth in the case of Huawei. Huawei is 100% private company owned by its employees and has no corporate links to Alibaba, Tencent, State Council, Ministry of State Security (MSS) or China Development Bank (CDB). Set up in 2014, the Cyberspace Administration of China (CAC; 国家互联网信息办公室), also known as the Office of the Central Cyberspace Affairs Commission (中央网络安全和信息化委员会办公室), is the central Internet regulator, censor, oversight, and control agency for the People's Republic of China.[14]

The Central Cyberspace Affairs Commission (中央网络安全和信息化委员会), formerly known as the Central Leading Group for Cybersecurity and Informatization (中央网络安全和信息化领导小组) is a policy formulation and implementation body set up under the Central Committee of the Communist Party of China to manage internet-related issues.

Ministry of Industry and Information Technology (MIIT; 中华人民共和国工业和信息化部[15]) of the Chinese government, established in March 2008, is the state agency responsible for regulation and development of the postal service, Internet, wireless, broadcasting, communications, production of electronic and information goods, software industry and the promotion of the national knowledge economy.[16]

---

[13]https://www.iqt.org/how-we-work/national-security/
[14]http://www.cac.gov.cn/
[15]http://www.gov.cn/fwxx/miit.htm
[16]http://www.gov.cn/english//2005-10/02/content_74176.htm

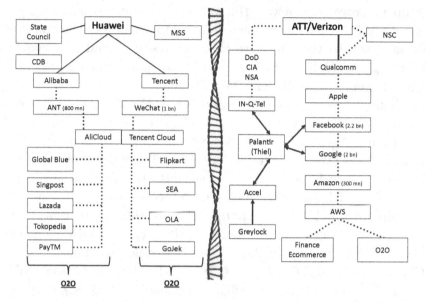

Figure 9.2: PRC vs the US.

Note: *Numbers are users.

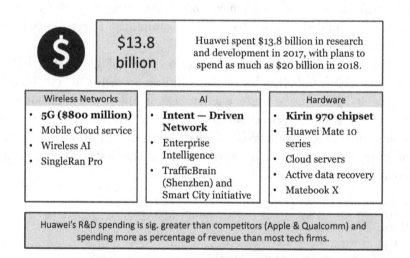

Figure 9.3: Huawei's research and development in 2017.

Figure 9.3 shows the R&D expenditure in 2017. Huawei's planned spending for 2018 R&D was US$20 billion, up from US$13.8 billion a year ago in 2017. 5G network, Intent-Driven Network, and Kirin 970 chipset are three main areas. Huawei's spending will soon catch up with Amazon which spent US$22.6 billion in 2017. Figure 9.4 shows the R&D expenditure for different tech companies.

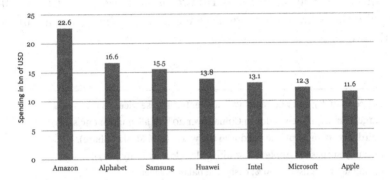

Figure 9.4: R&D spending in 2017: 4ᵗʰ among tech.

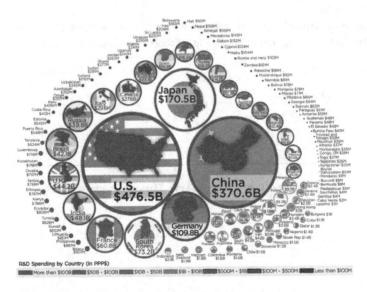

Figure 9.5: R&D spending in 2017: 2ⁿᵈ after the US.

The US still topped R&D spending at US$476 billion in 2017, followed by China's US$370 billion, Japan's US$170.5 billion and Germany's US$109.8 billion (see Figure 9.5). The fastest growth in Huawei came from Telecom (56%), followed by cell phones (34%) and enterprise services (98%). Huawei is China's largest smartphone vendor and third largest in the world. However, in July 2017, it surpassed Apple's in monthly sales globally. The profit margin remains strong at 7–9%, and Huawei is cash rich. Figures 9.6–9.10 describe the growth strategy, revenue, market share, and acquistions of Huawei.

- Telecom (56%), cell phones (34%), and enterprise services (8%).
- Largest smartphone vendor in China (over 20% of all smartphone sales).
- Third largest smartphone vendor in the world (11% Market Share).
- Recently surpassed Apple in monthly sales globally (July 2017).
- Strong net margins of 7–9%; cash-rich balance sheet.

Strategy
- Huawei's enterprise software in the cloud is now integrating into Huawei smartphones.
- Aggressive expansion into emerging markets with focus on SE Asia, India, and Africa.
- Competitive pricing and willingness to collaborate with local companies.

Revenue	Net profit	Cash flow from operating activities
603,621	**47,455**	**96,336**
CNY Million	CNY Million	CNY Million
↑ 15.7% YoY	↑ 28.1% YoY	↑ 95.7% YoY

Figure 9.6:   Summary: Huawei's growth.

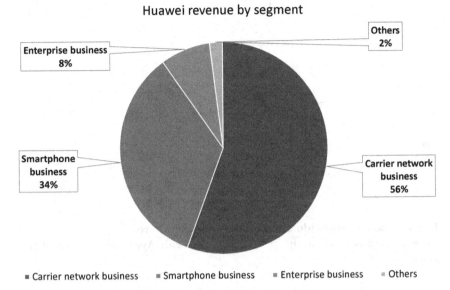

Figure 9.7: **Huawei revenue by segment: A telecom giant with fast growing smartphone business.**

Source: Huawei Annual Report, Schulte Research.

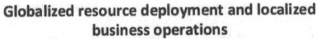

Figure 9.8: **Worldwide smartphone market share.**

Note: Huawei operating in 170+ countries, spreading rapidly in Eastern hemisphere.

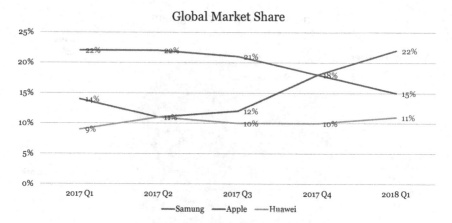

Figure 9.9:   Worldwide smartphone market share.
Note: Huawei is catching up. It is almost tied with Apple and it is number 3 globally.

Date	Acquired	Area
December 27, 2016	HexaTier	Database as a Service
December 7, 2016	Toga Networks	R&D: IT/Telecom
September 22, 2014	Neul	Cloud computing and IoT
November 14, 2011	Huawei Symantec	R&D: Security and Storage
February 19, 2011	3Leaf	Audio
September 11, 2006	Harbour Networks Holdings	SaaS

Figure 9.10:   Huawei's acquisitions.

9.2 Geographical Reach and Landscape

The contribution from Africa was around 15% of total revenue. The strategy of selling 5–15% cheaper than rivals such as Nokia and Ericson has gained the reputation as a "preferred low-cost, yet high-quality mobile network builder". Huawei has a 16% market share in South Africa and over 1 million P Lite Smartphones sold in the range of US$100–1200. With a new warehouse located in the free trade zone or Tambo International Airport in Africa, delivery time has been shortened from 3 weeks to 3 days. It made a US$100 million in a new R&D facility on campus. Figures 9.11 and 9.12 describe the activities in Africa.

Africa contributes 15% of Huawei's total revenue

Strategic pricing
- Reputation as a "preferred low-cost, yet high-quality mobile network builder"
- 5–15% cheaper than competitors Nokia and Ericsson

Growing Market Share with focus in South Africa
- 16% market share in South Africa, over 1 million P Lite smartphones sold
- Boosting $100–200 smartphones in Kenya

New warehouse and R&D Campus to be built in Johannesburg, first for Huawei in Africa
- Shortening delivery time of Huawei stock from 3 weeks to 3 days
- Warehouse located in free trade zone of OR Tambo International Airport
- $100 million investment to new campus

Figure 9.11: Africa: Growing presence in over 50 countries, among the top three telecommunications companies in the region.

Figure 9.12: Africa: Strong presence reflecting popularity as vendor.
Source: Apple, Huawei.

- Xpress Money and Huawei to bring **Huawei Pay**, targeting Africa's unbanked population

- **Huawei Music** to launch in Africa late 2018, with a catalogue initially made up of 80% South African music
  - Following Spotify's release in SA

- Plans to roll out Cloud Suite in Africa
  - App Gallery
  - Theme Store
  - HiCloud
  - Screen Magazine

- Currently has 5,800 Employees in African Region, over 60% are local hires

Figure 9.13: Africa: Collaborations to bring cloud technology to region.
Source: Apple, Huawei.

Figures 9.14 and 9.15 describe the activities in South East Asia. Figures 9.16 and 9.17 are for Latin America, and Figures 9.18 and 9.19 are for Europe.

Huawei to invest $81 million in Open Labs (ICT training centers) in SE Asia

Chinese smartphones are the leading mobile providers in the region
- Market traditionally dominated by Samsung, but Oppo, Vivo, and Huawei Technologies cornered a combined 29.6% market share over Samsung's 29.1% in 2017
- Oppo leading with 17.2%
- Huawei (5.1% market share) falling behind compared to other smartphone companies

Huawei strategy to enter affordable smartphone market through ICT collaborations with governments. See Thailand.

Figure 9.14: SE Asia: Thailand.

Note: Huawei strongly relied upon by Thailand gov to realize Thailand 4.0 vision.

Source: Apple, Huawei.

Thailand 4.0: Economic model aimed at pulling Thailand out of the middle-income trap, and push the country in the high-income range through cultivating a digital economy.

Huawei chosen by Thai gov. to help advance country's technology sector
- $15 million OpenLab Bangkok will offer
  - ICT training (800 persons/year)
  - ICT career certification (500 persons/year)
  - proof concept testing (150 persons/year)
  - host to a number of local and international collaborations, especially with the Thai Gov. Huawei strongly relied upon for Thai digital expansion.

Figure 9.15: SE Asia: More on Thailand.

Note: Huawei strongly relied upon by Thailand gov to realize Thailand 4.0 vision.

Source: Apple, Huawei.

- Seven offices, from Argentina to Mexico
- Over 12 million smartphones sold by 2016
- Huawei losing focus on Latam for SE Asia expansion

Telecom Argentina chose Huawei in 2016 to construct its all-cloud core networks

- Telecom Argentina now becomes the Latin American pioneer in cloudifying core networks
- Sparked more partnerships
- Spanish telecommunications Telefonica and Huawei to launch public cloud services in Chile, Brazil, and Mexico

Telefonica is Huawei's fifth largest customer

- Signed global alliance to develop Network Service Platform and CloudVPN for SMEs
- Huawei and Telefonica completed world's first Proof of Concept for 5G connection

Figure 9.16: Latin America: Strong presence, still engaging in enterprise but smartphone sales declining as focus is drawn to SE Asia.

Source: Apple, Huawei.

LATAM Smartphone Shipments (Millions Units)	2016	2017	YoY % Growth
Samsung	49.1	56.1	14%
Motorola	10.9	17.1	56%
LG	14.0	13.3	-5%
Huawei	11.5	11.1	-4%
Apple	7.1	6.1	-14%
Others	47.0	42.8	-9%
Total	139.5	146.5	5%

LATAM Smartphone Shipments (% Share)	2016	2017
Samsung	35.2%	38.3%
Motorola	7.8%	11.6%
LG	10.0%	9.1%
Huawei	8.3%	7.6%
Apple	5.1%	4.2%
Others	33.7%	29.2%
Total	100.0%	100.0%

Figure 9.17: Latin America: Huawei losing focus for SE Asia expansion.

> **Huawei is third most popular vendor, with 16.1% market share, 38.6% growth**
> - Huawei is top smartphone seller in Portugal and the Netherlands
> - Huawei — second biggest in Italy, Poland, Hungary, and Spain
> - Growing presence in European Countries
>
> **17 R&D sites**
> - Belgium, Finland, France, Germany, Ireland, Italy, Sweden, and UK

Figure 9.18: Europe: Growing presence, solid grounding in many countries.

Source: Apple, Huawei.

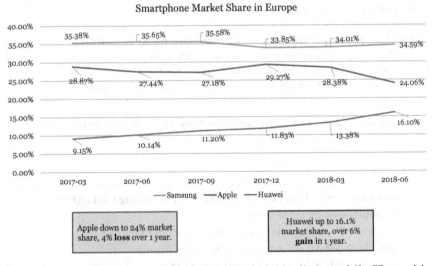

Figure 9.19: Europe market share: Apple is declining while Huawei is experiencing consistent growth.

Figures 9.20 and 9.21 compare the flagship chips of Huawei Kirin 970 with Apple's A11 Chips. Figure 9.22 gives an idea of the capabilities of HiSilicon, a subsidiary of Huawei. Figures 9.23 and 9.24 show that Huawei had already surpassed Apple phone sales in July 2017 and Huawei is leading in smartphone market share. In

Figure 9.25, Huawei is shown to be leading the telecom infrastructure sector and ahead of Ericsson from Sweden.

	Huawei Kirin 970		**Apple A11**
	8-core CPU — 4 Cortex-A73@2.4GHz, 4 Cortex-A53@1.8GHz		6-core CPU — 2 Monsoon, 4 Mistral
	12-core Mali-G72 MP12 GPU		3-core custom Apple GPU
	1.92 trillion operations per second		600 billion operations per second (1/3 of Huawei)

Tensor processing speed: Huawei is three times faster

Floating-point operation speed is benchmark for AI capacity
- Faster chips enables better AI operations for the customer
- Both chips have dedicated AI Neural Engines, but Huawei's is faster
- AI-powered products will be commercialized much faster for Huawei

Figure 9.20:   Flagship SoC (system on a chip) comparison.

Huawei Kirin 970 is first to contain a Neural Processing Unit (NPU)
- Dedicated processor for AI operations
- 1.92 TFLOPS — more powerful than a PS3's GPU (1.84 TFLOPS)
- Apple developed A11 which included a NPU after Huawei
- Supports mobile AI software (Tensorflow and Caffe)

Pioneer in mobile AI
- Instantaneous photo super-resolution
- Home mapping in 3D
- Real-time realistic AR in gaming

Figure 9.21:   Huawei and Apple comparison, Huawei ahead.
Source: Apple, Huawei.

Huawei phone chips produced by Huawei owned subsidiary **HiSilicon**

- Semiconductor and Integrated Circuits design.
- Gives Huawei in-house advantage for chip design.
- **Apple has had way too many problems trying to source its manufacturers in China.**
- Huawei has the sufficient resources, trend, and infrastructure to scale up.

Figure 9.22: Manufacturing: Huawei CPUs.

Note: Huawei has in-country advantage to make its own CPUs. HiSilicon the largest chip designer in China.

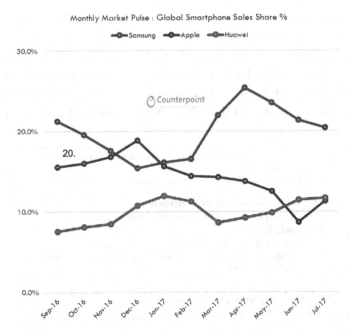

Figure 9.23: Huawei has already surpassed Apple in global sales shares in July 2017.

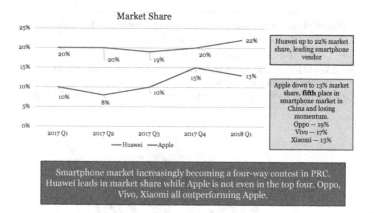

Figure 9.24: **PRC market share.**
Note: Apple is behind in one of its largest markets and saw growth SHRINK.
Huawei has momentum and leads the pack.

9.3　Telecoms Infrastructure: 5G

Figure 9.25 shows the latencies and download speeds. Figures 9.26–
9.30 look at the comparison of telecoms standard. Figure 9.31 compares Qualcomm x50 chip with that of Huawei's Balong 5G01 chip.
Figures 9.32 and 9.33 look at the telecoms infrastructure landscape.
Figures 9.34 and 9.35 raise the concerns regarding cybersecurity and
control. Figures 9.36 and 9.37 look at the competition in Korea.

> - Ericsson struggling in emerging markets while Huawei performed
>   well in Africa and SE Asia.
> - Huawei the only telecom vendor to increase market share in 2017.
>
> **Huawei is cheaper and more efficient**
> - Huawei cut industry profit margins from 50% to 30%.
> - Better technology
>   - SingleRAN : can handle multiple types of signals (2G, 3G,
>   WiMax, CDMA, GSM) on one box, freeing a carrier from building
>   separate networks.

Figure 9.25: **Huawei and infrastructure.**
Note: Huawei the leading telecom infrastructure vendor, beating Ericsson AB
(Sweden).
Source: Apple, Huawei.

Latencies: from 25 ms (4G) to 1 ms
- Military: autonomous driving, drone strikes, mission critical applications

Download speeds: from 1 gbps to 10—20 gbps
- Greater connectivity between devices
- Could be tool to actualize Internet of Things — theorized network of devices all connected to the internet
  - Smart homes
  - Autonomous cars
  - Smart cities
  - Smart farming

Figure 9.26: 5G Promises and Huawei.

Note: 5G promises faster connections, greater download speeds, and near 0 latency.

Source: Reuters, Huawei.

- Telecommunications standard governed by **3GPP** — a committee of members from all major telecommunications companies (Samsung, Huawei, China Mobile, Samsung, Nokia, etc.)
  - Chaired by Balazs Bertenyi (Nokia)
  - Vice-Chair Stephen Hayes (Ericsson), Xiaodong Xu (China Mobile), Satoshi Nagata (Docomo)
  - Decisions made are enforced as industry standard, and made by vote
- To be a member of 3GPP, you have to be a member of one of the seven major telecommunications standards development orgs. Ex. China has CCSA
  - These companies are representatives of their countries. Inherently political.
  - Standards are largely driven by which paradigm the company has more patents in.
- **Leading solutions: Polar (Huawei) vs. LDPC (Qualcomm)**

This is the first time a Chinese company has proposed a viable telecommunications paradigm that could be industry defining.

Figure 9.27: Standard: Governing board 3GPP defines the 5G Standard.

	Turbo	LDPC	Polar
Maturity	Proven 3G & 4G	Proven in Wi-Fi	Unproven
Age	30t+ years.	30+ years.	New (<10 years)
Flexibility	Flexible	Inflexible	Inflexible
Complexity	High	Low	Low
Backwards Compatible	Yes	No	No
Performance	Poor with small and large blocks	Poor with small blocks	High all around
Players	European Orange (FR) Ericsson (SE) LGE (SK)	North American Qualcomm (US) Intel (US) Samsung (SK) Nokia (FI)	Chinese Huawei (CN) China Mobile (CN)

Figure 9.28: Standard: Qualcomm leading push for LDPC; Huawei for Polar; scattered support for Turbo.

- Europeans pushed for **Turbo**, but were defeated and are now swing voters.

Voting Results of past meetings

PARADIGM	11.18.16 (Athens, GR)
LDPC Qualcomm-led	Qualcomm, T-Mobile, Verizon, AT&T (US) Samsung (SK)
LDPC and Polar (Huawei) HYBRID	Huawei, China Unicom, China Telecom, Motorola, Lenovo OPPO, Vivo, Xiaomi, ZTE, Alibaba *(CN)* Deutsche Telecom, Ericsson (EU)

Figure 9.29: Standard: Current debate revolving around the coding scheme for 5G — Heightening split between East and West.

- **June 2018 Release 15**: First 3GPP specs determined with hybrid solution
  - LDPC as data channels
  - Polar codes as control channels

- First time that Huawei's technology is incorporated in the standard
  - Polar technology will be used in all 5G development, huge win for China
  - **Huawei owns most polar encoding patents**

Figure 9.30: Standard: Ultimate decision with LDPC as data channels and Polar as control channels.

- **X50 chip demonstrated speeds of 4.51 gbps, the industry record**
- However, Apple and Samsung are not joining Qualcomm's 5G initiative (legal battles)

- **Huawei Balong 5G01 chip tested at 2.3 gbps**
- Mobile connection and remote driving demonstrated

Figure 9.31: Chipset: Qualcomm X50 outperforms Huawei's Balong with double download speed.

Huawei's advantage is its huge policy support from PRC
- Telecommunications is a **top-down industry**
  - 5G technology requires $200–300 billion of new infrastructure
  - 100 base stations minimum per city
  - Huawei has government support to expedite construction (China telecom)
  - **Huawei already controls 28% of global network equipment market**

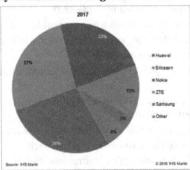

Figure 9.32: Telecoms infrastructure (1): Huawei already has leading market share with 28% of global market and PRC support.

Subscriber Data Management: Class of data of what users (subscribers) do when interacting with a network
- Overlooked wealth of information about every user
- Applications in Identity Management, Targeted Services and ads, Internet of Things
- Evolving from static integration of data to a dynamic personalized service enabler

In 5G network, services do not store data from one session to the next but instead rely on common external data management
- Instead of SDM, operators need cloud data management to store information

Huawei remains lead in field, which positions them well for 5G transition
- 2017 — Huawei leads with 33% market share in networks data management
- Most mature and reliable
- Services over 300 carriers in more than 130 countries

Figure 9.33: Telecoms infrastructure (2): Huawei has advantage in networks data management, essential for 5G.

Huawei has close ties to PRC, subject of investigation by the US
- Huawei sales already blocked in US and Australia
- Broadcom (Singapore) $117 billion buy of Qualcomm blocked by Trump citing security implications of Qualcomm's 5G R&D
- Ren Zhengfei began Huawei with a government loan after a career in the PLA
- Huawei growth aided by PRC support, incl. ~$100 billion credit line by CDB
- PLA is still Huawei's oldest customer
- Huawei and ZTE own 70% of China's telecommunications market

In leaked NSC document, 5G considered national security concern that necessitates nationalization
- Claim: Huawei is China's tool to nationalize 5G network
- US wants to do the same: nationalize a 5G network
- Fear that if China dominates 5G infrastructure, they will write backdoors into telecommunications for the entire world
- Claims are largely unproven

Figure 9.34: Implications (1): 5G Rollout will be defined by international cybersecurity concerns.

Source: Reuters, Huawei.

Whoever defines 5G standard will be industry leader and will define telecoms infrastructure for the world

- Governments are scared that telecoms infrastructure will be used as a method for spying
  - Huawei banned in US and Australia
  - Under attack & investigation for PRC government ties and breaking trade sanctions on Iran and North Korea
  - AT&T, Verizon, and BestBuy distribution deals blocked by Pentagon

Figure 9.35: Implications (2): 5G has massive applications, but the greater focus is who owns the infrastructure behind 5G tech.

Source: Reuters, Huawei.

- SK Telecom: Both a telecommunications provider and a lead in 5G R&D
  - Successful in innovative 5G applications (AI)
  - Acquired Quantum Computing cybersecurity company QKD for 5G security
- Together with KT and LG U+, they hold 99% of South Korean market share
- Korea vying to be first 5G country, with initial 5G telecoms infrastructure worth ~$9 billion
- Competition between **Samsung** and **Huawei** to bring infrastructure SK

Figure 9.36: Implications (3): Korea will most likely adopt 5G first, competition will define global 5G development.

Samsung long dominating player in Korea
- Long standing partnership with SK telecom to develop 5G technology
- Has always dominated SK market

After turned away from US, Huawei looking to expand to SK
- SK Telecom & KT Corp. in talks with Huawei, Samsung, Qualcomm for 5G contracts
- **LG U+ has already signed MoU with Huawei for their bid**
- Asserts Huawei growth and popularity in South Korea

> If Huawei defeats Samsung in SK, is significant financial win with broad symbolism. Chinese tech company unifying Asian market and capturing lead market share with 0 US support.

Figure 9.37: Implications (4): Huawei (28%) ahead of Samsung (3%) in telecoms infrastructure, high possibility to take SK.
Source: Reuters, Huawei.

9.4 Networks Infrastructure: Intent-Driven Network

Figure 9.38 describes the network infrastructure with Figure 9.39 looking into Huawei's flagship network technology. Figure 9.40 introduces the OpenLabs and finally, Figure 9.41 describes Huawei's rail cloud for Shanghai Metro.

Huawei growth attributed to expansion in Latin America, Europe, the Middle East, and Africa
- Competitive pricing and AI technology
- CISCO still holds massive market share (59%) because it dominates the profit-heavy US, where Huawei is banned
- Huawei is "China's Cisco" and is becoming more popular in emerging companies due to **Open Labs Initiative**
- Flagship's AI embedded network is called **Intent-Driven Network**

Figure 9.38: Networks infrastructure: Cisco (56.7%) dominates market, but Huawei (7.7%) beat Juniper (3.2%) for second place.
Source: Reuters, Huawei.

Intent-Driven Network (IDN) — introduces Big Data and AI technologies into All-Cloud Networks. Helps enterprises go digital.

- Campus Insights collects network data in real time to implement network fault analysis and proactive prediction
- Next-generation SD-WAN solution
- Highest-density data center switch
- Software-defined Security
- **All components rank either first or second in the Chinese market**

Huawei — Ping An partnership to bring IDN to the finance industry

- Ping An using Huawei's SD-WAN intent-driven interconnection to launch the first AI customer service
- Matching face recognition and voiceprint with Big Data to remotely verify customer identity information
- Solves the pain points of traditional insurance services, such as slow authentication and claim settlement

Figure 9.39: Intent-driven network is Huawei's flagship networks technology.

Source: Reuters, Huawei.

OpenLabs built to foster ICT industry in emerging countries and build collaborative solutions in developed economies

- Bangkok lab primarily training ICT workers
- Munich/Singapore labs focusing on bringing Huawei cloud technology to finance, transportation, energy sectors
- Ultimate goal of fostering smart cities, interconnected cloud infrastructures

120 cities in over 40 countries and regions implement Huawei smart city projects

- Plan to invest $200 million over 3 years to bring total labs from 12 to 20
- Suzhou, Munich, Mexico City, Singapore, Dubai, Johannesburg, Bangkok, Cairo, Istanbul

Figure 9.40: OpenLabs enable Huawei to market networks infrastructure to developing countries.

- Shanghai metro is second largest in world with 10 million riders each workday

- System's effectiveness has been hampered by legacy control center: three main issues
    1) Only supported basic rail network from years prior, not capable of controlling full system
    2) System lacked collaborative emergency response capabilities
    3) System unable to meet changing government requirements for operations

- Urban Rail Cloud is Huawei's application of their FusionCloud to Shanghai's metro; provides three primary advantages
    1) Strong cybersecurity protocols
    2) Multi-line information platform provides enormous data management improvements
    3) Expanded real-time data access allows better allocation of resources reducing costs and improving service

---

Figure 9.41: Huawei's rail cloud for Shanghai Metro.

9.5 Conclusion

1. On revenue, reach and R&D, Huawei is in top five telecom giants. This is why China is furious at US broadside.

   A. Huawei was started 31 years ago by a PLA major and has revenues of $100 billion with a presence in more than 170 countries. It also has R&D facilities in China, US, UK, Europe, India, Turkey, Latin America, and Israel. However, General Hayden of the NSA singled out Huawei as an "unambiguous national security threat" and its products have recently been banned in both the US and Australia. On the other side is a consortium controlled by Qualcomm, Apple, Google, Amazon, and Palantir with a "dotted line" to the NSC, NSA, and CIA. In fact, the *New York Times* today revealed that Secretary of Defense Mattis sent a memo to President Trump suggesting that the US implement a public–private program for technology for the US akin to China. The China 2025 national technology initiative was one of the reasons for the trade tariffs in the first place!

B. Huawei has a $14 billion annual R&D budget (4th largest globally) and has done advanced work in 5G, AI, Cloud, advanced chips, and quantum computing. China's national R&D budget is double that of Japan and almost on a par with the US. Its global market share for mobile phones is now greater than Apple, and its geographical reach is now truly global through massive outlays in internal growth, as well as M&A.

C. Tech specialists Kenta Iwasaki and Jason Kang from Schulte Research conclude the following when comparing the Huawei Kirin 970 and the Apple A11: (i) Kirin 970 is three times faster at 1.92 trillion per second and therefore is far more amenable for AI operations; (ii) Kirin's AI dedicated chip is better; (iii) Huawei developed the first Neural Processing Unit (NPU) and Apple's came later; (iv) Kirin has the newest instant photo super-resolution, home mapping in 3D and real-time AR; (v) Huawei's homegrown semi and IC design combines together cohesively; (vi) Huawei's massive R&D budget is 20% larger than Apple.

2. The race is on for telecom infrastructure and 5G, which is 10x faster. Huawei is cheaper and better, but politics allows them only to play second fiddle.

A. Huawei is the only vendor to increase telecom market share in 2017 with a dominant 28% market share. It cut industry margins from 50% to 30%. Its single RAN can handle multiple signals, nixing the need for separate networks. 5G is necessary for SMART everything: homes, drones, cars, cities, farms. The governing board is dominated by Nokia, Ericsson, DoCoMo, and China Mobile.

B. There is a definite problem of picking 5G. Polar is Huawei. Turbo is Ericsson. LDPC is Qualcomm. The final decision is a committee outcome: a muddled and inseparable hybrid combining Huawei and Qualcomm. Qualcomm's LDPC runs the data channels while Huawei's Polar runs the control channels. This is the first time Huawei has been included in all

5G equipment. The Qualcomm chip in this area of technology is superior. The rollout will be very expensive — north of $200 billion with 100 stations per city. Since it already has services to more than 300 carriers in more than 130 countries (with a 33% market share in network data management), this is a *fait accompli*.

C. The elephant in the living room is that Huawei's oldest customer is the PLA. Along with ZTE, it controls 70% of the Chinese market. It also controls 70 % of the software market for cell towers. The Qualcomm system, on the other hand, is linked to Google, Apple, Amazon, and US intel community. Both the US and China want to nationalize the 5G network, which is fundamentally incompatible with the unified nature of the system. This is highly problematic. Something will have to give. Korea will adopt 5G first, and then Africa will adopt it quickly. Huawei is present in 50 African countries.

3. Like 5G, the battle for the Cloud shows the absurdity (and impossibility) of bifurcating the data supply chain for smart cities, finance, payments, and banking).

A. The global market share of this $130 billion market is: (i) Alibaba, IBM, and Google are similar in mid-single digits; (ii) Amazon has 33% ($18 billion); (iii) Microsoft has 13% ($27 billion). Alibaba dominates in China with a 30% share followed by Microsoft, Tencent, and AWS. (Is Google coming into China to hook up with Tencent Cloud?) Amazon had a 4-year head start on everyone else.

B. Customers of all of these characters is a rich mélange of public, private, SMEs, social media, banks, manufacturing, and pharma. Moreover, they are utterly interlinked by geography as well as by industry. AWS and Microsoft are everywhere, whereas AliCloud's footprint is mostly in Asia. The great battle for this Cloud market will be seen in South and Southeast Asia with 1.5 billion people and millions of SMEs and thousands of conglomerates. Huawei is mostly restricted to China.

C. BOTTOM LINE: Huawei is an undeniable major player on the global stage now in R&D, market share, technological superiority, and key infrastructure for 5G. It dominates in Asia and Africa for telecom infrastructure. AliCloud dominates in Asia and significant parts of Europe. AWS has truly a global footprint as does Microsoft — inside and outside China. Google's re-entry into China may have a lot to do with a tie-up with Tencent Cloud.

Alibaba is the most undervalued of these companies and its deep and wide presence in fin tech (Paytm, Lazada, Tokopedia, Ant and many others) create a highly prized footprint. Lastly, this new 'cold war' language is perhaps a ploy to justify a large Pentagon budget. Trying to sever the telecom structure, chip technology, Cloud services, and 5G distribution is a fool's errand — it is impossible. It is likely more a show in a kabuki theater than for realpolitik. It is simply impossible to rend it in two because that would send us all back to the Stone Age and no one wants that.

Further References for Huawei are available on Wechat in Chinese.[17]

---

[17]https://mp.weixin.qq.com/s/QcI1T9ZuEmNTMPimnCzZkA
https://mp.weixin.qq.com/s/qLgevKaKvnocYYwITZi59g
https://mp.weixin.qq.com/s/SNqijP8NNNMA6W4ll0MHMA
https://mp.weixin.qq.com/s/5MZ5VEOhs1cdOne_ku_0g
https://bianjie.zuzitech.com/index/article/view/id/436720
https://mp.weixin.qq.com/s/N3GTZB2QoBLZT6qz279trg
https://mp.weixin.qq.com/s/1caO4pJv7ZI5jiz4uIljyg
https://mp.weixin.qq.com/s/a9lI8IzgSyHlQ-43gwmLaQ

Chapter 10

India: Walmart, Amazon, Alibaba, and Tencent in India: WHAT the Hell is Going on?

10.1 Summary

The problem is clear. Banks pay out way too much in dividends. They cannot make a return in the capital more than 2–3%. They do not invest nearly enough in the future. Moreover, their price/book ratios reflect a very definite structural problem. Why did banks not invest in cloud businesses? Why avoid e-commerce? Why is their SME business virtually non-existent? Why can't they process a credit application in 1–3 days? Why are their insurance products so minimal? If they know they are behind and need major investments, why do they continue to pay out 35–45% of profits in dividends? If they are consigned to a capital guzzling "holder of deposits" and only offer low margin loans, they have no one to blame but themselves. Why should banks get involved in the lifestyle, utilities, coupons, voice command and tickets? Now, it is clear. People say: Of course banks should have been involved in this. They know more about their customer than anyone else but never bother to collect all the data in one place and analyze it.

10.2 Amazon

1. Great potential in groceries; Fierce competition in other online retail; Heavily investing in financial services.

Amazon India is now focusing on grocery retail. It has a promising future in the online grocery market. While it is not the largest dominator in India yet, its fast delivery network, convenient online payment platform and reasonable prices are likely to help Amazon outperform its competitors in the future.

Amazon is performing well in Indian online cosmetic and smartphone markets, but not the largest dominator.

Amazon has been heavily investing in its financial services in India, including acquiring small companies and advertising its digital payment platform Amazon Pay. However, its financial services are not dominating the Indian market at the moment.

2. A leader in cloud services; Fierce competition in streaming media; Building fast and reliable logistics.

Amazon's Cloud Services being the global lead, see big potentials in India and focuses on the business of SMEs in the country. It also corporates with the government to provide cloud services for them.

Amazon's Prime Video performs well in India for its localized and original content. Although Hotstar is the giant in Indian video streaming market, Amazon succeeded in becoming the early subscription leader in the market.

Amazon is building its logistics network in India focusing on speed and reliability. Most of the logistics are for its e-commerce platform, but it also provides services for its partners.

BOTTOM LINE: Alibaba has unbeatable infrastructure and is neck and neck with Amazon in financial services, cloud, payments, food, and e-commerce. However, Alibaba surpasses in browsing, lifestyle, SME lending, public apps and its awesome integrated PRC apps, which it is simply inserting into Paytm. Amazon is creating a world-class 'farm to market' infrastructure which is a blueprint for 100 countries. The world has never seen such a thing. Walmart comes in third. Moreover, the incumbents, Commonwealth banks like HSBC and Stan C are rapidly being left in the dust. (India is Stan C's biggest market by revenue). These two banks will not BECOME the pipes for financial services — THEY ALREADY ARE. Firms like Bank of Baroda and Yes Bank got onboard with Amazon and Alibaba. HSBC and Stan C should have done the same — is it too late? Bank regulators prevent two things: collaboration and open systems. Without these two vital links, regulated banking entities can be transformed. We can blame the board of directors of these two banks, but is the fault with the bank regulators?

1. **Amazon:** Astonishing stealth-like buildout of financial infrastructure in 18 months — food, finance, fun, film, FX, Future... Cloud and more.

   A. There is one word for competition in the Indian online space: fierce. Amazon is pushing the envelope into uncharted waters faster than anyone. The online world is growing at 23% CAGR and 98% of India is offline. The extent of Amazon's penetration into all parts of Indian life: it is truly a lifestyle company... and a bank, insurance company, entertainment firm, publishing house, and an agricultural conglomerate.

   B. In a short period of time, Amazon can now challenge and beat Standard Chartered or HSBC in their Commonwealth backyard. AMZN is now far deeper and wider in the scope of products which it offers. It has Amazon Pay (wallet), Red Bus (transport), Faasos (food delivery), Niki (utilities), Book My Show (theater), Tone Tag (payments), Qwikcilver (gift cards), Emvantage (payments), Lending (corporate loans), Capital Float (SME loans), Prime (cards), Protect (insurance), Acko (insurance).

   C. *Retail, entertainment, cloud, logistics:* This is a company that can't stop.
      Retail development: Future (hypermarkets), Easy Day (local stores), Ezone (electronics), Shoppers Stop (department store), More (supermarkets), Pantry (delivery), Prime Now (fresh produce). Amazon Prime is new and has a 5% market share. Prime Video is also less than 1 year old. In the Cloud market, AWS is new but already has five centers with $100 million in revenue. Amazon in logistics is probably the most interesting in that it touches all parts of Indian life throughout the country. This includes Fulfillment (delivery

and warehouse management), Flex (warehousing), Easy Ship (local delivery), ATSPL (last mile delivery), IHS (crowd-sourced delivery), Prime Now (ultra-fast delivery). Morevoer, it is investing $700 million into a national infrastructure program for farm-to-market delivery of food. This is something that will offer tremendous productivity improvements, since 30% of food in India is wasted on the trip to the market.

2. **Alibaba:** Also on a hyperactive expansion binge in India with an equally impressive arsenal of tools with Paytm.

There is a lot of information in the 123 slides below (Figures 10.1–10.123) that lays out the differences in strategies of the Chinese and American companies in India.

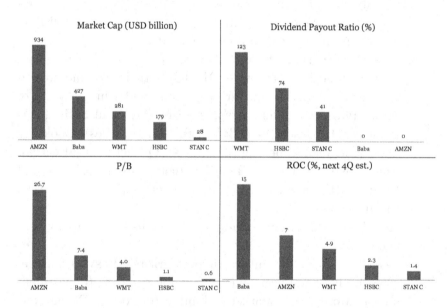

Figure 10.1:   **Banks do not invest enough for the future!**

Firm	Cloud	P2P	E-commerce	SME credit (in 1–3 days)	S/T instant credit	Insurance	Lending	FX
BABA	X	X	X	X	X	X	X	X
AMZN	X		X	X	X	X	X	X
WMT			X	X	X		X	X
HSBC						X	X	X
STAN C						X	X	X

Figure 10.2: They are consigned to a capital guzzling "holder of deposits" and only offer low margin loans, they have no one to blame but themselves.

Firm	Adver-tising	Food Delivery	Tickets/ Lifestyle	Streaming Music Content	Utilities	Coupons	Wallet	Voice Shopping	Groceries
BABA	X	X	X	X	X	X	X	X	X
AMZN	X	X	X	X	X	X	X	X	X
WMT	X	X	X	X		X	X	X	X
HSBC									
STAN C									

Figure 10.3: They know more about their customer than anyone else but never bother to collect all the data in one place and analyze it.

**INDIAN RETAIL SALES
IN 2017**

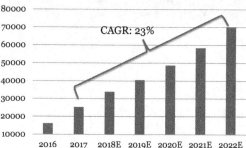

Although India's e-commerce makes up only 2% of the total retail sales in India, it may rise to 50% by 2023.

Indian e-commerce sales estimated growth rate in 2018 is 31%. The estimated CAGR from 2018 to 2022 is 23%. By 2022, India's e-commerce sector will be worth $70 billion.

Figure 10.4: **India e-commerce overview: Great growth potential.**
Source: eMarketer, Statista, Schulte Research.

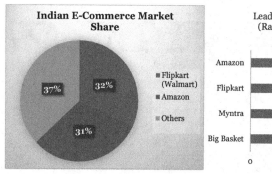

- Amazon and Flipkart (Walmart) are two dominant companies in Indian e-commerce market.
- Besides Amazon and Flipkart, Alibaba-backed Big Basket is also trying to get a share in the market.

Figure 10.5: **Amazon in India: Fierce competition with Flipkart.**
Source: Statista, Schulte Research.

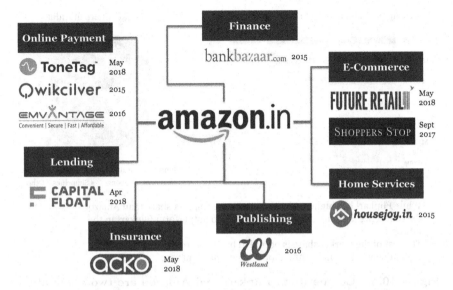

Figure 10.6: Amazon's key acquisitions in India.

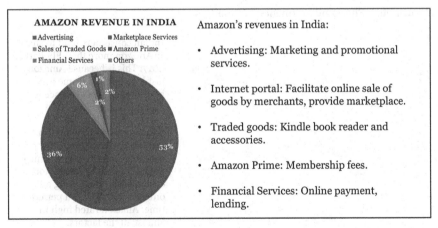

Figure 10.7: Amazon's revenue in India breakdown: Advertising and marketplace sales make up a large proportion of revenues.

Source: Trak.in, Schulte Research.

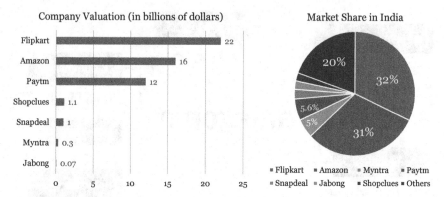

- While Flipkart has higher valuation and larger market share than Amazon, the difference is small and Amazon is likely to outperform Flipkart in the future.
- The rest of the market share is occupied by many small companies. They are unlikely to become a threat for Flipkart and Amazon.

Figure 10.8:   Competitors: Flipkart and Amazon are two dominators in India.

Source: The Economic Times, Schulte Research.

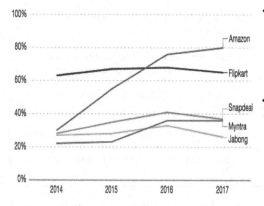

Consumer's Shopping Frequency in Retailers

- Indian consumers tend to shop in Amazon more than Flipkart in 2017. This is because Amazon's fast logistics and Prime service have brought convenience to customers.

- Shipping cost, shipping time, product review ratings, low price guarantee, and retailers return policy are the key factors for customers while picking an online retailer. Over a period of time, Amazon rated high in almost all the factors.

Figure 10.9:   Competitors: Amazon attracts most of the customers.

Source: Forrester Research, Schulte Research.

Figure 10.10:   Amazon India vs Standard Chartered India.

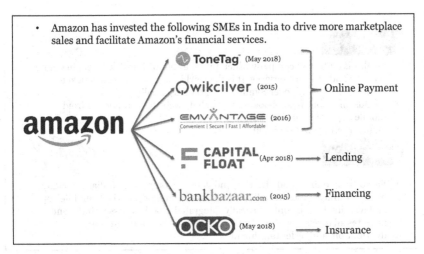

Figure 10.11:   Fintech investments in India: Three acquisitions in 2018.

Source: EconomicTimes, Schulte Research.

amazon pay

- Amazon opened digital wallet for India in 2016.
- Amazon Pay has evolved to include a digital wallet for customers and a payments network for both online and brick-and-mortar merchants.

Partnership
- Amazon Pay's partners will stimulate more transactions executed on Amazon Pay.

Figure 10.12: Online payment: Amazon Pay — Still new in India but has lots of partners.

Source: Crunchbase, Schulte Research.

redBus
- redBus is an Indian company which provides online bus ticket booking, hotel booking and bus hire services. It is the world's largest online bus ticket booking service with over 8 million customers globally.
- redBus offers bus ticket booking through its website, iOS, and Android mobile apps for all major routes.
- Number of Users: 7.5 million

Faasos
- Faasos is an online food ordering and delivery company in India. It operates in the 16 largest cities in India and takes customer orders via its mobile app and website. It is the only vertically integrated food business in India and operates all three stages of a "food on demand" business: ordering, distribution, and order fulfillment.

Figure 10.13: Amazon Pay partnership: redBus and Faasos — Online ticketing and food delivery.

Source: redBus, Wikipedia, Schulte Research.

Niki.ai

- Niki is an artificial intelligence company which helps users book cabs & hotels, order food, recharge, pay utility bills, and book movie and flight tickets via online chatting.
- It is India's first fully-automated chat bot app with no human intervention.
- Number of Users: 2 million

BookMyShow

- BookMyShow offers showtimes, movie tickets, reviews, trailers, concert tickets, and events booking. It is India's largest entertainment ticketing website.
- It covers 250 cities in India, and has partnerships with all major Indian production houses, studios, and cinema with real-time ticketing services.
- 85% of all cinema tickets sold online are booked on bookmyshow, resulting in sales peaking up to 7 million tickets per month.

Figure 10.14: Amazon Pay partnership: Niki.ai and BookMyShow — Chat Bot app and tickets booking.

Source: redBus, Wikipedia, Schulte Research.

Profitability

- Amazon Pay India has reported net loss of Rs 177.8 crore (US $26 million) on total revenues of Rs 7.4 crore (US $1 million) in 2017.
- The company's losses in 2016–2017 increased significantly from the Rs 7.8 crore in the previous financial year.
- Total revenues also increased during the period. Its overall expenses for FY17 jumped from Rs 8.2 crore in FY16 to Rs 185.3 crore in FY17.

Reason

- Amazon was spending heavily on promotions and advertising in India to publicize its newly-launched Amazon Pay.

Figure 10.15: Amazon Pay — Huge loss in 2017 because of heavy advertisements.

Source: Tofler, The Economic Times, Schulte Research.

Market Share

- Paytm is the dominator in Indian digital payment sector. It has a market share of 68%. Paytm is even accepted in shabby shops on the street.

- Amazon Pay has a market share of less than 5% in Indian digital payment industry.

- Paytm is backed by Alibaba. Alibaba has invested $700 million in Paytm. It now holds 62% stake in Paytm.

- Amazon Pay will have no competitive advantage unless Amazon manages to boost the sales on its online retail platform.

Digital Payment Market Share in India

10.20%
5%
5.40%
11.40%
68%

▪ Paytm ▪ Freecharge ▪ AirtelMoney ▪ Mobikwik ▪ Others

Figure 10.16:   Amazon Pay competitors: Paytm dominates the market with 68% market share.

Source: Financial Express, Schulte Research.

Introduction

 **ToneTag**

- ToneTag is a sound-wave communication tech platform which enables payments and proximity customer engagement services on any device, independent of the instrument or the infrastructure. It transfers data into sound waves via mobiles and payments can be executed once the payment machine receives the sound waves. With its simple SDK installation, ToneTag can seamlessly work inside mobile apps and devices.

Services

 • Encodes data into sound waves which is then transmitted between devices making it highly interoperable.

 • Enables contactless digital transactions on mobile phones, card swiping machines, automated teller machines, and other payment-enabling devices.

Figure 10.17:   Online payment investment: ToneTag — Execute digital payments through sound waves.

Source: ToneTag, Deal Street Asia, Schulte Research.

Introduction   **Qwikcilver**

- Qwikcilver is the world's first and only end-to-end Gift card solution provider. It has business across 16 countries. It helps companies to issue gift cards to improve customer loyalty and track transactions globally.

Services

 • Gift Card Solutions: Due to regulations in India, Amazon cannot issue gift cards by itself. Qwikcilver's license and prepaid payment technology can help Amazon issue gift cards.

 • Gift Card Digitalization: Based on Amazon's need, Qwikcilver can digitalize gift cards on websites and mobile apps.

Figure 10.18: Online payment investment: Qwikcliver — World's first gift card issuer.

Source: EconomicTimes, Schulte Research.

Introduction

- Emvantage offers an online payment gateway platform compatible with credit card, debit card, and net banking. The platform's MIS tool allows users to extract daily, weekly, monthly, and quarterly merchant and bank reports.
- Emvantage's platform includes a payment gateway for online transactions made using credit or debit cards, mobile payment tools that integrate into merchant apps, and a prepaid wallet.

Services

 Online Payment Gateway   Prepaid Digital Wallet

 Mobile Payment Tools Integration   Technology Support

Figure 10.19: Online payment investment: Emvantage — Online payment gateway for credit and debit cards.

Source: EconomicTimes, Schulte Research.

- Amazon.in announced an alliance with the India's second largest public sector bank — Bank of Baroda — to extend loans to its sellers.

- Under this alliance, Bank of Baroda will support micro, small, and medium businesses on Amazon.in to get fast and easy access to working capital at highly competitive interest rates in a short span of 3 to 5 days.

- Bank of Baroda will offer ECOM loans (BEL) up to Rs 25 lakh ($36,541) to sellers that have been selling on Amazon India marketplace and performing well consistently.

Figure 10.20:　Amazon lending — Partners with Bank of Baroda to issue loans to Amazon's online merchants.

Source: EconomicTimes, Schulte Research.

- Amazon India is launching its seller lending network which will offer merchants on its platform loan options from multiple third party lenders. This will help sellers get better loan rates from lending startups like Capital Float and established banks including Bank of Baroda and Yes Bank.

- Bank of Baroda is the second largest public sector bank in India. It offers internet banking services, mobile banking services, corporate banking services, etc.

- Yes Bank is India's fourth largest private sector bank. Yes Bank offers personal banking, corporate banking, and internet banking services including accounts, deposits, credit cards, home loan, personal loans, etc.

Figure 10.21:　Amazon lending — Seller lending network: to lend loans to online merchants.

Source: Times of India, Yes Bank, Bank of Baroda, Schulte Research.

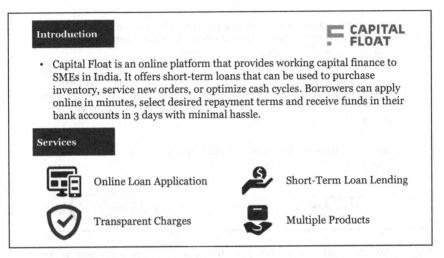

Figure 10.22: **Amazon India online lending investment — Capital float: Provides working capital finance to SMEs in India.**

Source: EconomicTimes, Schulte Research.

- Amazon offers Amazon Prime credit cards to help serve two broader corporate goals: grow Prime customers and increase marketplace sales. To attract card customers, Amazon has been adding perks exclusive to Prime members. Cardholders are likely to spend more on Amazon than non-cardholders, which also benefits Amazon's marketplace and boosts customer loyalty.

Figure 10.23: **Amazon consumer lending: Amazon Prime stimulates online sales in India.**

Source: CBINSIGHTS, Schulte Research.

- Amazon has not formally launched an insurance business in India but has shown interest in issuing insurance products. It invested $12 million in Indian car and bike insurer Acko in May 2018.

- Acko offers traditional car and bike insurance policies, but is increasingly focused on "internet economy" deals, which primarily consist of e-commerce, travel, and ride hailing-focused products such as an in-trip insurance program with Ola.

- Acko may partner with Amazon India to issue an Amazon Protect Product (Amazon Insurance) in India.

Figure 10.24:   Insurance: No product yet but has invested in Acko.
Source: Insurance Times, Schulte Research.

Introduction

- Acko General Insurance is a digital insurance company in India that redefines insurance for the consumers. It implements technology into insurance buying process to make it more convenient. It provides customers with affordable and convenient insurance and claims for car, bike, and mobile.

Services

  Car Insurance and Claim   Mobile Protection and Repairs

  Bike Insurance and Claim   Trip Insurance and Claim

Figure 10.25:   Insurance investment — ACKO: Provides convenient and affordable insurance.
Source: Acko, Schulte Research.

10.3 Groceries

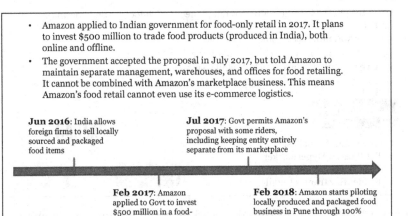

- Amazon applied to Indian government for food-only retail in 2017. It plans to invest $500 million to trade food products (produced in India), both online and offline.
- The government accepted the proposal in July 2017, but told Amazon to maintain separate management, warehouses, and offices for food retailing. It cannot be combined with Amazon's marketplace business. This means Amazon's food retail cannot even use its e-commerce logistics.

Jun 2016: India allows foreign firms to sell locally sourced and packaged food items

Jul 2017: Govt permits Amazon's proposal with some riders, including keeping entity entirely separate from its marketplace

Feb 2017: Amazon applied to Govt to invest $500 million in a food-only retailing venture

Feb 2018: Amazon starts piloting locally produced and packaged food business in Pune through 100% subsidiary Amazon Retail India

Figure 10.26: Food retail regulations: Amazon has to separate food retail from marketplace business.

Source: The EconomicTimes, Schulte Research.

- Amazon is in talks to acquire a 10% stake in **Future Retail Limited**, a company which owns brick-and-mortar retail shops like Big Bazaar and Easy Day, in a bid to **expand its offline presence**.

- The deal values Future Retail at about $6 billion. Amazon might end up spending about **$500–600 million.**

- This is not the first time that Amazon has bought stakes in an offline venture in India. In October 2017, Amazon had bought 5% stake in **Shoppers Stop** for $26 million. With Future Retail, Amazon will move the needle forward in its online–offline strategy.

Figure 10.27: Amazon's acquisitions: Future Retail and Shoppers Stop.

Source: Factor Daily, Schulte Research.

Introduction

- Future Retail is India's offline retail pioneer which focuses on hypermarket, supermarket, and home solutions.
- Over 500 million customers every year.
- Covers more than 250 cities across India.
- Over 12 million square feet of retail space.
- Products supplied by over 30,000 entrepreneurs and manufacturers across India.

Services

Offline Retail Branches:
- Big Bazaar: Hypermarket
- Easy Day: Neighborhood Store
- Ezone: Electronics Retail

Figure 10.28: **Future Retail — Offline retail dominator which focuses on hypermarket and supermarket.**

Source: Future Retail, Schulte Research.

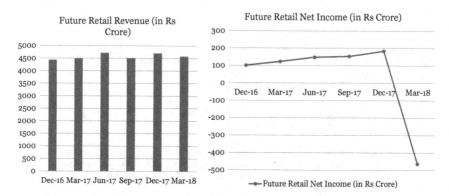

- Future Retail's revenue and profitability have been steady.
- The loss in March 2018 results from an exceptional expense of Rs 603.87 crore caused by the demerger of Hypercity Retail (India).

Figure 10.29: **Future Retail — Revenue and net income: Steady revenue and profitability.**

Source: ASIA TIMES, Schulte Research.

Market Share

OFFLINE RETAIL MARKET
SHARE IN INDIA

■ Reliance Retail ■ Future Retail ■ D-Mart ■ Others

- There are only three big retailers left: Reliance Retail, D-Mart, and Future Retail.

- Future Retail spearheaded a series of acquisitions to consolidate leadership in India's offline retail. Nilgiris, HyperCity, Aadhaar from Godrej Group, etc.

Figure 10.30: Future Retail competitors — Future Retail has the second largest market share.

Source: ASIA TIMES, Schulte Research.

Benefits

- Future Retail can learn from Amazon's global sourcing business. Amazon India can use the help of Future Retail's supply chain. Future Retail has a fully-owned subsidiary called Future Supply Chain Solutions. It has a network of 46 distribution centers, including four temperature-controlled centers, which spread out in a 'hubs and branches' model and cover 11,228 PIN codes.

Global Sourcing   Supply Chain

Figure 10.31: Why Future Retail? — Amazon can utilize Future Retail's supply chain.

Source: Factor Daily, Schulte Research.

Reasons of Acquisition

Amazon already sells a large number of phones, books, clothes, and electronics online in India. Now it is aiming at:

(1) Selling fresh produce and groceries;

(2) Being present on shelves in brick-and-mortar stores;

(3) Having access to a large cache of customer data that Future Retail has built over time, to glean insights about Indian buyers.

Online Retail　　　　　　　　　Offline Physical Stores

Figure 10.32: Why Future Retail? — Amazon wants to expand its offline groceries business.

Source: ASIA TIMES, Schulte Research.

Introduction　　　　　　　　　　　　SHOPPERS STOP

- Shoppers Stop is an Indian department store chain. There are 83 stores across 38 cities in India, with clothing, accessories, handbags, shoes, jewelry, fragrances, cosmetics, health products, home furnishing, etc.

Services

Online & Offline Retail:

- Online: Same products, prices, and services as offline physical stores. Provides delivery services to more than 750 cities in India. Provides Express Store Pickup service, where customers can shop online and collect the products at the nearest physical store.

- Offline: Clothing, accessories, shoes, jewelry, beauty products, etc.

Figure 10.33: Shoppers Stop — Offline retailer with online shopping and delivery services.

Source: Shoppers Stop, Schulte Research.

- Amazon acquired 'More' retail chain with $580 million investment for 49% ownership in September 2018.
- 'More' has over 540 supermarkets and hypermarkets across India.
- The investment is a major step for Amazon's offline expansion in India.
- The combination of online and offline data will build a full ecosystem for Amazon to compete with Walmart in India.

Figure 10.34: More: Amazon's offline expansion in India.

Source: CNN, Reuters, Schulte Research.

Introduction

- Amazon Pantry, a grocery and household item delivery service from Amazon India, has expanded its services to a total of 34 cities.

Services

- Amazon Pantry service allows users to fill a box with a maximum capacity of 15 kg, with items from across six categories: cooking essentials, personal hygiene, household supplies and pets, snacks and beverages, beauty products and baby products. It guarantees next day delivery and charges Rs 20 per box as delivery charges. Delivery is free for Amazon Prime members.

Figure 10.35: Amazon groceries — Amazon Pantry: Convenient next-day delivery.

Source: Medianama, Schulte Research.

Introduction

- Amazon has another service called Kirana Now (Amazon Now) for grocery delivery. Through KiranaNow, the company has tied up with various Kirana stores as well as grocery retail chains such as Big Bazaar, Spar Market, Hypercity, etc, that will deliver the perishable items ordered within two to four hours. Amazon Now is available in Delhi, Mumbai, Bangalore and Hyderabad.

Services

 Two to four hours Delivery   Order Real-Time Tacking

 Fresh Produce Delivery from Local Whole Foods Market   Restaurant Takeout Delivery

Figure 10.36:   Amazon groceries — Amazon Now: Selling fresh produce with 2–4 hours delivery.

Source: Prime Now, Schulte Research.

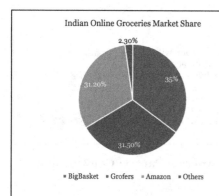

- Alibaba backed BigBasket (BigBasket received $200 million from Alibaba in 2018) is the largest dominator in Indian online grocery retail market.
- However, Amazon is still the third largest dominator and occupies considerable amount of market share.
- While Flipkart currently has no place in Indian online groceries retail business, it might push more aggressively with Walmart's entry.

Figure 10.37:   Amazon's Grocery retail — Competitors: BigBasket is still the largest dominator.

Source: Inc42, Schulte Research.

10.4   Other Online Retail

Competition

- Flipkart is still ahead of Amazon in 2017 online smartphone sales.

- However, Amazon has the upper hand in the premium segment in the category, which is in the price range above Rs 30,000.

- Amazon contributed 63% to online smartphone premium segment shipments in 2017 driven by sales of OnePlus and Apple. Flipkart's contribution was 31% of online smartphone premium segment.

INDIA ONLINE SMARTPHONE SHIPMENT MARKET SHARE

- Flipkart - Amazon - Xiaomi - Others

6%
10%
51%
33%

INDIA ONLINE PREMIUM SMARTPHONE MARKET SHARE

- Amazon - Flipkart - Others

6%
31%
63%

Figure 10.38: Mobiles: Flipkart is still ahead of Amazon.

Source: Entracker, Schulte Research.

Competition

- Flipkart owns Myntra's and Jabong's fashion unit. This gives Flipkart a combined market share of 60% in Indian online fashion market.

Action

- Amazon India unveiled the Amazon Fashion Studio in Gurugram called BLINK, a facility that offers allied services such as high-quality catalogue imaging to fashion sellers on Amazon.in.
- BLINK, a 44,000 square feet digital imaging studio, will allow Amazon's fashion brand partners to produce and shoot over 250,000 images, offer video-editing facilities, creative space, including editorial content.

INDIAN ONLINE FASHION MARKET SHARE

- Flipkart+Myntra+Jabong - Amazon - Others

11%
29%
60%

Figure 10.39: Fashion: Flipkart dominates online fashion market.

Source: Livemint, The Times of India, Schulte Research.

Market Share

Beauty Product Market Share in India

Online Cosmetic Market Share in India

- Amazon India is already selling 2 million beauty products across 19,000 brands, but India's online beauty market is dominated by Nykaa, which commands a 33% market share. 5-year-old Nykaa closed 2017 with gross merchandise volume, or gross sales, of Rs 1,000 crore ($150 million).
- Purplle, which has a foothold in about 13% of the market, plans to open five stores in Mumbai this year. It also plans to introduce its own brands within sub-segments of makeup and skincare in 2018.

Figure 10.40: Beauty products: Nykaa dominates online cosmetic market.

Source: Redseer, Schulte Research.

- Amazon India is preparing to launch its own brand of beauty and personal care products, as the country's top two digital retailers chase a fast-growing cosmetics market dominated by much smaller companies.
- Amazon India and Flipkart have increased focus on the high margin cosmetics category as growth in the larger online apparel market decelerates but experts caution they need to establish themselves in the category before introducing their own brands.

Who else is launching their own products?

Myntra

- Myntra (owned by Flipkart), which sells beauty products across more than 100 brands, is expected to launch its beauty and personal care brands this year in addition to retailing through offline stores in collaborations with international brands.

purplle

- Purplle also plans to introduce its own brands within sub-segments of makeup and skincare in 2018.

Figure 10.41: Beauty products: Amazon launching its own brand of beauty products.

Source: The Economic Times, Schulte Research.

10.5 Streaming Media

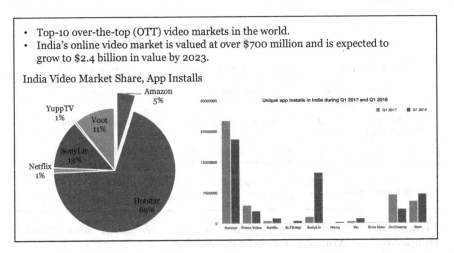

- Top-10 over-the-top (OTT) video markets in the world.
- India's online video market is valued at over $700 million and is expected to grow to $2.4 billion in value by 2023.

India Video Market Share, App Installs

Figure 10.42: India online video market: Great growth potential.
Source: Media Pratners Time of India, CNBC, AsiaVenturebeat, Schulte Research.

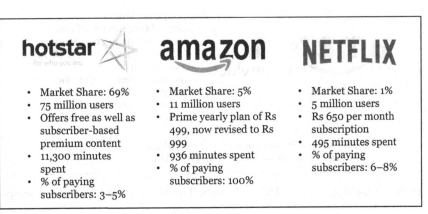

- Market Share: 69%
- 75 million users
- Offers free as well as subscriber-based premium content
- 11,300 minutes spent
- % of paying subscribers: 3–5%

- Market Share: 5%
- 11 million users
- Prime yearly plan of Rs 499, now revised to Rs 999
- 936 minutes spent
- % of paying subscribers: 100%

- Market Share: 1%
- 5 million users
- Rs 650 per month subscription
- 495 minutes spent
- % of paying subscribers: 6–8%

Figure 10.43: India online video market: Amazon and competitors — Hotstar has 69% market share while Amazon has 5%.

Source: App Annie, Media Partners, The Quint, The Times of India, Venturebeat, Schulte Research.

- Amazon is the early leader in the subscription race, but Netflix is the revenue leader

	amazon	**NETFLIX**
India launch	Dec 2016	Jan 2016
India subscribers (end of 2017)	610, 000	520, 000
Original Hindi and regional content	18 shows planned	4-5 shows planned
Languages	English, Bangla, Hindi, Tamil, Telugu, Marathi	English, Bangla, Hindi, Tamil
Deals with local talent and production houses	25+	4+

Figure 10.44: India online video market: Amazon and Netflix comparison — Amazon outperforms Netflix.

Source: Forbes, Livemint, HIS Markit, Quartz, The Economic Times, Schulte Research.

- **Subscription Model**
  - Low subscription cost: Costs Rs 129 ($1.90) a month or Rs 999 ($14.50) for a year in India($119 a year in the US)
  - Includes access to Prime Video and Prime Music
  - The early subscription leader in India

- Possible integration of some of its other offerings such as its e-commerce marketplace and Amazon Music with Prime Video

Figure 10.45:   Amazon Prime Video: The early leader in the subscription race.

Source: Economics, Forbes, Livemint, HBR, Schulte Research.

- Committed 300 million investment in India market
- Hindi content is around a quarter of all titles
- Over 10% regional content
- Launched its **Indian studios** in 2017
  - 17 original programs
  - Shows in Hindi, Tamil, and Telugu
  - Additional Amazon programming available in other regional languages like Bengali and Marathi
  - Four "Prime Original" shows already released:
    Music competition — The Remix
    Fiction — Inside Edge
    Crime thriller — Breathe
    Reality series — Comicstaan
  - Planned to launch as many as 10 Originals in 2018

Figure 10.46: Amazon Prime Video — Adaption to India market: Launched Indian studios in 2017.

Source: Amazon, Forbes, Gadgets 360, Television Post, Schulte Research.

10.6   Web Services

- India's public cloud market is likely to rise by more than 53% to $4 billion by 2020 as the world's fastest growing economy becomes more digitized.
- The cloud services market in India is forecast to grow at over a CAGR of 22% during 2015–2020.
- **Hybrid Cloud solution** drives overall cloud adoption.

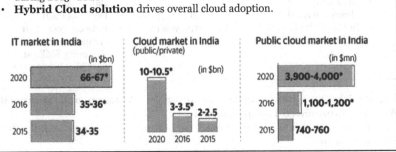

Figure 10.47:   Cloud market in India: Great growth potential.

Source: Forbes, First Post, Zinnov Analysis, Gartner, Synergy Research Group, Schulte Research.

Competitors in cloud service market:

Global majors
   Google, Alibaba, Amazon and
   Microsoft

Google Cloud

All opened data centers in India

Indian local players

 NxtGen:
Pushing AI development

ESDS: Smart
city hosting
managed
services

enabling futurability

CtrlS:
Pushing India on
the global map

CtrlS | Asia's Largest
Tier IV Datacenter

Figure 10.48:   Cloud market in India — Amazon and competitors.
Source: Forbes, First Post, Gartner, Canalys, Schulte Research.

C-] Alibaba Cloud

- First in India market
- Data Centers: 5
- Heavy advertising
- Simpler adoption and
  architecture

- Second in India
  market
- Data Centers: 3
- Easy deployment

- Announced joining
  India market in
  December 2017
- Data Centers : 1(in
  January 2019)
- Targeting SMEs and
  possibly government

**Figure 10.49:   Cloud market in India — Amazon and competitors:
Amazon dominates Indian market.**
Source: Business Line, Forbes, First Post, International Trade Administration,
Schulte Research.

- Amazon Internet Services Pvt Ltd(AISPL): Indian subsidiary of the Amazon Group
- Undertakes the resale and marketing of AWS Cloud services in India
- Revenue: $99 million
- Estimates 5 billion worth with accelerating growth
- Sees big potential in India: has the chance to be one of the biggest regions of AWS
- Offers more than 70 services for compute, storage, databases, analytics, mobile, Internet of Things (IoT), and enterprise applications
- Most popular among SMEs
    India is a very heavy base of SMEs and startups
    AWS has simpler adoption and architecture
- 5 data centers in India
    Opened its first data center in 2016

Figure 10.50: Amazon Internet Services Pvt Ltd (AISPL): Great growth potential.

Source: Forbes, Financial Express, First Post, Schulte Research.

- The first global cloud service provider (CSP) to achieve full empanelment for delivering Public Cloud services to government customers in India.
- Provides broad suite of cloud services, and a transformational approach to accelerate the deployment of digital services.
- Increase agility and speed innovation with AWS's flexible and highly secure "pay-as-you-go" Cloud services.
- Reasons of the government choosing AWS:
    - Massive economies of scale
    - Support for open standards
    - Secure infrastructure
    - Reduced cost
    - Run high performance computing (HPC)

Figure 10.51: Web services for Indian government: First global cloud service provider to deliver public cloud services to government.

Source: First Post, SCMP, Schulte Research.

> • **Cloud Research Lab**: provide students with opportunities to use AWS Cloud technology to pursue research initiatives that focus on AI and ML innovation.
>
> • Help students innovate via the same AWS platform which all AWS customers have access to.
> • Students can receive hands-on training and access to content prepared by some of the top computer science institutions globally.

Figure 10.52: 'AWS Educate': Provides cloud technology education for students.

Source: Forbes, SCMP, Schulte Research.

10.7 Logistics

> • India's logistics industry which is worth around US $160 billion.
> • Is likely to touch US $215 billion in the next 2 years.
> • Expected to grow at a CAGR of 8.6% between 2015 and 2020.
>
> **Segmentation of Logistics in India**
>
>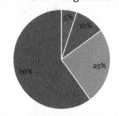
>
> ▪ Value Added Logistics ▪ Freightforwarding
> ▪ Warehousing   ▪ Transportation
>
> Logistics costs are high in India:
> • Underdeveloped infrastructure: 67% of the population lives in rural areas
> • High dwell time at ports
> • Low levels of containerization
> • Multi-layered tax system: interstate tax system contributing to significant delays at state border crossing points

Figure 10.53: India shipping market — High logistics costs and underdeveloped and great growth potential.

Source: Bloomberg, Medium, Economic Survey, Schulte Research.

- Expanding its fulfillment network
- Builds fulfillment centers
- Focus on speed and reliability
- FBA (Fulfillment By Amazon)

- Investment in Logistics startup **»XPRESS BEES**
  XpressBees
  - An e-commerce logistics company
  - Offering logistics services and solutions and on-time delivery through extensive network across India
  - More than 100 million investment
- Caters to the services such as last-mile delivery, reverse logistics, payment collection, drop shipping, vendor management, cross-border services, fulfillment services, and tailored software solutions.

Figure 10.54: Amazon and Alibaba Logistics in India comparison: Amazon focuses on speed while Alibaba makes more investments.

Source: Forbes India, Inc42, YourStory, Schulte Research.

Figure 10.55: Fulfillment by Amazon: Amazon's flagship product.

Source: Inc42, Efficient Era, Schulte Research.

- Heavy investment in infrastructure, especially in stocking products in its warehouses.
- 56 FCs across 13 states in India, with a storage capacity close to 13.5 million cubic feet.
- Specialized fulfillment network building
    - 15 grocery service
    - 25 large appliances and furniture
- Plan to double storage capacity
- Will have 67 operational fulfillment centers in 13 states across the country with an overall storage space of 20 million cubic feet.

Figure 10.56: Fulfilment by Amazon: Expansion of Fulfilment network.

Source: Amazon, Bloomberg ,Inc42, YourStory, Medianama, Schulte Research.

Advantages

- Save time and resources
- Quick delivery and great customer satisfaction
- Trustworthy
- Eligible for free delivery
- Increase in sales

Disadvantages

- Not preferable for low margin items
- Determining inventory volume
- Shipping products to Amazon warehouse
- Only for domestic orders

Figure 10.57: Fulfillment by Amazon: Pros and cons for sellers.

Source: Amazon, Inc42, Seller Hub, Schulte Research.

Vendors designate a section of their own warehouses for products to be sold on Amazon and coordinates the delivery logistics.

amazon easy ship

Pick up packaged goods from a seller's place of business and deliver them to consumers.
Enables its sellers to choose their courier partners for better operations and proceed to ship products on the same day.

Figure 10.58: Fulfillment platform localization: Easy Ship and Seller Flex.

Source: Amazon, Bloomberg ,Inc42, YourStory, Medianama, Schulte Research.

Cost of Transport Comparison

Air Cargo
Rs 40/kg

Amazon
Rs 25/kg

Surface Transport
Rs 18/kg

Amazon Transportation Services Private Limited (ATSPL):

- A subsidiary company to offer delivery services to its online merchants
- Incorporated in May 2012
- Has put extra resources (primarily drivers) to ensure that the distance is covered in 24 hours
- Utilizes bicycle and motorbike couriers for last-mile deliveries in both urban and rural communities
- One of the logistic partners for Amazon's Indian marketplace

Figure 10.59: Amazon building a nationwide surface transport network.

Source: Amazon, India Times, Schulte Research.

"I Have Space" (IHS):
- Crowdsourcing platform where local entrepreneurs/stores partner with Amazon for delivery and pickup services within an area of 2–3 kilometers.
- Launched in 2015
- 17,500 stores across 225 cities
- Problem in India: Hard to find precise addresses in the last miles

 Amazon assigns packages based on user's location

 Store owner delivers packages to the customers

 Payment amount is calculated based on the number of packages delivered

 Amount credited to the store owner's account by the first week of every month

Figure 10.60: "I Have Space" (IHS): Helping with last-mile delivery.
Source: Amazon, Zauba Corp, YourStory, Medianama, Schulte Research.

Amazon Prime Now:

- An app-only fast delivery service
- Launched in March 2015

- Customers can shop for products from the Now store, which is fulfilled by Amazon.
- "Speed app" across Mumbai, Delhi, Bengaluru, and Hyderabad.
- **Ultrafast delivery:** will provide exclusive Express two-hour and same-day delivery to Prime members anytime between 6 AM and midnight.
- They can buy over 10,000 products across categories such as fresh produce, groceries, and home and kitchen items on the app.
- Largest seller: Clow

Figure 10.61: Amazon Prime Now — the 'Speed App': App-only fast delivery service.
Source: Economic Times, First Post, Amazon, Inc42, Schulte Research.

10.8   Alibaba in India

Figure 10.62:   Alibaba has four key assets in India.

Figure 10.63:   Alibaba is a major player in India with 3 years of experience.

Source: Times of India, Statcounter, Schulte Research.

Two factors have increased use of mobile payments in India
1) Policy: India's 2016 demonetization move and creation of a national digital payment architecture has incentivized use of digital payments.
2) Smartphone penetration: 25% of Indians currently use smartphones, up from 15% in 2015.

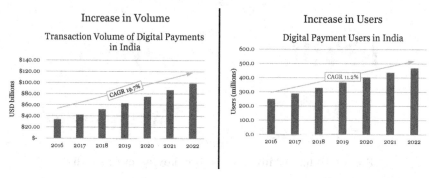

Figure 10.64: **Mobile payments are growing fast in India driven by policy changes and smartphone penetration.**

Source: Statista, Schulte Research.

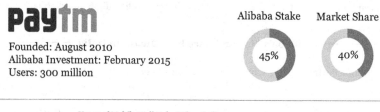

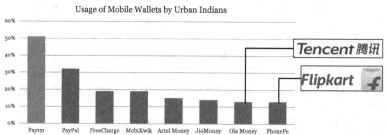

Figure 10.65: **Alibaba dominates Indian mobile payments via Paytm investment.**

Source: eMarketer, Paytm, India NPCI, Schulte Research.

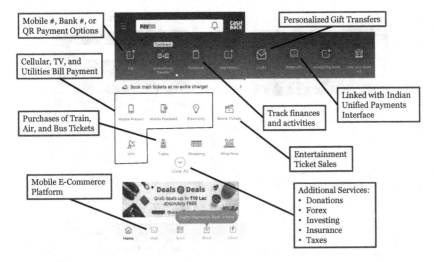

Figure 10.66: Paytm homepage.

Source: Paytm, Schulte Research.

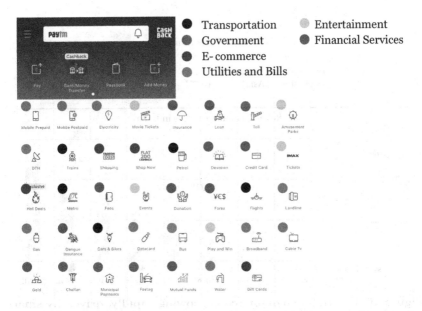

Figure 10.67: Paytm app covers a comprehensive array of needs for the Indian consumer.

Source: Paytm, Schulte Research.

Figure 10.68:　**Alipay and Paytm have nearly identical user interfaces.**

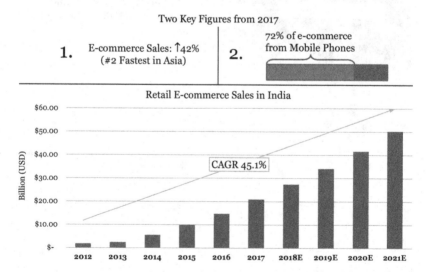

Figure 10.69:　**India e-commerce is growing rapidly, driven by smartphone penetration.**

Source: eMarketer, Schulte Research.

Alibaba has invested primarily in two Indian e-commerce ventures, Big Basket and Paytm Mall, which had a combined net sales of US $240 million.

Despite Amazon and Walmart's leading position, Alibaba's e-commerce sales more than doubled in 2017.

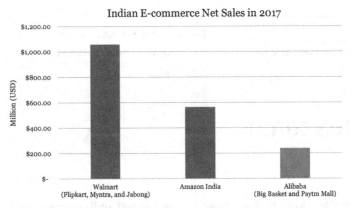

Figure 10.70: Alibaba is behind the curve with Indian e-commerce, but is making up ground fast.

Source: Indian Times, Statista, Schulte Research.

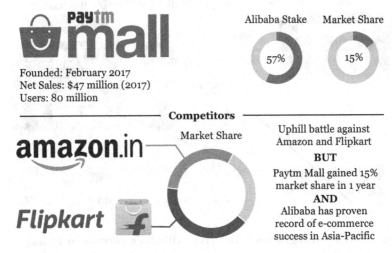

Figure 10.71: Paytm is bringing to India the same successful strategies used by Taobao.

Source: Bloomberg Quint, Indian Times, Forrester, Schulte Research.

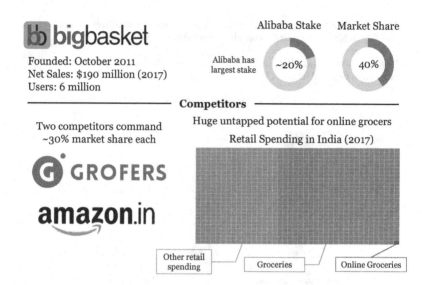

Figure 10.72:   Alibaba has stake in the biggest online grocer in India.
Source: Statista, Reuters, Schulte Research.

Figure 10.73:   Integration will drive Alibaba's success in India.
Source: Paytm, Schulte Research.

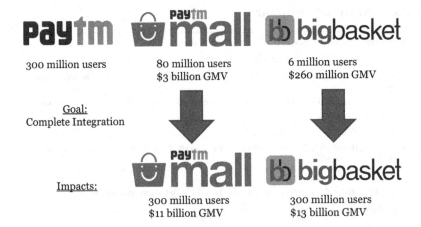

Figure 10.74: Integration solves user disparity between Alibaba's Indian ventures, enables market leadership.

Source: Statista, Reuters, Schulte Research.

- Alibaba's UC Browser has 50% market share of India mobile browsing market.
- India contributes 100 million MAU among UC Browser's 430 million global MAU.
- UC Browser also has a 41% market share in Indonesia.
- UC Browser is winning because its data compression and small app are more suitable for budget phones.
- Alibaba and UC Web is aiming to expand the user-generated content part in the future.

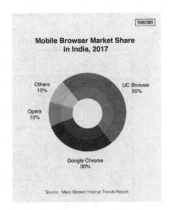

Figure 10.75: Mobile internet browser: Alibaba's UC web is dominating.

Source: yourstory.com, Mary Meeker, Schulte Research.

10.9   Walmart–Flipkart

- Walmart is not doing anything new in the O2O or analytics space — but they're doing a lot of it.
- Bringing their offline customer base online, investing in Big Dataand analytics.
- Huge customer base and large market cap. ensures that they will be competitive with Amazon and co.
- Outside of acquisition of Flipkart — non-existent in India (FDI regulation prevented expansion).
- Struggling in China with failure to understand consumers and inability to cope with supply chain differences.
- Showing strong commitment to India and China — exited British, Brazilian, and Japanese markets (rumored).

Figure 10.76:   Assessment of Walmart's global position.

Source: Schulte Research.

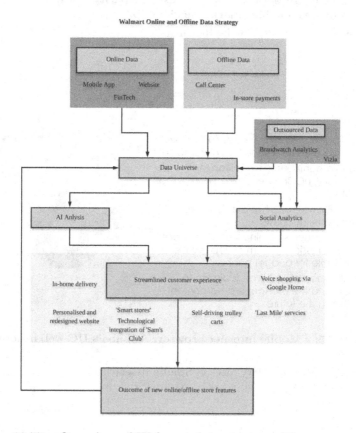

Figure 10.77:   Overview of Walmart data strategy (1).

Source: Schulte Research.

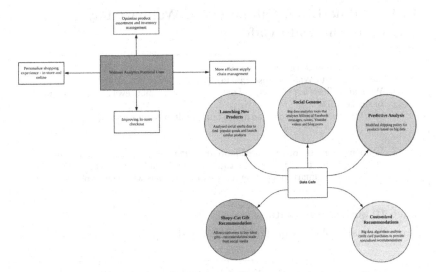

Figure 10.78: **Overview of Walmart data strategy — Improving O2O connection + experience.**

Source: Schulte Research.

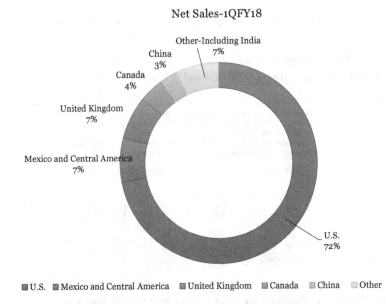

Figure 10.79: **Walmart revenue breakdown by region.**

Source: Walmart, Schulte Research.

10.10   Online Data, Offline Data, Walmart Pay and the Data Café

- Walmart.com has 100 million AMU
- Online sales for Walmart rose 33% in 1QFY18
- Walmart collects 2.5 petabytes of data from its customers every hour — 10 km of stacked CDs
  - Walmart now has data on approx. 145 million Americans

- Walmart's mobile app registered more than 27 million users as of June 2016
  - Mobile app users visit store locations 2x more and spend 40% more
  - 55% of Walmart shoppers enter in-store locations with a smartphone

Figure 10.80:   Online data.

Source: Walmart, Bloomberg, Forbes, Schulte Research.

- December 2015: Released Walmart Pay. In-store payment service — available on iOS and Android
  - Walmart Pay: Scanned QR code on pin pads at checkout
  - Unlike many other mobile payment methods (Apple Pay), Walmart Pay does not use (NFC) to transmit data
  - Access to one's mobile pay account requires a Touch ID or passcode

- Walmart Pay is now available in 4,774 stores (41% of worldwide stores)
  - 2/3 of users use the service more than 2x/month
  - 5.1% of customers now use Walmart Pay to complete in-store purchases

Figure 10.81:   Walmart Pay.

Source: Walmart, Bloomberg, Forbes, Schulte Research.

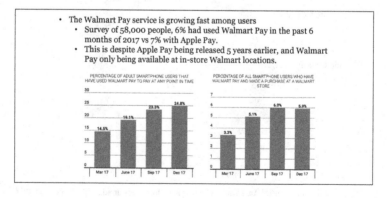

Figure 10.82:   Walmart Pay cont. — Growth and users.

Source: Walmart, Bloomberg, Forbes, PYMNTS, Schulte Research.

- Walmart runs 12,000 stores worldwide — more than 4,500 in the US
  - Collects data on in-store customers through connection to Walmart Wi-Fi connection spots
  - Tracks and stores data from its customers

Figure 10.83:   Offline data → Online data.

Source: Walmart, DeRyze, Schulte Research.

- Walmart announced the 'Data Café' analytics hub at the company HQ in Bentonville, Arkansas to deal with the immense amount of data the company receives.
  - The Café: the world's largest private cloud built to process the 2.5 PB/hr.
  - 200 external and internal streams of data modelled and examined and 40 PB of transactional data — 250 billion rows of data.
  - Sources of data include: economic, meteorological, telecom, social media, gas prices, local events.

- Walmart also tracks social media traffic through third party companies like Brandwatch Analytics, a social listening platform, that gathers data and responds to consumer trends.

Figure 10.84:   Analytics.

Source: Walmart, Brandwatch, Schulte Research.

10.11   Case Studies for Analytics Application and Streamlined Customer Experience

- Analytics conducted by Walmart discovered that the #1 reason consumers download retail applications in-store was to compare prices.

- In 2014 Walmart released 'Savings Catcher': Scans the QR codes of products into the app directly.
  - Downloads increased by 10 million in the week following the release of this feature.

Figure 10.85:   **Case 1: Savings catcher.**

Source: Digital Turbine, Walmart, Schulte Research.

- In 2017, Walmart announced its new feature, voice shopping, that would allow customers to shop on Walmart.com using voice commands via Google Home and Google Home Mini devices.
  - Benefits the online shopping experience at Walmart.com as it guarantees convenience when combined with their 'Easy Reorder' feature that expedites the delivery of previously ordered items.

Figure 10.86:   **Case 2: Voice shopping.**

Source: Walmart, Schulte Research.

- New Walmart services that come as the result of social analytics are imperative for Walmart to continue to grow its revenue and customer base.
  - 'In-home delivery' — that allows couriers to bring deliveries into your house, like groceries into your fridge.
  - 'Last Mile' service that allows company employees to make deliveries on their way home from work.

- Important for Walmart as it has more than 20% of the market share in US groceries (online and offline).

Figure 10.87: Case 3: Last mile service and in-home delivery.

Source: Forbes, Bloomberg, Schulte Research.

- Walmart appears to be acquiring Flipkart for market share primarily and technology.
  - Walmart COO (McKenna): advantages of Flipkart are data, AI, and mobile payment (PhonePe).
  - Did not acquire 100% so that they could work with existing investors (Microsoft, Tencent, etc.).

- Can take advantage of Tencent/Microsoft/eBay expertise in technology in developing their O2O strategy and advancing their online presence.

- Access to 17% of the global population and one of the fastest growing emerging markets.
- India is the fastest growing economy at 7.4% in 2018.
- Indian retail market is estimated to be worth US $650 billion.

Figure 10.88: Walmart's India/Flipkart strategy — 17% of global population potential customers.

Source: Fool, Schulte Research.

- Access to the Indian e-commerce market — Flipkart has 10 million active monthly users.
  - Breakthrough after failed venture with Bharti Enterprise and tight FDI regulation regarding multi-brand retail.
  - Walmart has 21 in-store locations in India — cash-and-pay locations.

- India is the only e-commerce market available with incumbent firms dominating China.

Worrying Sign: Flipkart grew GMV 43% FY17 while Amazon India GMV grew 67%.

Figure 10.89:   Walmart's India/Flipkart strategy.

Source: IMF, Schulte Research.

10.12   Flipkart Pvt. Ltd.

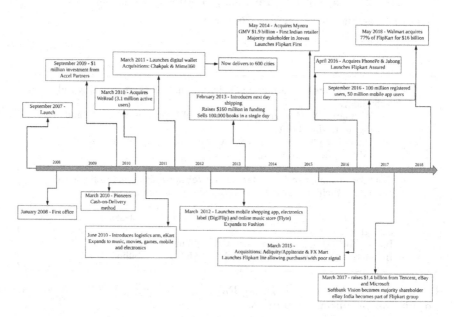

Figure 10.90:   History of Flipkart (1).

Source: Flipkart, Schulte Research.

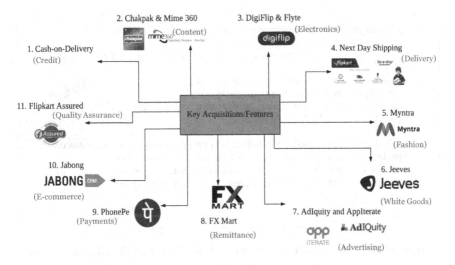

Figure 10.91: History of Flipkart (2).

Source: Flipkart, Schulte Research.

Figure 10.92: Cash-on-Delivery (March 2010).

Source: Flipkart, Medianama, Hackernoon, Business Insider, Schulte Research.

- Flipkart made two acquisitions in 2011 — Chakpak and Mime360
  - Chakpak — Bollywood news site
  - Mime360 — Platform connecting content owners with content publishers

- Chakpak — 5 million AMU at time of acquisition
  - Flipkart received access to digital catalogue
  - Co-founder, Sachin Bansal, said it was a part of their digital strategy
  - Also funded by Accel Partners

- Mime360 — Mime360's API (Application Programming Interface) helps to prevent piracy as content is delivered directly to the consumer.
  - More significant of acquisitions as it eliminates infrastructure costs for partners and allows content owners to set regional pricing.

Figure 10.93:   Acquisition of Chakpak and Mime360 (March 2011).
Source: Flipkart, NextBigWhat, Schulte Research.

- DigiFlip (Private electronics label): launched in March 2012 to break into the electronics market
  - In June 2016 DigiFlip launched its own tablet (DigiFlip Pro XT712) to compete with Google and Amazon.
  - Tablet failed due to aging software, poor audio, and low battery life.
  - DigiFlip was discontinued in December 2016.

- Flyte, Flipkart's digital music store, aimed to capitalize on the recent acquisition of Mime360 whose API technology would help the arm of the company to deal with piracy issues.
  - Shut down a year after launch due to piracy issues and inability to compete with iTunes and free online streaming services.

Figure 10.94: Acquisition of DigiFlip and Flyte launch (March 2012) — Disasters.
Source: Flipkart, Economic Times, Medianama, Schulte Research.

- Flipkart launched its 'In-a-Day Guarantee' in 2013 to compete with other major e-commerce business in India (Amazon).
  - For an additional fee of Rs 90 (less than US $2) Flipkart will guarantee next day delivery provided proximity to a major city.
  - Most e-commerce business in India (aside from Flipkart and Amazon) take 3–7 working days to deliver.

- Later that year Flipkart expanded this service to 'Same-day Guarantee' delivery in 10 cities (Bangalore, Delhi, Mumbai, Kolkata, Noida, Gurgaon, Faridabad, Manesar, and Thane) if order is placed before 12 PM.

Figure 10.95:   Next day shipping (February 2013).

Source: Flipkart, TechinAsia, The Hindu, Schulte Research.

- In May 2014, Flipkart acquired Indian e-commerce fashion company, Myntra.
  - 100% of Myntra was acquired for US $300 million.

  - One of Flipkart's biggest successes — claims fashion accounts for approx. US $1 billion in sales — one of their largest source of revenue (with smartphones and appliances/electronics).

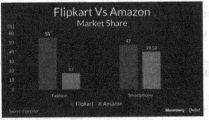

- Indian apparel market expected to grow at a CAGR of 9% to US $102 billion by 2023.

- ½ of the customers who visit Flipkart shop for fashion items.

Figure 10.96:   Myntra acquisition (May 2014).

Source: Flipkart, YourStory, Bloomberg, Inc42, Schulte Research.

- Jeeves Consumer Services Pvt. Ltd: After-sales service provider on large home appliances and electronics —
- Jeeves operates in more than 225 cities and is India's largest electronics service provider.

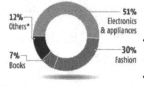

BREAK-UP OF INDIAN E-COMMERCE

51% Electronics & appliances
30% fashion
7% Books
12% Others*

- The 3 largest components of e-commerce in India currently are smartphones, fashion, and large appliances
  - Approx. 75% of market

- Partnership with Jeeves gives Flipkart credibility in launching its private brands.

- Eases pressure from retailers threatening to stop offering guarantees on products sold online.

Figure 10.97:   Majority shareholder — Jeeves (May 2014).

Source: Flipkart, Jeeves, Livemint, Trefis, Morgan Stanley Schulte Research.

- AdIquity — Mobile app network that allows app developers and mobile publishers to earn revenue from their mobile inventory.
  - 25 million ad impressions/month
  - Ads in 200+ countries
  - 15,000 app developers/publishers using this platform
  - Uses 'Geo-sense' to determine what ads are most popular by location
  - Allows Flipkart to generate revenue by selling online advertising

- AppIterate — Mobile engagement and marketing automation company
  - Uses A/B testing tool for mobiles — tests two variants for popularity
  - Delivers 100 million + personalized mobile notifications/month
  - Integrated into Flipkart's mobile app to increase ad revenue

---

Figure 10.98:   AdIquity and AppIterate acquisitions (March 2015).

Source: Flipkart, YourStory, Schulte Research.

- FX Mart: Provides payment services that include electronic payments, remittances, and FX.
- Majority stake acquired by Flipkart in 2015 worth US $6.5 million.
- Owns a prepaid licence issued by the Reserve Bank of India (RBI).
- Offers a digital wallet through its mobile app and avoid paying external e-wallet providers.
- Can use e-wallet to make transactions with third-party sellers.

Figure 10.99:   FX Mart (March 2015).

Source: Flipkart, Bloomberg, Livemint, Schulte Research.

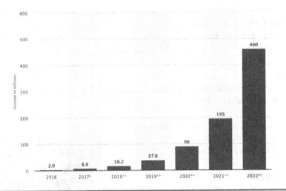

Number of mobile payment transactions in India (billions)

Figure 10.100:   Number of mobile payment transactions in India.
Source: Statista, Schulte Research.

- Electronic to make transactions online or in-store
  - In April 2016 Flipkart acquired the mobile payments app — PhonePe
  - PhonePe operates on a UPI which acts as an email ID for money
  - UPI does not require credit card or bank details and runs on IMPS

- Late arrival to online payments market with previous ventures failing and rival Snapdeal launching a similar service in 2015
  - Since the acquisition of PhonePe, Flipkart has partnered with Yes Bank to allow P2P transactions
  - PhonePe currently has over 75 millions downloads

Figure 10.101:   Launch of digital wallet — PhonePe (April 2016).
Source: Flipkart, Economic Times, Schulte Research.

- TPV (Total Value of Payments):
  - Paytm — US $29 billion
  - PhonePe — US $20 billion (4x increase from March 2017)

- Total Users:
  - Paytm — 300 million
  - PhonePe — 100 million

- UPI Users (Real time payments service):
  - Paytm — 62.9 million
  - PhonePe — 42.4 million

- Total Merchants:
  - Paytm — 7 million
  - PhonePe — Approx. 300,000

Figure 10.102: Launch of digital wallet — PhonePe (April 2016) cont. — Paytm vs PhonePe.

Source: YourStory, Schulte Research.

- Acquired Indian e-commerce fashion portal, Jabong, through its subsidiary, Myntra, for US$70 million.

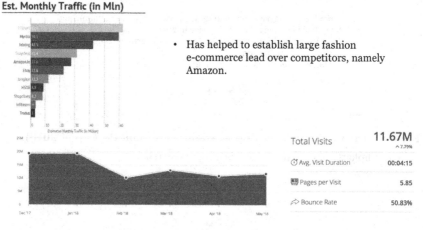

- Has helped to establish large fashion e-commerce lead over competitors, namely Amazon.

Total Visits	**11.67M** ^7.79%
⏱ Avg. Visit Duration	00:04:15
📄 Pages per Visit	5.85
↗ Bounce Rate	50.83%

Figure 10.103: Jabong (April 2016) — e-commerce fashion portal.

Source: Flipkart, SimilarWeb, Quora, Schulte Research.

- Flipkart Assured, launched in 2016, is India's first quality and speed assurance program
  - The Flipkart Assured badge on delivered products guarantees that the product has gone through six levels of quality control checks and at an expedited speed.
  - Flipkart Assured products are sourced only from the highest ranked sellers.
  - Available to all customers via the desktop and mobile app.
  - Response to Amazon's 'Fulfilled by Amazon' quality assurance.

Key Statistics

I.	65% of first time Flipkart customers use the service
II.	60% of products sold are Flipkart assured
III.	50% fewer returns on Flipkart assured items

Figure 10.104: Flipkart Assured (April 2016).

Source: Flipkart, Schulte Research.

Time	Investor	Investment Amount (USD)	Valuation (USD)
2007	Sachin/Binny Bansal	$6000	N/A
2009	Accel Partners	$1 million	$50 million >
2010	Tiger Global	$10 million	$50 million >
2011	Tiger Global	$20 million	$50 million >
2012	Naspers Group and ICONIQ Capital	$150 million	$1 billion
2013	Existing Investors	$200 million	$1.6 billion
May 2014	DST Group/Tiger Global/Naspers/ICONIQ	$210 million	$2.6 billion
June 2014	Tiger Global/Morgan Stanley/Accel/GIC	$1 billion	$7 billion
May 2015	Ballie Gifford/Greenoaks Capital/Steadview Capital/T. Rowe Price Associates/Qatar Investment Authority/Existing Investors	$1.25 billion	$11 billion
April 2017	eBay/Tencent/Microsoft	$1.4 billion	$11.6 billion
August 2017	Softbank Vision Fund	$2.5 billion	$15 billion

Figure 10.105: Valuation-Funding.

Source: Schulte Research.

- $2.23 billion USD total revenue FY16
- $2.9 billion USD total revenue FY17
  - Statistic includes subsidiaries – no clear topline
  - Breakdown of revenue by region/service unclear
  - 29% revenue increase from FY16 to FY17

Myntra

- Flipkart's most successful products are fashion (comes from the acquisition of Myntra), smartphones and electronics

Figure 10.106: Revenue breakdown.

Source: Flipkart, Economic Times, Schulte Research.

- 1.28 billion USD costs in FY 17 – 68% increase in costs from FY16 primarily caused by 5x increase in financial costs driven by valuation cuts during the year
  - Finance costs FY 17: $630 million USD (434% increase from FY 16)
  - Advertisement expenses FY 17: $173 million USD (9.4% increase from FY16)
  - Employee expenses FY 17: $300 million USD (9% increase from FY 16)

- Estimated profit FY 17 - $1.62 billion USD
- Estimated profit FY 16 - $1.487 billion USD
  - Approx. 8.9% increase

Figure 10.107: Profit breakdown.

Source: Flipkart, Economic Times, Schulte Research.

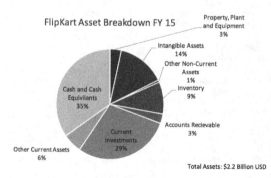

- No available figures for FY16/17

Figure 10.108: Asset breakdown.

Source: Flipkart, Schulte Research.

10.13 Flipkart Impact on Walmart

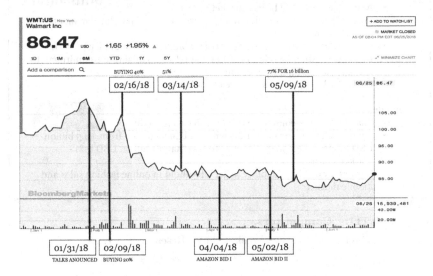

Figure 10.109: Timeline of Walmart–Flipkart talks.
Source: Bloomberg, Schulte Research.

- E-commerce in India still accounts for less than 5% of total retail
- Flipkart has 40% of the e-commerce market share – it's main competitors being Amazon and Snapdeal

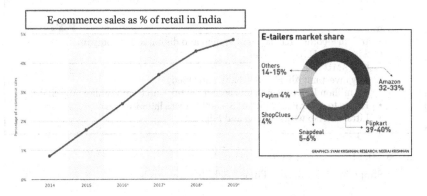

Figure 10.110: E-commerce in India.
Source: The Week, Statista, Schulte Research.

Amazon:
  - Amazon entered the Indian market in 2014
  - Spent $5 billion–$750 million for Amazon Pay India (online payment processing service)
  - In May of 2018, Amazon made a formal bid for 60% of Flipkart for $12 billion (including $2 billion USD breakup fee) highlighting commitment to the region – 20 billion valuation
  - Currently has 33% market share of Indian e-commerce

Snapdeal:
  - Snapdeal is an e-commerce company based in Delhi
  - Sharp decline in recent months with a fall in valuation from $6.5 billion USD (approx. 26% market share) to less than $1 billion USD (5%> market share)
  - Factors for decline: Inability to get a foothold in online fashion sales and mobile market

Figure 10.111:   Competitors.

Source: VCCircle, PrivCo, The Week, Schulte Research.

ShopClues:
  - An e-commerce company owned by Clues Network Pvt. Ltd, established in Silicon Valley and based in India
  - Valued at $1.1 billion USD
  - Key investors: Tiger, Helion and Nexus
  - 70% of GMV comes from 'Tier II' and 'Tier III' cities – smaller/rural – less overlap with Amazon and Flipkart

Paytm:
  - Indian e-commerce payment system and digital wallet company
  - $10 billion USD valuation
  - 250 AMU
  - Key investors: Alibaba, Softbank and One97
  - 7.5 million Indian merchants accept Paytm – up from 1 million March '17
  - Hopes to enter PoS (Point of Sales) business later this year
  - Little overlap with Amazon and Flipkart

Figure 10.112:   Competitors cont.

Source: ShopClues, Financial Times, Schulte Research.

10.14   Walmart and Tencent

- Walmart dropped Alibaba in March of 2018 and the Alipay service in all of its stores across China
- Walmart now working with Tencent to advance 'Walmart Scan and Pay' mobile payment service
  - Using WeChat Pay technology
  - Working on facial recognition payment technology
  - Found in all 443 Walmart China stores by the end of the year

Figure 10.113:   Background.

Source: Pandaily, Schulte Research.

- Walmart and Tencent are both investors in JD.com
  - Walmart recently merged its membership program with JD.com
  - Walmart launched logistics system that allows JD.com to use Walmart's inventories
  - Walmart currently owns 12.1% of JD.com ($4.87 billion USD)
  - Tencent merged WeChat data with JD.com's customer shopping data
  - Tencent owns 15% of JD.com

- Tencent is one of the earliest investors in Flipkart

- Alibaba a threat to Walmart's O2O PRC strategy due to greater overlap
  - Alibaba is already Tencent's main rival

Figure 10.114:   Why Tencent? (1)

Source: The Motley Fool, Schulte Research.

- Tencent and Walmart collaborating on digital and smart retail
  - Targeting marketing and payment services

- Plan to integrate Tencent technology into Walmart stores
  - Walmart staff to interact through WeChat
  - Walmart Scan and Go technology + facial recognition advanced
  - Self service checkout through WeChat Pay

- Ads for Walmart on WeChat app

Figure 10.115:   Why Tencent? (2)

Source: Yicaiglobal, Schulte Research.

- Walmart entered the Chinese market in 1996 under a joint venture
- Since 1996, Walmart has opened 424 retail stores in China and 15 wholesale
  - Walmart China accounts for only 3% of total revenue

So what are the reasons for the sluggish growth?

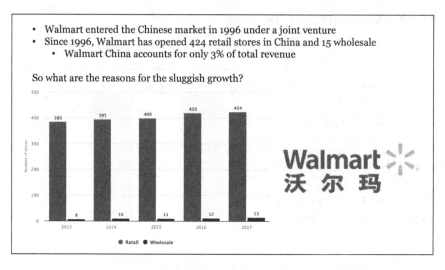

Figure 10.116: Walmart in China so far.

Source: Walmart, Statista, Schulte Research.

- Sun-Art Retail Group Limited is an investment holding company based in Hong Kong that operates through brick-and-mortar super and hypermarkets in China
- Despite having fewer in store locations than Walmart, Sun-Art has a higher % of the market share due to it's understanding of Chinese consumers
  - Sun-Art provides an outdoor market experience by giving stores a local street look which Chinese consumers are more accustomed to, as appose to the large box format of Walmart locations

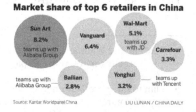

Figure 10.117: Dominance of Sun-Art — Alibaba owns 36% of Sun-Art.

Source: Walmart, Forbes, China Daily, Schulte Research.

10.15 Tencent India

- Flipkart is India's leading e-commerce marketplace (40% market share)

- Tencent participated in Flipkart's $1.4 billion funding round (with Microsoft and eBay in April 2017)

- In the same month, Alibaba invested $45 million more in Paytm.

- Tencent continues to be an investor in Flipkart (6% shareholder post Walmart deal)

Figure 10.118: Flipkart (April 10, 2017) — $1.4 billion for Amazon of India.

Source: Flipkart, Reuters, Techcrunch, Schulte Research.

- Ola is India's leading ride-hailing service, operating in 110 cities
  - Uber operates in 30 cities

- Ola FY17 Total Revenue – $110 million USD

- Tencent lead Ola's $1.1 billion funding round (with Softbank in October 2017)

- Ola money, is Ola's own payment system

- Ola is also investing in AI and machine learning as well as in-car entertainment (Ola Play)

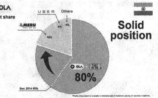

Figure 10.119: Ola (October 11, 2017) — $1.1 billion for ride sharing app.

Source: Ola, Reuters, Medianama, Schulte Research.

- Hike is a messaging app with over 100 million users
- Tencent led Hike's $175 million funding round in August 2016

- In 2017, Hike launched Hike Wallet – mobile payments service
  - Hike Wallet is now compatible with Ola Cabs

- However, Paytm (Alibaba) and PhonePe (Flipkart) crowd the mobile payments market – 300 and 100 million respectively
- WhatsApp, Facebook and Gmail dominate messaging

Figure 10.120: Hike (August 16, 2016) — $175 million for mobile payments.

Source: Techcrunch, CNN, Schulte Research.

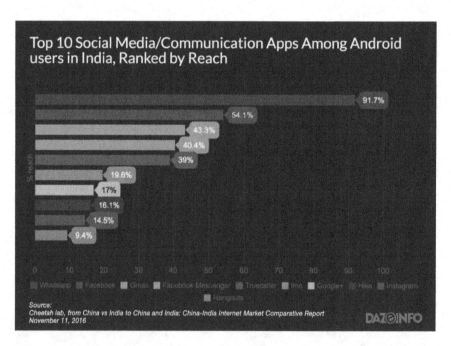

Figure 10.121: Hike cont.

Source: DazeInfo, Schulte Research.

- Music streaming service launched by Times Media

- Tencent lead Gaana's $115 million funding round in February2018

- Crowded Market – Saavn (Tiger Global); Hungama (Xiaomi); Apple Music; and Spotify (Tencent)

- Total revenue Indian digital music - $126 million USD

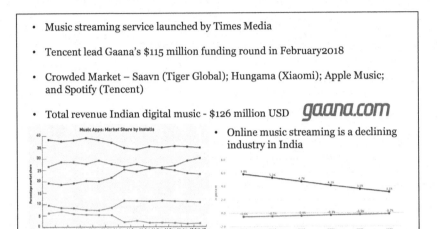

- Online music streaming is a declining industry in India

Figure 10.122: Gaana (Februrary 20, 2018) — $115 million for music streaming (Spotify).

Source: Techcrunch, Statista, Medium, Schulte Research.

- NewsDog is an India-focused vernacular news aggregator with 50 million readers

- Tencent lead NewsDog's $50 million funding round in May 2018
  - Tencent's first investment in the Indian news industry

- Tencent missed the rise of news aggregators in China (Bytedance and its news aggregator app Toutiao)
  - Bytedance backs one of NewsDog's competitor – Dailyhunt
  - UC News, another competitor, is owned by Alibaba.

Figure 10.123: NewsDog (May 22, 2018) — $50 million for news aggregator (Daily Mail of India).

Source: NewsDog, Techcrunch, Schulte Research.

PART 4
How New Technologies Will Revolutionize the Infinity of Digital Signals Out There

Chapter 11

Quantum Computing: It is Around the Corner — Get Ready!

Jason Kang

Harvard University

11.1 Introduction

Quantum computing is perhaps the most expensive innovative technology and the most difficult to understand because it has been on the Gartner Hype Cycle up-slope (Figures 11.1 and 11.2) for more than 10 years! Since 2005, quantum computing has been mentioned as an emerging technology and in 2017, it is still considered emerging.[1]

The foundation of quantum computing is the understanding that theoretical methods of computing cannot be separated from the physics that govern instruments of computing. Specifically, the theory of quantum mechanics presents a new paradigm for computer science that drastically alters our understanding of information

---

[1]https://www.gartner.com/smarterwithgartner/the-cios-guide-to-quantum-com puting/

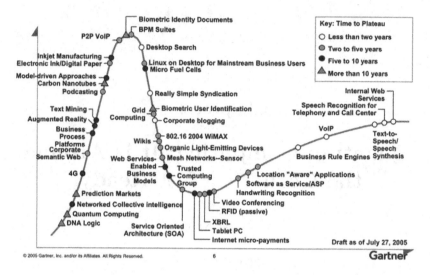

Figure 11.1:   Emerging technologies hype cycle 2005.

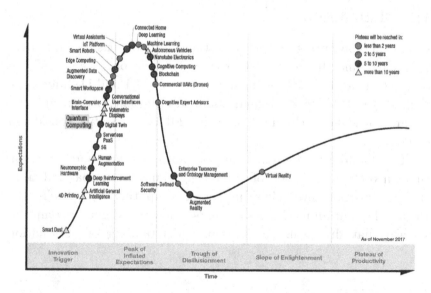

gartner.com/SmarterWithGartner

Figure 11.2:   Gartner hype cycle for emerging technologies.

Source: Gartner (November 2017). © 2017 Gartner, Inc. and/or its affiliates. All rights reserved. PR_338248.

processing and what we have long assumed to be the upper limits of computation. If nature is ruled by quantum mechanics, we should simulate it through the creation of QCs. Figure 11.3 is the executive summary of this chapter and Figure 11.4 gives an idea of the new generation of computing.

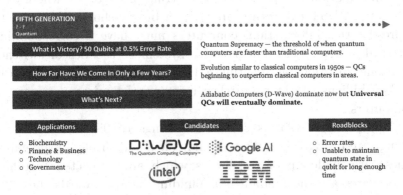

Figure 11.3: Executive summary: Early days, but quantum computers are already surpassing computers.

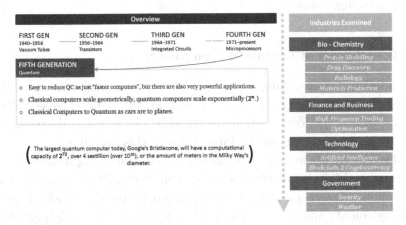

Figure 11.4: Applications: The beginning of a new generation of computation.

11.2　The Journey from Classical Computers to Quantum Computers (QCs)

The development of the computer starting in the 1930s has allowed us to create economic, social, and technological models for all walks of life. These computers are based on a binary system. This means that information is represented as a string of 0's or 1's where each character MUST **unambiguously** be a binary choice of 0 or 1. To represent this information, computers must have a corresponding physical system. Let's imagine this system as a series of switches with one direction representing a 1 and the other direction as a 0. Today, billions of these switches exist on the microprocessors in computers.

While information is stored as strings of 0's and 1's, these strings are processed, evaluated, and calculated through logic gates. These are made up of transistors which are connected together. Logic gates are the fundamental building blocks towards the large computations we ask of our computers, and when chained together hundreds of millions of times, they build up the capacity to perform advanced algorithms.

11.3　Physics of Quantum Computers

As advanced as computers have become in the past century, they still depend on binary choices between 0 and 1 to make sense of the chaos around us. Yet as our understanding of the world deepens, so are we more conscious of the limitations in this paradigm.

Developments in quantum mechanics are continuing to remind us of the unexplained complexities of our universe. At the core of this expanding branch of physics are the theories of superposition and entanglement. Simply put, this is the idea that subatomic particles like electrons can exist at different places at once (**superposition**) and also appear to affect one another across seemingly empty space (**entanglement**). These phenomena present a unique physical system to analyze and store information at speeds which are faster by a large order of magnitude compared to classical computers. QCs, first imagined in 1980, are now championed as the technology to fulfill this purpose. Figure 11.5 shows the potential of QCs.

	Classical Computer	Quantum Computer
Information	o Represented by binary bits o 0 or 1	o Represented by quantum bits (qubits) o 0 or 1, **superposition** of 0 & 1 (infinite)
Physical Systems	o Silicon based switches o Transistors	o Single Electrons, Diamonds o Must engage in quantum phenomena o **Superposition** and **Entanglement**
Calculations	o Deterministic	o Probabilistic

Quantum Advantage
o Qubits can store more information.   o Computational capacity increases exponentially (2^n).   o Qubits can interact with one another (entanglement) and effect computation more efficiently.

Figure 11.5: Introduction: Quantum computers have exponential advantage.

The computational advantage of QCs is derived from the idea that quantum computer bits (or **qubit**) can represent information not only as 0's or 1's, but as a superposition of both 0 or 1 — potentially infinite variations of numbers between 0 and 1 (Figure 11.6). So, each quantum-bit is empowered with phenomenal amounts of information. If computers today can already accomplish so much with just two states, imagine the possibilities of a machine that can access millions of superpositions between 0 and 1. QCs will be able to calculate information exponentially quicker and will shatter our current limits of information processing. They are the vehicle to artificial intelligence, risk analysis, optimization and the litany of technologies we have long imagined. For many new tasks, they are the natural successor to the modern computer that has defined the information age. This has important implications for understanding brain degenerative diseases, energy, agriculture, finance, biochemistry and many other branches of science. Figures 11.7 and 11.8 show the challenges.

Analogy

- o Consider electron spin to be the direction of an arrow.
- o Whereas classical bits can only be 0 or 1 (polar), qubits have an "in between phase".
- o The more possible configurations, the more possible information you can store.
- o Instead of strings of 0's and 1's, you would compute with "decimals", which can be stored in infinite states.
- o Instead of representing 5 as "101" (classical, requires 3 bits) you represent it as 0.5 (a physical state of 1 qubit).

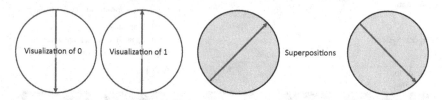

Figure 11.6: Qubits: How does it represent information?

- o Quantum Computing progress is like the development of the light bulb.
- o The logic exists, but finding the right material to execute the logic is challenging.
- o The most pressing problem is qubit coherence, when qubit information is corrupted.

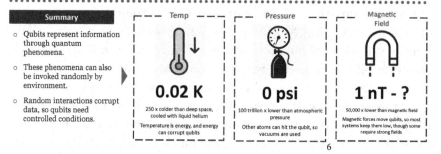

Figure 11.7: Roadblocks: Qubit coherence.
Source: Google, D-Wave, University of Edinburgh, Massachusetts Institute of Technology.

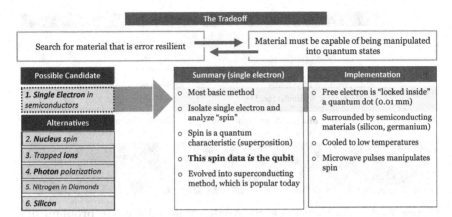

Figure 11.8: Qubits: What are Error Resilient candidates for Qubits?
Source: Google, University of Edinburgh, Massachusetts Institute of Technology.

11.4 The Hardware of Quantum Computing

11.4.1 *History and Overview of Quantum Computers*

Long before computers were miniaturized into the MacBooks and PCs that flood commercial use today, they were a jumbled mess of wire, tubing and, metal that weighed tons and occupied large rooms. They began as calculators designed for specific tasks. These computers, called *analog computers*, vary from tools as rudimentary as abacuses to devices with greater resemblance to modern computers. They could calculate the range, trajectory and deflection data of gun fire, and, for example, automate temperature and pressure flow in factories and airplanes.

Analog (Adiabatic)	Digital (Universal)	
Examples	o Abacus o Film, Tape Records o Flight Control	o Laptops, Smartphones o Supercomputers o Data Centers
Information	o Represented **Physically** o Mechanical, hydraulic backing	o Represented **Symbolically** o Bitwise, electrical backing
Application	o Solves only **one** specific problem	o Solves different problems

**Analog :: Digital
as
Adiabatic QC :: Universal QC**

Figure 11.9: Classical analogy: Analog tailored for one problem like Adiabatic; digital solves many like Universal QC.

The fundamental difference between what is considered an analog computer and a digital one is the way information is processed (Figure 11.9). Analog computers represent information by using a physical model that mimics the problem it is meant to solve. Because the problem is hardwired into the design of the machine, analog computers are limited to single tasks. On the other hand, digital computers represent quantities and information symbolically. Because that symbolic nature is flexible, it can constantly be restructured for different problems. It is the difference between an abacus, which represents numbers with beads and slides, and the calculator application on a smart phone, that crunches numbers as binary values sent through a processing chip. It is the difference between a music record, (where sounds are engraved onto the disk) and the music application on a smart phone (where data is encoded as binary values).

It is important to note that digital computers weren't immediately superior to their analog counterparts. Indeed, digital computers are the standard today and are inherently designed to have far more potential than the many restrictions that come with analog computers. However, until that potential was fully reached, analog computers were considered a competing alternative to digital computers in many fields, especially industrial process control. Both technologies were constantly at a race for improvement, and before digital computers were advanced enough to outpace analog, the frontier of technology was based on digital–analog hybrid systems like those implemented in the NASA Apollo and Space Shuttle programs. It wasn't until the 1980s when the discovery and subsequent mass production of the silicon transistor and microprocessor that the digital revolution was finally ignited. This process from pure analog to pure digital took 25 years.

Today, the journey to develop the first functional quantum computer mimics a similar evolutionary narrative. The analog equivalent of QCs is a paradigm known as **Adiabatic QCs** (or AQC), with research and development lead by a Canadian company D-Wave Systems and the United States Intelligence Advanced Research Projects Activity (IARPA). On the other end are computers that, like digital computers now, employ logic gates on different qubits to effect computation. These are so named **Universal QCs** (or UQC).

11.4.2 *What is a Qubit?*

While the notion of a qubit has already been touched upon, it is important to understand that the fundamental technology of any quantum computation paradigm — whether adiabatic or universal — is this concept of a qubit. A qubit is a physical system that serves as the most basic memory block for a quantum computer. They are the quantum equivalent of classical bits (transistors) used in today's computers and smartphones. The information in both bits and qubits share a common goal: to physically capture the information that each computer is processing. As information is changed during computation, the bit or qubit must also be manipulated to represent that change. This is the only way that the computer can keep track of what is happening. Because quantum computers store information in quantum states (superpositions and entanglement states), qubits themselves must be able to physically represent these quantum states, and so in turn qubits must be quantum by nature. This is challenging, largely because quantum phenomena only occur in extremely fringe conditions. To worsen the problem, even given the right environment, quantum phenomena are natural events. Anything from a ray of light to a change in pressure or temperature can invoke such phenomena, and in turn excite the qubit into a different quantum state than intended, therefore corrupting the information that qubit was meant to hold. To address these issues, researchers place quantum computers in extremely controlled conditions, with temperatures held at no more than 0.02 Kelvin — 20,000 colder than outer space — in nearly an empty vacuum — 100 trillion times lower than atmospheric pressure   and in either extremely light magnetic fields or extremely strong ones, depending on circumstances. Ultimately, all this trouble goes towards enabling such a qubit candidate to engage primarily in superposition states. This event, which enables qubits to hold not only 0 or 1 but also a superposition between 0 and 1, is the crux behind quantum computing. By enabling multiple states — possibly infinite states — for each qubit, these blocks of memory can hold much more information than their binary cousins (classical bits). And in turn, quantum computers can effect computation much faster.

11.4.3   *Adiabatic Quantum Computers*

We showed above how analog computers were designed specifically for certain tasks. Adiabatic QCs (AQC) can be considered as an analog equivalent to general quantum computers (Figure 11.10). AQCs are particularly suited for optimization algorithms where they calculate the best choice out for all possible solutions in a specific, given scenario. The promise of efficient optimization is particularly rewarding since it is both necessary in nearly every industry and one of the most noticeable limitations in modern computation. Complex optimization problems include: (1) Deciding on fuel-payload ratios for space flight; (2) energy conformations for protein-folding in mad-cow and sickle-cell disease; (3) mapping flight routes for an airline company; and (4) legislating economic policy and predicting market volatility. Current technology requires us to approximate because often times, the vast number of possibilities in a given scenario may take years or may be too difficult to analyze. But it is this very niche in which AQCs may excel in. An announcement from Google in 2016 had promising predictions that the D-Wave system may be able to perform specific computations 10^8 times faster than modern supercomputers — though realizing these gains is still years away.

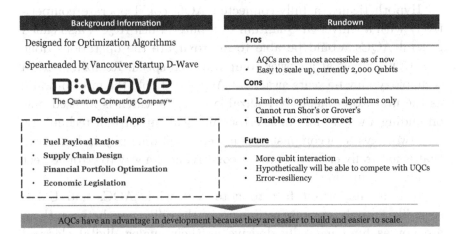

Background Information	Rundown
Designed for Optimization Algorithms	**Pros**
Spearheaded by Vancouver Startup D-Wave	• AQCs are the most accessible as of now • Easy to scale up, currently 2,000 Qubits
D::WƏVE The Quantum Computing Company™	**Cons**
	• Limited to optimization algorithms only • Cannot run Shor's or Grover's
- - - - - Potential Apps - - - - -	• **Unable to error-correct**
• Fuel Payload Ratios	**Future**
• Supply Chain Design	
• Financial Portfolio Optimization	• More qubit interaction • Hypothetically will be able to compete with UQCs
• Economic Legislation	• Error-resiliency

AQCs have an advantage in development because they are easier to build and easier to scale.

Figure 11.10: Adiabatic quantum computers (analog equivalent in the quantum world).
Source: D-Wave.

While this paradigm for calculation works well in certain situations, it isn't able to complete many of the algorithms that are envisaged for Quantum Computation. Put another way, they may not be able to perform algorithms as they are currently written. More pressingly, AQCs suffer from qubit de-coherence and lack any infrastructure for error correction (see Section 11.4.4). Similarly, analog computers suffered from this same notion of errors and the inability to autocorrect. This is precisely why their digital counterparts eventually outpaced them. (The analogy, as we will see, is that universal QCs will have auto-correct functions and should, therefore, overtake adiabatic QCs).

Another issue AQCs face is that the nature of these computations require multiple interactions between qubits occurring simultaneously. However, the more interactions there are between qubits, the more vulnerable the qubits are to errors and the harder it is to keep track of each interaction. The problem is two-fold. To reduce noise, interactions between qubits must be limited. But limiting interactions in turn limit what the particular AQC can accomplish.

Hypothetically, a fully-connected AQC could operate competitively with a Universal Quantum Computer. Such AQCs, also called general AQCs, would be able to approximate the operations done in a UQC, albeit at a slower computational speed. Realizing such a technology may be years away, but AQCs still hold the most promise as the first experimentally realizable QCs. Research is focused now on finding a way to minimize errors and scale up current technology. The two ways to accomplish this are either finding physical material that is naturally error-tolerant or conceiving of a way to error correct during computation.

It is an important fact to note that while AQCs and UQCs seem like entirely separate technologies, the differences between them are not as fundamentally divisive as their analog–digital classical computer counterparts. Both technologies are different paradigms that utilize the same core phenomena to effect computation. One may not be fundamentally better than the other. Rather, each has its strengths in solving certain problems, and weaknesses in others.

11.4.4 *Universal Quantum Computers*

Opposed to Adiabatic Quantum Computers, which specialize only in optimization calculations, Universal QCs are a form of quantum computation that hold greater promise and potential (Figure 11.11). In this paradigm, computation is realized by sequences of logic operators acting on different qubits. Because of its flexibility, UQCs are the system upon which many quantum algorithms are intended to operate. This means that there is already much academic infrastructure surrounding UQCs and their potential applications.

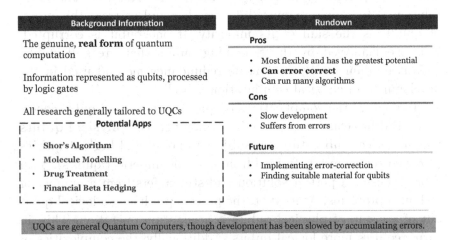

Background Information	Rundown
The genuine, **real form** of quantum computation	**Pros**
	• Most flexible and has the greatest potential
Information represented as qubits, processed by logic gates	• **Can error correct**
	• Can run many algorithms
	Cons
All research generally tailored to UQCs	• Slow development
Potential Apps	• Suffers from errors
• Shor's Algorithm	**Future**
• Molecule Modelling	
• Drug Treatment	• Implementing error-correction
• Financial Beta Hedging	• Finding suitable material for qubits

UQCs are general Quantum Computers, though development has been slowed by accumulating errors.

Figure 11.11: Universal quantum computers (digital equivalent in the quantum world).
Source: Google, IBM, Intel, Mermin "Quantum Computer Science".

While this technology is the more flexible of the two options, UQCs also face the most problems in terms of development. These systems are fundamentally reliant on their qubits to store information while logic gates access that information and draw conclusions from that information. Thus the preservation and processing of information as a whole is the fundamental building block of UQCs. UQCs, struggle with this the most. Because quantum superposition is a delicate phenomenon, the physics of maintaining multiple qubits in independent superpositions is challenging. This is aggravated by the fact that quantum mechanics isn't just a laboratory-generated phenomena. All of nature operates by quantum mechanics, so any interaction of the qubits with their environment can corrupt the qubit and therefore the information they represent.

The seeming impossibility of preventing these kinds of errors, called **error correction** or **qubit coherence**, has long been the greatest argument against the feasibility of QCs. However, rapid developments in this study of quantum error correction are helping realize the possibility of a UQC that is error-resilient. The basic solution is to couple many qubits together and to treat this group of qubits as the sum of its majority. If one qubit is corrupted, it is overshadowed by the remaining qubits that are still stable. This is fundamentally the failsafe redundancy method used in the development of classical computation as well.

To capture the complications of qubit quantity and qubit coherence, IBM has classified such a distinction between **physical qubits** (the imperfect qubits that are subject to noise) and **logical qubits** (error-corrected qubits formed from multiple imperfect qubits). Such a project isn't a perfect solution and still suffers from a number of technical problems. At its root, the process producing a logical qubit out of a group of physical qubits is a quantum operation. This, in turn, requires more logical qubits. Additionally, the complexities of quantum interactions impose a significant barrier on fabricating and operating qubits in close proximities to one another. Nonetheless, the very fact that there is infrastructure for error correction is encouraging and empowers UQC to potentially be the most viable paradigm for quantum computation.

11.4.5 *Hybrid*

Classical computers eventually evolved into a hybrid analog–digital system to solve problems of both. In a similar fashion, we think that adiabatic and universal QCs can eventually morph into a hybrid Quantum Computer. Imagine there are two classmates Bob and Paul who are both listening to a lecture. Paul is sitting in the front row, while Bob is in the back row, and both are trying to record the professor's talk. In this scenario, both Bob and Paul are using a tape recorder which takes in the voice from a microphone and lays it onto a tape (an example of analog technology at its finest). Because of his close proximity to the professor, Paul's recording is crisp and clear, whereas Bob's recording in the back is riddled with coughing noises and movement throughout the classroom. This interference is emblematic of the "noise" that affects analog technology, and the very kind of "analog error" that disrupts calculations within AQCs.

So, instead of both recording, Paul offers to digitize his recording, encode it in an mp3 file, and send it to Bob who then reconverts the file to a tape. Now both Paul and Bob have nearly the same, clear recording of the lecture. The only difference between the two are the errors that may have occurred between digitizing and un-digitizing the tape.

This is the fundamental logic for analog–digital hybrid technology. So long as the errors involved in conversion are less than the errors otherwise incurred by environmental noise, an analog–digital hybrid is able to combine the advantages of both technologies. Such a hybrid Quantum Computer would run computations with qubits operating as an adiabatic system, but with connections between qubits controlled by a digital, error correcting, network. This offers the flexibility and scalability of AQCs while still maintaining a degree of error correction during computation. However, this kind of hybrid quantum computer is still largely hypothetical. It requires technology from both AQCs and UQCs that aren't available yet, and academic papers on this topic are scarce as well.

That isn't to say that hybrid solutions to quantum hardware are non-existent. There is a significant amount of literature, and field-tests, for hybrid classical-adiabatic quantum computers. The foundation of this paradigm is to use adiabatic quantum computers solely in the specific fields that they excel in. In a large scale computation, certain calculations can be offloaded to an adiabatic quantum computer while a classical supercomputer performs the computations for everything else. This approach is promising, because it offers computational speedup without actually having to perfect adiabatic quantum technology, or build a universal quantum computer. So long as the computations are specific and small, such a hybrid solution could very-well offer significant boosts in computational speed, and be conceivably tested and used in industry within the next 5 years.

11.4.6 Summary

Currently, research is divided into two camps. Adiabatic QCs are analog systems thought to **only** be able to run **one** algorithm at a time, namely an optimization calculation. Luckily, optimization has tremendous applications. The largest argument against AQCs is its lack of infrastructure to address errors and that there is no base theory for quantum advantage on an AQC system. Without addressing errors, it would be impossible to scale AQCs to a point where they would compete with classical computers. That being said, if researchers are able to design an AQC that is error-resilient and fully-connected, it would enable AQCs to run a wider range of algorithms, potentially making it competitive with the Universal model.

The digital quantum computer is a system called Universal QC. These systems operate just as a classical computer does. It processes information as bits and binds different logic gates together in order to effect computation. This is the most advanced form of computation because in theory it would act as a platform that could run any number of algorithms, especially algorithms like Shor's and Grover's that first instigated popular interest in QCs (see 11.5.2). Furthermore, UQCs have the infrastructure to error-correct, and though its implementation still faces problems, it remains the most viable platform to realize sufficient qubit coherence.

It is important to understand that with all considerations of quantum computation, divisions are never binary. Quantum technology is still in its developmental infancy, and different problems are handled best by different kinds of quantum computation. There isn't a quantum computer that is completely superior to all others, and progress in hardware will only progress by acknowledging that different paradigms each have their own contributions. That they are all pieces of a puzzle that will ultimately converge upon what we imagine to be the quantum computer. In fact, while these are the two main divisions in quantum hardware, they aren't the only two. Research spread across different universities and companies are continuing to imagine new possibilities to solve our questions. As a third option, Microsoft has announced another method of achieving UQCs with what they call a topological quantum computer, and it will only be through a combined effort that we will be able to bridge the gap between hardware and software.

11.5 Algorithms

As with most new technology, the physical hardware of the system is a much more complicated and multidisciplinary field — and the slowest to develop. On the other hand, algorithms are the theoretical proofs of what we *could do* with this technology once we have it. In short, algorithms carry the promise of quantum computing. Here, we will highlight two of the most fundamental algorithms in the study of quantum computing: Grover's and Shor's (Figure 11.12).

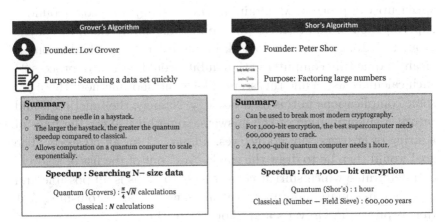

Figure 11.12: Quantum algorithms: Theoretical algorithms prove exponential speedup.

Source: Grover, Shor, Mermin "Quantum Computer Science".

11.5.1 *Quantum Parallelism*

Before diving into these two algorithms, it is interesting to get an intuitive sense of how quantum algorithms are able to achieve so much so efficiently. The core of this phenomena is the idea of quantum parallelism.

As an analogy, consider that you have 49 dirty shirts and one clean shirt, all mixed together in a pile. The goal of your algorithm is to search this pile of laundry until you find the clean shirt. A classical computer would approach this problem by blindly reaching into the pile, checking if it is clean, discarding the shirt if it is not and then moving on. Even though today's computers are able to work through data sets this small at astonishing speeds, it is still fundamentally a redundant process since there is no intelligent design in the blind choice of a shirt. In the worst-case scenario, this method would require 50 iterations before arriving at the clean shirt.

Quantum parallelism promises a method that would dramatically speed up the process. Maintaining this analogy, one could build a superposition of all 50 shirts (metaphorically lining them up on the ground in front of you) and check in one calculation whether one is clean. A quantum computer with n qubits would be able to process 2^n such calculations. While this seems like a one-shot method, there are some complications in this methodology. Namely, the function would return a superposition of the answers, something along the lines of a string of answers "no, no, no, yes, no, no..." for all 50 shirts. The goal would then be to draw the single "yes" out of the list of answers and determine which shirt it corresponds to. This is all possible and only requires marginally more calculations — an astonishing feat that is proven in Grover's and Shor's algorithm.

11.5.2 *Grover's Algorithm*

In 1996, this very challenge of searching for a superposition of "results" from a quantum calculation was solved by Lov Grover, an

Indian–American computer scientist from Bell Laboratories. Sparing the mathematical details of the algorithm, Grover's is an iteration that, when applied to a superposition of output states, helps to express the desired output with greater probability while eliminating undesired outputs. Especially with large data sets, Grover's algorithm can search the data set for the desired result at an exponential speedup from its classical counterpart.

The first important use of Grover's Algorithm occurs in the case when n = 4, or when there are four inputs. In this case, it takes exactly one invocation of Grover's algorithm to determine the solution, while a classical computer can do no better than to test each of the four possible solutions in random order, solving it in a mean time of 2.25 iterations. This simple example highlights the capacity for Grover's to shorten run times and provide efficient calculations.

Scaling such a demonstration to a more meaningful size, Grover's algorithm is expected to produce the solution with no more than $\sqrt{N}$ where N is the size of the input. Conversely, a classical computer needs N iterations to do the same thing. While this may seem like a small improvement, industry calculations often consider millions of data points, and such an algorithm is not only valued for its efficiency, but it can also raise the upper-limit of how many data points we can handle. This provides more efficient and accurate machine learning procedures as the data from AI proliferates.

11.5.3 *Shor's Algorithm*

In 1994, Peter Shor published a paper and coined Shor's Algorithm (Figure 11.13). This would spark the entire industry of quantum computation. It was designed to reduce numbers into their prime factors, a pursuit that may seem trivial but is actually the foundation of a wide range of cyber-security protocols (see 11.7.4.1). Today, Shor's algorithm is still considered the landmark discovery in the field of quantum computation and is frequently considered a benchmark for quantum computation.

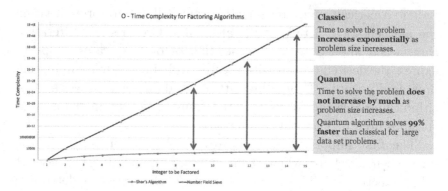

Figure 11.13: Shor's algorithm: Computation does not increase sig. as data set increases.

Shor's algorithm is especially relevant in the context of cryptography. Today, the gold-standard used by the private and public sectors is RSA encryption, a protocol that is formed by using two large prime numbers and their product as keys. To break the encryption, a third-party would have to factor that large product into the two prime factors. However, even the best classical algorithm to do so, called the *number-field sieve*, would take millions of years due to the sheer size of the prime numbers used in such protocols. The number-field sieve isn't especially efficient either, and has a run-time complexity of $O\left(\exp\left(\sqrt{\frac{64}{9}n(\log n)^2}\right)\right)$ where n is the length of the bit-representation of the particular prime number. For context, if someone used a 300-digit key, the fastest supercomputer in the world performing over 10^{15} operations a second around the clock and would still take over 10^{30} years — more years than there are atoms in this universe — to break the encryption. Considering that today's modern RSA encryption doesn't even use 300 digit keys, but opts for 1024 or 2048 digit keys, breaking RSA with even an army of supercomputers is considered impossible.

Yet give this problem to a quantum computer, and the results are much different. With Shor's algorithm, factoring a number is expected to be done in $O(n^3)$. To translate this back to our original problem, researchers think a 300-digit key would be solved between 1 second to 3 hours. The ability to break RSA encryption is not a trivial feat, and the race to do so is precisely why QCs have received so much attention since the publication of Shor's paper. Even the government has a stake in this technology, with the Snowden leaks in 2014 revealing an NSA program to develop quantum cryptography machines.

11.5.4 *Summary*

The power of Shor's Algorithm is derived from a particular blackbox of mathematics called the Quantum Fourier Transformation. It is a landmark discovery of superfast calculations that take advantage of superposition and entanglement, and has since paved the way for a number of other algorithms. Today, these algorithms are all publicly available on the Quantum Algorithm Zoo by the US National Institute of Standards and Technology. These algorithms are generally designed for UQCs that have tens of thousands of qubits, are noiseless, and are far more advanced than the types of devices that we have access to today.

Still, the study of algorithms does matter. Both Shor's and Grover's algorithms are heralded because they are a theoretical testament to what QCs *can* do. Both were developed in the late 1990s, and 20 years later we are still struggling to build the hardware that could support such ambitious pursuits. There is a chase going on between algorithm research and hardware development — and it can be frustrating. At the same time, it is this very pursuit that reminds us what we are developing this technology for, and the incredible possibilities that await us on the other end.

11.6 Software

After an algorithm is derived, and a quantum computer is built which is capable of running such an algorithm, the final part is the deliverance of these sets of higher-level instructions to the computer. This is the role of quantum software, to design a language that translates what the algorithms intend to do into a set of commands that the hardware can understand and execute.

The software of classical computers engages in this same fundamental pursuit, to translate mathematics into actionable programs, but there are a few problematic distinctions that alter how quantum software must be implemented. These distinctions began at the instruction level. Classical computers use boolean logic on their memory, or bits, to effect computation. Operations in UQCs also hope to model their computation through logic gates acting on different qubits, but unlike classical computers, qubits are not deterministic. That is, by the very nature of quantum behavior, qubits cannot be cloned between different processor registers, and reading the state of any one quantum register alters the information stored within it.

Furthermore, computation with QCs isn't deterministic either. When a classical computer processes information, the information represented is always in a single form, e.g., 1 + 1 is always 2. The quantum behavior of qubits means that all calculations done with it are probabilistic. 1 + 1 will almost always be 2, but there is a chance it could also be 3. The goal of a good algorithm is to produce the desired result *with a high degree of probability,* but there is still always a degree of uncertainty. As we shall see later, QCs operate in a very delicate ecosystem determined by extremely low temperatures and air pressure. Any software must be careful not to disrupt these delicate systems.

There is no consensus yet as to how programmers will be able to handle all the nuanced behaviors of a quantum computer, but progress will likely result in a new programming paradigm that is probabilistic as well. In this, QCs draw another parallel with the development of their classical counterparts, whose programming was once run on a probabilistic object-oriented paradigm.

11.6.1 *IBM-Q Experience*

Easy, accessible programming interfaces are necessary to encourage user usage, which in turn provides valuable data on system performance. The poster-child of these interfaces is the IBM-Q experience, a drag and drop quantum computing platform that actually takes the circuits users provide and runs them through the cloud to the physical machine (at the Thomas J. Watson Research Center in Yorktown Heights, New York). As of right now, there are three processors that users can use: two 5-qubit processors and one 16-qubit processor. There are also simulators you can run that give you mock results, including a 20-qubit simulator. As the company describes it, the IBM-Q experience is a "living experiment" that is constantly building towards new updates, calibrations, and progress in developing the technology as a whole.

Behind the IBM-Q experience is a delicate double layering of software that translates what the users input into the programs the system actually reads (Figure 11.14). The first layer of this is the **Open Quantum Assembly Language** or OpenQASM which functions as an intermediary for quantum instructions. These instructions are then passed onto **Quil**, which functions as the actual 'instruction set' architecture. The fundamental contribution of Quil to quantum software is that it allows for a shared quantum/classical memory model. A shared memory architecture is the infrastructure that allows for many quantum phenomena such as quantum teleportation, quantum error correction, and so on. More sophistication is planned for the future, including processor optimization and a higher-level language for writing and compiling these programs (Figures 11.14 and 11.15).

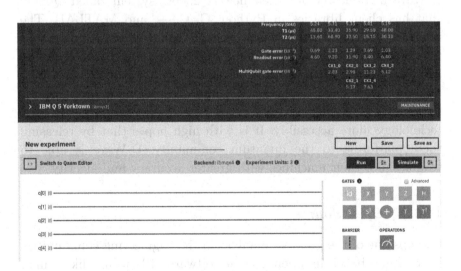

Figure 11.14: IBM-Q.

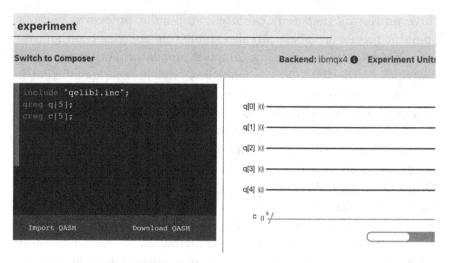

Figure 11.15: **OpenQASM assembler (bottom left) and the Quantum Information Software Kit (bottom right) with access to IBM's 5-qubit Machine.**

11.6.2 *QbSolv*

The equivalent platform for Adiabatic QCs is the standard API that D-Wave has released for their newest 2000Q system called QbSolv that has client libraries in Python, C, C++, and MATLAB. The interface can be accessed both over the cloud or as an integrated part of a High Performance Computing System (HPC), which is just a synonym for a supercomputer, or a data center.

Just as how the OpenQASM functions as a higher-level translator for Quil, so does **QSage** and **ToQ** help users interface with the technology more accessibly. It is with high hopes that by releasing these platforms to the quantum community, D-Wave is able to troubleshoot their machines with greater accuracy and alacrity.

11.6.3 *Open Source*

Perhaps one of the greatest mantras in the engineering communities is its rigid belief in open source software. Platforms like Linux have demonstrated how crowdsourcing innovation is the vehicle to a more successful and powerful technology, and as QCs develop, so are platforms adopting this paradigm. They rely on a community rather than a company to generate advancement. Whereas IBM has implemented the IBM-Q experience to interface with their machines, and D-Wave with their QbSolv, the pursuit of Open Source technology is to remove these higher-level interfaces and allow the community to directly send their code to the language interpreter. By removing the intermediate representatives, users will be able to write directly to the system interface and control. This is then translated into instruction and fed into the system for computation. In layman terms, it is empowering anybody to access their systems with the hope that they will be able to test for and troubleshoot unexpected cases that are otherwise not generated within the parameters of their higher-level interfaces.

While complex in theory, open-source platforms are more or less playgrounds for software developers, there are no shortages in quantum platforms to support the community. Among the list include **Regitti Forest**, a python-based suite that builds instructions on Quil and runs them on a 26-qubit simulator called the Quantum Virtual Machine. **Project Q** is an open-sourced version of the IBM-Q experience implemented in Python. Microsoft's **LIQUi |⟩** is a platform implemented in F#, **Scaffold** in C++, and **Chisel-Q** in Scala. Needless to say, there is no shortage for different developers wanting to experiment in different languages.

More tailored platforms are available for researchers interested in specific applications of quantum computing. For example, **QuTiP** simulates open quantum systems like superconducting circuits and **Open Fermion** is a chemistry platform used to generate and represent different chemical and material systems. Open Fermion is also connected with the Regitti Forest initiative so that researchers can run their equations on QCs. In this, Open Fermion and Regitti exemplify how collaboration within different developer communities are tied into the larger narrative of quantum computation's pursuit into reality.

11.7   Applications

Discussions on quantum computing often devolve into rebranding the technology into "faster computers". Indeed, this is marketable and true — QCs would be able to engage our current computational needs with more speed and accuracy. But the merits of a quantum age are of a qualitative kind. As Richard Feynman, a quantum-theorist who earned the 1965 Nobel Prize in Physics, said "Nature isn't classical, dammit, and if you want to make a simulation of nature, you'd better make it quantum mechanical, and by golly it's a wonderful problem, because it doesn't look so easy."

11.7.1   *Biochemistry*

11.7.1.1   *Modelling*

Perhaps one of the most tangible goals QCs can realize is the understanding and production of better biochemical processes. It is important to realize that molecular and chemical interactions are quantum by nature. If researchers want to model any molecule, they need to account for the push and pull between every subatomic particle. This is phenomenally laborious, especially when taking into consideration the exotic nature of quantum interactions like superposition, entanglement, and indistinguishability. In this, we are drastically limited by our current computational ability and as a result, researchers have to make approximations by using classical mechanics. In other words, rather than considering molecules as they are, as living quantum systems, they are reduced to a science project of balls and sticks, connected by springs, suspended by their triviality.

For a long time, these models were sufficient because we assumed that many of these quantum phenomena are only exhibited in temperatures near absolute zero. The study of this branch of physics — where electrons are not governed by quantum interactions — is called chemistry. However, as research deepens our understanding of these quantum interactions, we can no longer simply ignore such a class of interactions. When subatomic particles exhibit quantum phenomena, they alter the shape of the atom, which in turn changes how it interacts with neighboring atoms. And understanding how atoms interact with one another is the very basis for grasping chemical reactions. If we could understand these interactions, and perhaps harness them for our own benefit, we are empowered to capture and revolutionize an entire branch of biochemical reactions. Figure 11.16 shows the potential of modelling quantum chemistry.

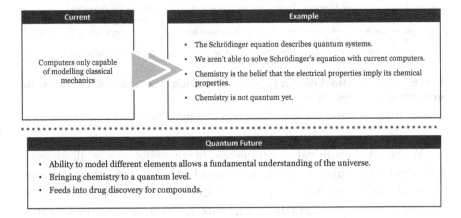

Figure 11.16: Biochemistry: Modelling quantum chemistry.
Source: IBM, D-Wave, Mermin "Quantum Computer Science".

11.7.1.2 *Drug Discovery*

The implications of understanding how molecules interact with one another — and their behavior with the environment — have the power to unlock an understanding of a new generation of medicine. Proteins, for example, are able to perform their role in biological processes through their shape. These shapes, in turn, are governed by the interactions between different parts of their amino acid chain. They fold over on themselves to form three-dimensional structures. This then enables them to perform a number of duties such as carrying isotopes around cells, blocking harmful infections, and so on. The destructive side of this "folding" is what leads to malfunctioning proteins. This creates conditions and diseases ranging from general allergies to Alzheimer's and Mad-Cow (Figure 11.17).

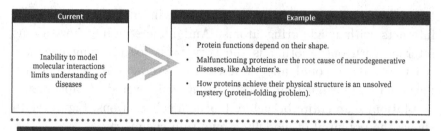

Figure 11.17: Biochemistry: Understanding the role of proteins in neurodegenerative diseases.

Source: IBM, D-Wave, Mermin "Quantum Computer Science".

While the chemical components of proteins are well known, the way they interact with one another and how they achieve their physical structure is poorly understood. Having access to a computer that can simulate the minute interactions within proteins that may cause these diseases would inform a generation of medical researchers and enable a revolution in drug discovery and patient treatment. Researchers are now proving just that. In 2012, Harvard Professor Alan Aspuru-Guzik published a paper predicting the lowest-energy configurations of folded proteins by using the D-Wave One computers.

On the corporate side, Accenture Labs' researchers collaborated with Biogen, the third largest biotechnology company, and 1QBit to prove that a quantum-enabled molecular comparison method was just as good or better than existing methods. (Molecular comparison is one of the first steps in developing a new drug). Their quantum method was able to provide more contextual information about shared traits between compared molecules vs the traditional method, which only infers such trait matches. Furthermore, it enabled researchers to see exactly how, where, and why molecule bonds matched, offering hope for more expedited drug discovery, trials, and effectiveness.

In the future, QCs are predicated to be capable of sequencing entire genomes, allowing drug treatment to be individually tailored, and perhaps even allow us to understand hereditary disease and how we can best combat them before they are even expressed.

11.7.1.3 *Radiology*

To date, radiation therapy remains one of the toughest and most effective treatments against cancer. Radiation beams are targeted at cancerous cells to stop them from multiplying, but determining the intensity and location of such beams must be optimized so that they minimize the size effects of killing healthy cells as well (Figure 11.18). Currently, medical dosimetrists leverage medical software to make these complex calculations, and the production and sale of these technologies will constitute a $1 billion market by 2024.

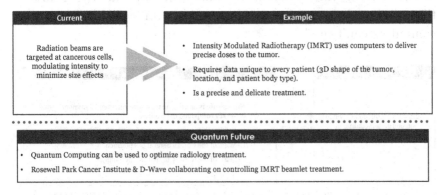

Figure 11.18: Biochemistry: Radiology.
Source: IBM, D-Wave, Mermin "Quantum Computer Science".

The reason why this is so expensive is that the best radiation isn't delivered as a brute-force treatment, but through what is called Intensity Modulated Radiotherapy (IMRT). In this high-precision treatment, a computer controls linear accelerators to deliver precise doses to the tumor, conforming around its shape by modulating the intensity of the radiation beam in small volumes. Optimizing and controlling such an operation demands constant evaluation of variables specific to every patient. The promise of a quantum computing infrastructure could provide these calculations more efficiently and accurately. Even slight improvements in these fields could affect significant improvements in the fight against cancer. An initial study on optimizing beamlet intensity for IMRT treatment has already been conducted between the Rosewell Park Cancer Institute and D-Wave.

11.7.1.4 *Materials Production*

One of the most tangible applications of QCs is to benefit bio-materials production (Figure 11.19). Currently, fertilizer production hinges on the production of ammonia from atmospheric nitrogen. Production depends on bacteria, which uses a particular enzyme called nitrogenase to operate this chemical reaction. However, the mechanism of this reaction is unknown. And at the heart of this reaction is an iron-molybdenum co-factor called FeMoco that computers can't model.

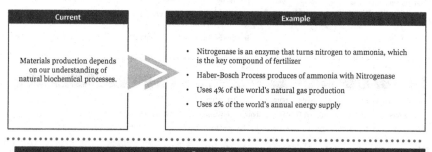

Figure 11.19:   Biochemistry: Materials production.

Source: IBM, D-Wave, Mermin "Quantum Computer Science".

This reaction to create ammonia-based fertilizers derived from nitrogen for the world's food supply has been industrialized as the Haber–Bosch process, which uses high temperatures, high pressures, and metal catalysts to force production. This is a laborious task that consumes 4% of the world's natural gas production and 2% of the world's annual energy supply. The promise of QCs is that they could bring about methods to better produce ammonium, and ultimately reduce annual energy usage. This is just one example in a long list of materials production. A better understanding of living systems around us can unlock new levels of efficiency, reduce energy consumption, and create cures for disease.

Another example of this is Photosystem II. It is a large enzymatic complex that carries the first steps to photosynthesis. A deeper understanding of its manganese center could enable artificial photosynthesis. Yet another revolves around solar cells. They are built with multi-crystalline silicon production, and understanding the reactions in its production could enable for fewer impurities and higher levels of efficiency. The list is endless, and it is this promise of application that engages so many companies in the push towards development.

11.7.2 Finance and Business

11.7.2.1 High-Frequency Trading

In the meager 36 minutes between 2:32 and 3:08 pm on May 6, 2010, millions of people around the globe were confounded as they watched the New York Stock Exchange hemorrhage $1 trillion in market value. Later, this day would be named the Flash Crash of 2010 and regulators would trace the root cause to a spoof in the algorithms that many firms had depended on for trading — a dependency that, once the stock market began to fall, ignited a massive selling panic in the marketplace.

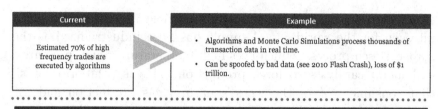

Figure 11.20: Finance and business: High frequency trading.
Source: IBM, D-Wave, Mermin "Quantum Computer Science".

With researchers estimating today that algorithms account for up to 70% of the high-frequency trading on Wall Street, algorithms have become invaluable commodities. The best algorithms achieve results by monitoring the thousands of transactions that happen each second, and then analyzing their variables for possible profit through arbitrage. But accounting for *every* variable is challenging, especially when it comes to random human behavior. This is especially challenging in the context of economics because unlike any other scientific field, economics is unique in its lack of a controlled environment where researchers can run experiments and test hypothesis. As a result, researchers can only grasp at models for the future based on what has happened in the past. Attempts to predict the future are governed by probability and a huge number of *past* variables. Dictating this exact pursuit is the class of algorithms known as Monte Carlo Simulations. Because of their computational weight, it is a field that molds particularly well around the advantages of a quantum computer. The ability to process more variables and draw conclusions from larger data sets could provide more accurate accounting of projected returns, risk assessments, and other factors necessary in evaluating baskets of investments.

Research engaged in this particular study has already proven quantum speedup. In 2018, a team in Toronto presented a quantum algorithm for the Monte Carlo pricing of trading derivatives that demonstrated a $O(\sqrt{N})$ run time as opposed to the classical $O(N)$ run time. Different teams around the world are investigating and discovering new parts of what is now considered "quantum finance" (Figure 11.20).

1QBit recently published two papers. One explored the calculations of optimal arbitrage opportunities using a Quantum Annealer like the D-Wave system. Another analyzed the impact of Brexit on financial markets. Investment firms are in a kind of financial arms race as well. DE Shaw, Renaissance Technologies, Two Sigma, and JP Morgan are all adding quantum computing to their quantitative investment arms.

So long as the financial industry depends on computation to aid or even power investments, quantum computing will stand as one of the most important factors in the development of future financial technology and, ultimately, ways to beat the market.

11.7.2.2 *Optimization*

While Adiabatic QCs may never be able to run Shor's Algorithm, their ability to compute optimization algorithms has merit. Nearly every industry has to deal with a notion of optimization: (1) what are the best flight routes for optimizing passenger seat miles; (2) what is the most economical sourcing of materials for manufacturing cars; (3) what are the best communication routes to maximize advertising? The list is endless. Virtually anything that depends on a choice that will minimize loss and maximize gain requires optimization. Figure 11.21 shows the possible future.

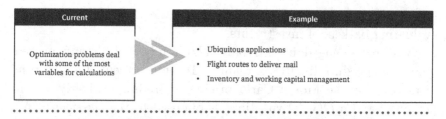

Figure 11.21:   Finance and business: Optimization.
Source: IBM, D-Wave, Mermin "Quantum Computer Science".

The implications are tremendous if this technology is industry-scalable, but especially so for retail companies like Amazon whose slim profit margins depend entirely on supply chain efficiency. The promise of even a 1% improvement can translate to millions of dollars in gained profit. Researchers examining the role of quantum computing in transportation operations are already making progress, with Volkswagen leading a study on how to tackle traffic flow in Beijing by examining data from 10,000 taxis. The sheer amount of movement data, destination points, and alternative routes can lead to what is known as a "combinatorial explosion" for traditional computers, and is precisely why QCs are necessary as an industry tool.

11.7.3   *Technology*

11.7.3.1   *Artificial Intelligence*

Imagine you are given a digital image consisting of different colored shapes. It would be a rather menial task for you to sort them by shape or by color, but for a computer, this is a phenomenally challenging notion to grasp. For one, the computer has to *define* the notion of a shape, and then the notion of a color. The study of building computers capable of doing so without human assistance is the branch of artificial intelligence known as unsupervised machine learning.

While researchers debate about the varying methods for reaching a functional AI system, everyone agrees that artificial intelligence is a technology built on massive amounts of data. Luckily, our society in the information age produces on average 2.5 quintillion bytes of data every single day (This is the number of neurons in every single adult brain in Europe). Massive amounts of information are traversing our countries and seas, and the ability to learn from this data is the burgeoning interest of the information age.

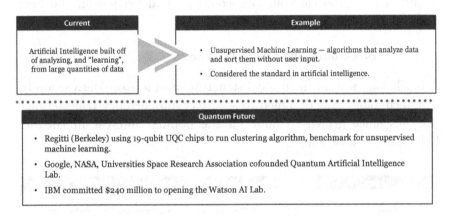

Current

Artificial Intelligence built off of analyzing, and "learning", from large quantities of data

Example

• Unsupervised Machine Learning — algorithms that analyze data and sort them without user input.

• Considered the standard in artificial intelligence.

Quantum Future

• Regitti (Berkeley) using 19-qubit UQC chips to run clustering algorithm, benchmark for unsupervised machine learning.

• Google, NASA, Universities Space Research Association cofounded Quantum Artificial Intelligence Lab.

• IBM committed $240 million to opening the Watson AI Lab.

Figure 11.22: Technology: AI.
Source: IBM, D-Wave, Mermin "Quantum Computer Science".

Advances are being made in laboratories around the world, and none are more impressive than the team at Regitti Computing that is successfully using one of its 19-qubit UQC chips to run a clustering algorithm. This is a method of sorting different things into similar groups that operate as a benchmark for unsupervised machine learning.

AI has always been the natural candidate positioned for a quantum revolution, and many experts consider quantum computing to be the ark upon which such a technology could be delivered. So, naturally, research and funding for this growing field of "Quantum Machine Learning" isn't lacking. In 2013, Google, NASA, and the Universities Space Research Association launched the Quantum Artificial Intelligence Lab to explore the D-Wave AQC.

Similarly, IBM has committed $240 million to opening the Watson AI Lab in Cambridge, MA that will study the possibilities and applications of quantum machine learning (Figure 11.22). Interest is especially concentrated in a possible hybrid classical-quantum solution, where the computationally heavy problems are offloaded to a quantum computer while the classical computer analyzes the data. This may be promising because short calculations may evade dealing with errors still prevalent in current QCs, and such quantum — classical solutions could be realized within the next decade.

11.7.3.2 *Blockchain and Cryptocurrencies*

A blockchain is a mathematical structure used to store data securely. The technology was designed to secure cryptocurrency transactions, and has since grown to fame off the popular interest in Bitcoin, Ethereum, and others.

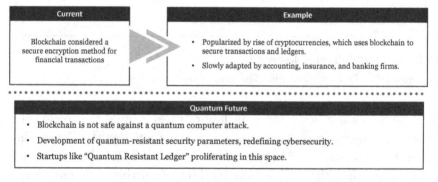

Current	Example
Blockchain considered a secure encryption method for financial transactions	• Popularized by rise of cryptocurrencies, which uses blockchain to secure transactions and ledgers. • Slowly adapted by accounting, insurance, and banking firms.

Quantum Future
• Blockchain is not safe against a quantum computer attack.
• Development of quantum-resistant security parameters, redefining cybersecurity.
• Startups like "Quantum Resistant Ledger" proliferating in this space.

Figure 11.23: Technology: Blockchain.
Source: IBM, D-Wave, Mermin "Quantum Computer Science".

However, blockchain can be used to store any type of data, and its utility is already being noticed by governments, accounting firms, and so on. Although such a technology is safe by today's standards, the blockchain technology is not quantum-safe and could theoretically be broken by a sufficiently large functional UQC. Namely, cryptocurrencies use what is called ECC-encryption to secure their public/private wallets. Security of this level would not withstand quantum decryption algorithms, and on top of that, QCs could also mine cryptocurrencies exponentially faster than their classical counterparts.

That being said, new security parameters are emerging that are quantum-resistant. Startups like "Quantum Resistant Ledger" are already offering their titular product and new methods to defend against such attacks will only continue to develop (Figures 11.23 and 11.24).

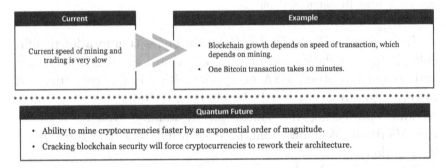

Figure 11.24: Technology: Cryptocurrency.
Source: IBM, D-Wave, Mermin "Quantum Computer Science".

11.7.4 *Government*

11.7.4.1 *Security*

As explained in Section 11.5.3, one of the benchmarks defining the progress of quantum computing is the successful execution of Shor's Algorithm — an algorithm that would, in theory, enable a quantum computer to return the prime factors of any number in polynomial time (Figure 11.25).

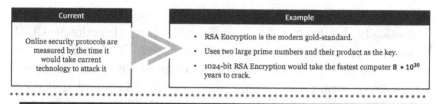

Current	Example
Online security protocols are measured by the time it would take current technology to attack it	• RSA Encryption is the modern gold-standard. • Uses two large prime numbers and their product as the key. • 1024-bit RSA Encryption would take the fastest computer $8 * 10^{30}$ years to crack.

Quantum Future

• Quantum Computer can break 1024-bit RSA Encryption in 1 day.
• Emerging "Quantum Encoding" market **now**, emphasis on Quantum Key Distribution (QKD).
• Cybersecurity firm ID Quantique (Korea), startup using QKD to protect data ($65 million by SK Telecom).
• Cybersecurity marketspace expected to be worth $25 billion in 2025.

Figure 11.25: Government: Security.
Source: Market Research Mediam.

Large-number factorization, because of its computational difficulty, is the foundation for security procedures like ECC encryption and its older brother, RSA encryption. RSA is the gold-standard in cryptography and is used ubiquitously, including securing online banking, protecting file transfers, fire-walling systems software like Microsoft Windows and in encrypting sensitive government information.

The promise of QCs poses a dangerous threat to the validity of these encryption procedures. Concurrently, "Quantum Encoding" is emerging to combat this future and proposals like Quantum Key Distribution (QKD) are providing alternatives that are Quantum-safe. The fear of decryption is revitalizing the cybersecurity industry, and startups are already proliferating this market space — a market space that is expected to be worth $25 billion by 2025, according to Market Research Media. ID Quantique is one such example that was founded in Switzerland and uses QKD to protect data. It was acquired in early 2018 by SK Telecom for $65 million. SK Telecom, the largest cell carrier in South Korea, plans to use the technology to guarantee security for a 5G broadband expected to roll out as early as March 2019.

11.7.4.2 *Weather*

While not as publicly known as their financial industry counterparts, weather agencies are also among the top users of supercomputers (Figure 11.26). The National Weather Service processes almost all of its observational data from buoys, satellites, and weather balloons through 160 feet of supercomputers housed in Reston, Virginia, and Orlando, Florida. These computers help different governmental agencies to predict the timing, location, and structure of different weather patterns and allows the close monitoring of up to eight storms at any given time.

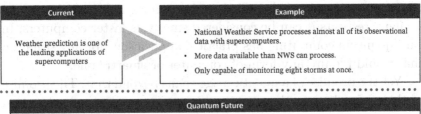

Figure 11.26: Government: Weather.
Source: IBM, D-Wave, Mermin "Quantum Computer Science", National Weather Service, UK Met Office.

Despite the combined computational capacity of 5.78 petaflops, more data is coming into these weather computers than they are capable of processing. Current computers are capable of predicting regional weather events, like snowstorms and hurricanes, but they are incapable of localizing those predictions. Because 30% of the US GDP ($6 trillion) is directly or indirectly affected by weather, such predictions are invaluable for food production, supply-chain optimization, air traffic control, and transportation. The United Kingdom is already considering the weather-implications of quantum computing with the MET office publishing a report on the practical necessity of such a technology for the needs of a post-2020 timeframe.

Hartmut Neven, the Director of Engineering at Google, also highlighted how better climate models can lend insight to how humans are influencing the environment and what steps need to be taken to prevent future calamities.

11.7.5 *Summary*

Quantum computing is heralded as the fifth computational age, and in many respects it can be considered simply as faster computers. In this, quantum computing has practical applications for any situation that would require any form of computer or smartphone.

Yet the promise of a quantum age spans beyond that. The promise of this technology is the next step in our response to the question that asks us the depth of our technological capacity. Whether that is breakthrough medical treatment, artificial intelligence, or synthetic material development, quantum computing may be the key to unlocking a new age of human advancement. It seems farfetched, but so was the idea of a computer only 40 years ago. After all, the idea of grains of sand being capable of performing advanced mathematics is a preposterous notion!

11.8   Current Development and the Future

While quantum computing holds immense promise for the future, the technology is still far from being actualized for commercial use. The leading companies in this field include many that have been named in the paper. D-Wave is working on producing quantum annealers while Google, IBM and Intel are pushing the frontier on

Universal QCs. Startups like 1QBit and QxBranch[1] are focused on the applications of this technology to industries while the academic world makes headway on potential algorithms that could be run on such a device. Finally, other startups like Rigetti are trying to bridge all teams by building processors, designing software for the processor, and applying it to industries.

The ultimate goal of all these computers is the point when QCs can surpass classical computers, a finish-line that is aptly named **Quantum Supremacy**. It is estimated that 50 **logical** qubits on a UQC operating at below a 0.5% error rate are all that's required, but reaching such a goal may be 5–10 years away. The most significant roadblock on this path is the notion of environmental noise and errors, which is explained in the following section.

11.8.1 *Error Correction*

Much of the quantum computing landscape today resembles the discovery of the filament for the lightbulb. The general logics of the technology exists, but we still struggle to find the materials that can deliver that technology.

As has been repeated, the biggest challenge facing QCs is retaining information. QCs operate using quantum phenomena as a source of data and storage. A change in temperature, a magnetic flux, or a random ray of light can also invoke these kinds of quantum behaviors, and in turn corrupt the information stored in the qubit. Furthermore, the possible physical candidates for qubits are limited. Whatever is used must (a) be capable of being manipulated into quantum states like superpositioning and (b) be able to be stored and have that information accessed. Leading candidates include electrons, photons, superconductors, and even diamonds, but none have proven to be the perfect fit.

---

[1]Paul Schulte is an investor in Q Branch.

Aside from a candidate qubit that is relatively error-tolerant, researchers are also working on ways that a computer can actively perform error correction. The leading solution was proposed by Peter Shor (who also discovered Shor's Algorithm). This involves coupling multiple qubits together with ancillary qubits to have a fault-tolerant redundancy within the design. If one qubit is corrupted, its ancillary would hypothetically be uncorrupted and could restore the corrupted qubit back to its original state. Such a solution has drawn the distinction between **physical qubits** and the operational **logical qubits** (see Section 11.4.3).

Of course there are still issues with this design. Namely, an incredible number of physical qubits would need to contribute to a logical qubit, with estimates as high as 10,000 of today's physical qubits for a single logical qubit, according to Alan Aspuru-Guzik of Harvard University. Moreover, the overhead computational costs of error-correcting requires almost as much power as today's QCs can handle in total, thus leaving little room to run the actual algorithm.

It's not just a race for adding new qubits, but a race for adding error-tolerant qubits and then evaluating the progress of technology requires benchmarking both metrics. The current goal is a 0.5% coherence rate. At those levels, computational power is expected to increase exponentially. The growth of performance decreases significantly at higher error rates, and beyond 1% error rates adding more qubits does not add any computational power at all.

This cannot be a race to add qubit quantity. IBM has proposed a metric called **Quantum Volume** that scales with both error tolerance and qubit count (Figure 11.27). It reduces errors as qubits are added. In other words, the reduction is errors is as important as the addition of computing power.

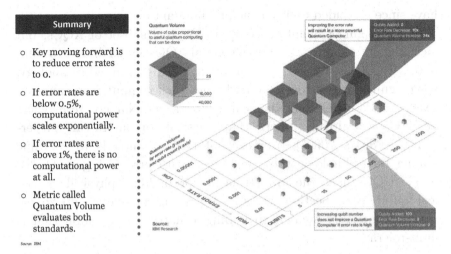

Summary

o Key moving forward is to reduce error rates to 0.

o If error rates are below 0.5%, computational power scales exponentially.

o If error rates are above 1%, there is no computational power at all.

o Metric called Quantum Volume evaluates both standards.

Source: IBM

Figure 11.27: Quantum volume: Error rates and qubit quantity must be considered.

Source: IBM.

11.8.2 *D-Wave 2000Q*

D-Wave is the leading producer of QCs, with their most recent model housing an impressive 2000 qubits. Their numerous collaborations with research institutes, companies, and universities are testament to their progress in the field, but their 2000 qubits are still far from Quantum Supremacy. Most importantly, their system aims at AQC, not UQC, and thus doesn't qualify for the 50-qubit benchmark. Furthermore, there is controversy on whether D-Wave computers are actually engaging in quantum calculations, or whether they are simply performing calculations with a potentially quantum infrastructure. The main point is that even though D-Wave originally planned to move towards Adiabatic Quantum Computation, they have since grounded their technology in quantum annealing systems. This paradigm of quantum computation is a subset of AQCs, but lacks the potential to scale into a general AQC that enables it to compete with UQCs. Rather, quantum annealing is a limited technology, with significant controversy on whether such computation can even reach speeds faster than classical computation. Namely, researchers at USC have already proved that classical computers can outperform D-Wave's 2000Q. Moreover, if D-Wave intends to continue scaling its technology, they must deal with errors (Figure 11.28). While UQCs have the infrastructure for building physical vs logical qubits (see Section 11.4.4), AQCs are unable to do so, so research is largely focused on finding a material for a qubit that is fault-tolerant.

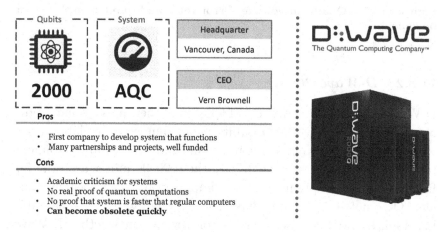

Figure 11.28: D-Wave: 2000Q.
Source: D-Wave.

Despite all the skepticism, D-Wave is engaging in numerous projects and collaborations. Even if they aren't producing profoundly advanced QCs, their \$200 million funding is at least a testament to their potential to do so.

11.8.3   *Google Bristlecone*

In March 2018, Google announced its newly minted 72-qubit quantum computer that they are "cautiously optimistic" can demonstrate quantum supremacy. The technology comes after Google's trial run with their 9-qubit device, which demonstrated only 0.6% error rates, just 0.1% from the generally accepted threshold of 0.5% error rates. Their 72-qubit device is a scaled up version of their 9-qubit device and they hope to preserve such error rates, though benchmarks for this new chip have not been released yet.

Google has also released plans that they are considering hybrid Adiabatic-Universal QCs, though no system has been produced as of yet (Figure 11.29).

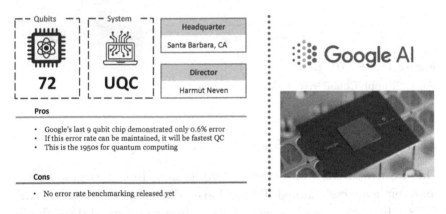

Figure 11.29:   Google: Bristlecone.
Source: Google.

11.8.4   *IBM*

In November 2017, IBM released a 50-qubit quantum computer that can preserve its quantum states for 90 milliseconds — an industry record. In addition to its 50-qubit system, it also has a fully functional

5-qubit system and 20-qubit system both functioning at low enough error rates that researchers can access and use them on the cloud. Benchmarking information on the error rates for their 50-qubit computer has not been released (Figure 11.30).

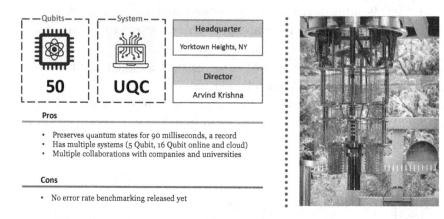

Figure 11.30: IBM.

Source: SupplyChainDrive, IBM, Azure, Asia Times.

11.8.5 *Intel Tangle Lake*

Despite a relatively late start in production, Intel has produced a 49-qubit chip codenamed Tangle Lake which was displayed at the 2018 CES. While such a chip doesn't bring any breakthroughs in quantum computing technology, Intel is researching a kind of qubit that could be built out of single electrons in silicon. Technical details aside, such a qubit is beneficial because it can be mass-produced with the same fabrication methods as transistors today. Unsurprisingly, this is a similar fabrication method that Intel has become particularly good at after mass-manufacturing the microprocessor for the past few decades (Figure 11.31).

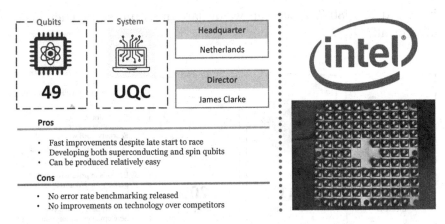

Figure 11.31: Intel: Tangle Lake.
Source: Intel.

11.8.6 *Global Race*

As different private companies quickly advance toward quantum supremacy, a similar race is taking hold of countries around the world. In 2015, China's largest retailer Alibaba teamed up with the state-backed Chinese Academy of Sciences to conduct research in the field. In February 2018, the first prototype 11-qubit chip became available for cloud testing, and the Chinese government has since pledged $10 billion for research and development in a new national quantum lab. Similarly, the European Union is planning a $1.1 billion investment in research (see Figure 11.32).

11.8.7 *The Complicated Reality*

With the birth of every technology, the most pressing question is how far we have to go in the foggy and seemingly endless road towards development. Where is this point of critical mass point? When will we see the practical proof of the quantum computer that was first imagined in 1980s? When can we declare quantum supremacy once and for all?

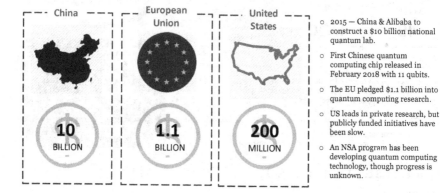

Figure 11.32: Global race.
Source: Bloomberg.

It's a fundamental point to understand that while quantum supremacy is a good benchmark, it isn't concrete by any means, and is constantly shifting as the technology we develop for QCs benefits classical computers as well. Just this last year, the newest supercomputer tested at 200 petaflops, doubling last year's record of 93 petaflops. Furthermore, as faster and more efficient classical algorithms are developed, so does the goalpost of quantum supremacy shift ever further.

Indeed, quantum computing is not a single sweeping victory. It's just one technological camp, fighting multiple battles, across multiple battlefields. Quantum computing hardware will slowly develop until it's able to run the algorithms that have already been written for them. Some of these algorithms will be faster than their classical counterparts. The ones that aren't faster will be tailored until they are, and the hardware will continue to develop to meet these new needs. It is progress that is made by the inch, across different territories, with years of testing and academic review.

Yet the promise of a quantum information age does exist, and a future where everything from wristwatches to vehicles of space exploration are powered by quantum technology is burgeoning within sight. After all, in 1998, the best quantum systems were 2 qubits large and 20 years later, applications with D-Wave's 2000-qubit AQC and Google's 72-qubit UQC are published. In fact, our progress now strongly resembles the state of classical computers in 1950. We are getting close to the first devices that are able to perform calculations which classical computers can't do.

Quantum computing can materialize within the next 30 years, in a form that is applicable, and with applications that will dramatically alter the undercurrents of our culture. It will be a long wait, but the road will be exciting, and the progress immense.

Bibliography

Ball, B. (n.d.). The Era of Quantum Computing is Here. Outlook: Cloudy. Retrieved August 6, 2018, from https://www.quantamagazine.org/the-era -of-quantum-computing-is-here-outlook-cloudy-20180124/

Barends, R., Shabani, A., Lamata, L., Kelly, J., Mezzacapo, A., Heras, U. L., & Martinis, J. M. (2016). Digitized adiabatic quantum computing with a superconducting circuit. *Nature, 534*(7606), 222–226. https://doi. org/10.1038/nature17658

Bleicher, A. (2018, February 19). The Ongoing Battle Between Quantum and Classical Computers. *Wired.* Retrieved from https://www.wired.com/story/ the-ongoing-battle-between-quantum-and-classical-computers/

Buchanan, W., & Woodward, A. (2017). Will quantum computers be the end of public key encryption? *Journal of Cyber Security Technology, 1*(1), 1–22. ht tps://doi.org/10.1080/23742917.2016.1226650

Devoret, M. H., Wallraff, A., & Martinis, J. M. (n.d.). Superconducting Qubits: A Short Review, 41.

Giles, M. (n.d.). Google thinks it's close to "quantum supremacy." Here's what that really means. Retrieved August 6, 2018, from https://www.technology review.com/s/610274/google-thinks-its-close-to-quantum-supremacy-heres- what-that-really-means/

Grassl, M., Langenberg, B., Roetteler, M., & Steinwandt, R. (2015). Applying Grover's algorithm to AES: Quantum resource estimates. *ArXiv:1512.04965 [Quant-Ph].* Retrieved from http://arxiv.org/abs/1512.04965

Hemsoth, N. (2016, June 28). Novel Architectures on the Far Horizon for Weather Prediction. Retrieved August 6, 2018, from https://www.nextplatform.com /2016/06/28/novel-architectures-far-horizon-weather-prediction/

Ilievski, E. (n.d.). Adiabatic Quantum Computation, 17.

Jackson, M. (2017, June 25). 6 Things Quantum Computers will be Incredibly Useful for. Retrieved August 6, 2018, from https://singularityhub.com/2017 /06/25/6-things-quantum-computers-will-be-incredibly-useful-for/

Jordan, S. P., Farhi, E., & Shor, P. W. (2006). Error correcting codes for adiabatic quantum computation. *Physical Review A, 74*(5). https://doi.org/10.1103/P hysRevA.74.052322

Kendon, V. M., Nemoto, K., & Munro, W. J. (2010). Quantum analogue computing. *Philosophical Transactions of the Royal Society A: Mathematical, Physical and Engineering Sciences, 368*(1924), 3609–3620. https://doi.org/1 0.1098/rsta.2010.0017

Kissell. (n.d.). *An Update on the Google's Quantum Computing Initiative.*

Knight, W. (n.d.-a). IBM announces a trailblazing quantum machine. Retrieved August 6, 2018, from https://www.technologyreview.com/s/609451/ibm-ra ises-the-bar-with-a-50-qubit-quantum-computer/

Knight, W. (n.d.-b). Serious Quantum Computers are Finally Here. What are We Going to do With Them? Retrieved August 6, 2018, from https://www.technologyreview.com/s/610250/serious-quantum-comp uters-are-finally-here-what-are-we-going-to-do-with-them/

Lekitsch, B., Weidt, S., Fowler, A. G., Mølmer, K., Devitt, S. J., Wunderlich, C., & Hensinger, W. K. (2017). Blueprint for a microwave trapped ion quantum computer. *Science Advances, 3*(2), e1601540. https://doi.org/10.1126/sciad v.1601540

Mermin, N. D. (2016). *Quantum Computer Science: An Introduction.* Cambridge University Press, New York, US.

Moore, S. K., & Nordrum, A. (2018, June 8). Intel's New Path to Quantum Computing. Retrieved August 6, 2018, from https://spectrum.ieee.org/nano clast/computing/hardware/intels-new-path-to-quantum-computing

Moses, T. (2009). Quantum Computing and Cryptography, 12.

Proos, J., & Zalka, C. (2003). Shor's discrete logarithm quantum algorithm for elliptic curves. *ArXiv:Quant-Ph/0301141.* Retrieved from http://arxiv.org/ abs/quant-ph/0301141

Radiology Information System Market Size Worth $980.2 Million By 2024. (n.d.). Retrieved August 6, 2018, from https://www.grandviewresearch.com/press-release/global-radiology-information-system-ris-market

Rebentrost, P., Gupt, B., & Bromley, T. R. (2018). Quantum computational finance: Monte Carlo pricing of financial derivatives. *ArXiv:1805.00109 [Quant-Ph].* Retrieved from http://arxiv.org/abs/1805.00109

Roell, J. (2018, February 1). The Need, Promise, and Reality of Quantum Computing. Retrieved August 6, 2018, from https://towardsdatascience. com/the-need-promise-and-reality-of-quantum-computing-4264ce15c6c0

Simonite, T. (2018, May 19). Google, Alibaba Spar Over Timeline for "Quantum Supremacy." *Wired.* Retrieved from https://www.wired.com/story/google-a libaba-spar-over-timeline-for-quantum-supremacy/

Soon, W., & Ye, H. Q. (2011). Currency arbitrage detection using a binary integer programming model. *International Journal of Mathematical Education in Science and Technology, 42*(3), 369–376. https://doi.org/10.1080/0020739X. 2010.526248

US Department of Commerce. (n.d.). About Supercomputers. Retrieved August 6, 2018, from https://www.weather.gov/about/supercomputers

Williams, P. Collin. (2011). *Explorations in Quantum Computing.* Springer-Verlag, London, UK.

Wolf, R. de. (2016). Quantum Computing: Lecture Notes. University of Amsterdam, Amsterdam, Netherlands.

Young, K. C., Sarovar, M., & Blume-Kohout, R. (2013). Error suppression and error correction in adiabatic quantum computation I: techniques and challenges. *Physical Review X, 3*(4). https://doi.org/10.1103/PhysRevX.3.0 41013

Chapter 12

Cloud Wars: Alicloud vs AWS and Azure (and What Huawei Wants to Do About It)

Cloud wars are out there and Figures 12.1–12.11 give an overview and comparison.

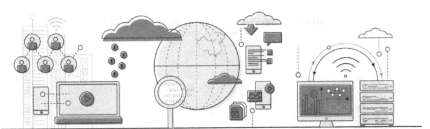

Global Cloud Market: Amazon, Microsoft, Alibaba, Huawei Cloud Comparison.

12.1   Cloud Market Overview: Global and China

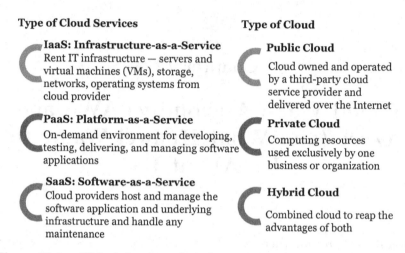

Type of Cloud Services

IaaS: Infrastructure-as-a-Service
Rent IT infrastructure — servers and virtual machines (VMs), storage, networks, operating systems from cloud provider

PaaS: Platform-as-a-Service
On-demand environment for developing, testing, delivering, and managing software applications

SaaS: Software-as-a-Service
Cloud providers host and manage the software application and underlying infrastructure and handle any maintenance

Type of Cloud

Public Cloud
Cloud owned and operated by a third-party cloud service provider and delivered over the Internet

Private Cloud
Computing resources used exclusively by one business or organization

Hybrid Cloud
Combined cloud to reap the advantages of both

Figure 12.1:   Global cloud market share: Overview of cloud computing structure.

Source: Microsoft, Forbes, Synergy Research Group, Schulte Research.

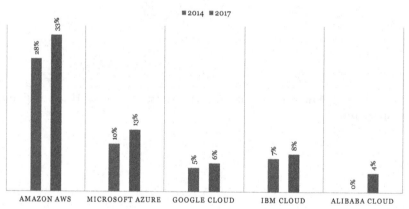

Global Cloud Services — Market Share 2014 and 2017
(IaaS, PaaS, Hosted Private Cloud)

■2014 ■2017

AMAZON AWS: 28%, 33%
MICROSOFT AZURE: 10%, 13%
GOOGLE CLOUD: 5%, 6%
IBM CLOUD: 7%, 8%
ALIBABA CLOUD: 0%, 4%

Figure 12.2:   Global cloud market share: AWS being the growing global leader.

Source: Forbes, Synergy Research Group, Schulte Research.

Size of the Cloud Computing and Hosting
Market Worldwide, 2011–2019
(in billion US dollars)

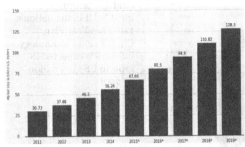

- Cloud computing to increase from $67 billion in 2015 to $162 billion in 2020 at CAGR 19%.
- Infrastructure as a Service (IaaS) to grow 36.6%, reaching $34.7 billion.
- SaaS revenue is expected to grow 21% reaching $58.6 billion by the end of this year.

Figure 12.3: **Global cloud market trends: Great increase of the cloud market.**
Source: Forbes, Gartner, Insight, 451 Research, Synergy Research Group, Schulte Research.

- By 2019, more than 30% of the 100 largest vendors' new software investments will have shifted from cloud-first to cloud-only.
- 81% of enterprises have a multi-cloud strategy.
- Amazon Web Services has been adopted by 64% of firms in 2018, while Microsoft Azure has been adopted by 45% of firms .
- AWS is the global leader by holding 33% of the market share.

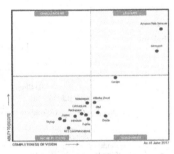

Figure 12.4: **Global cloud market: Cloud services performance breakdown.**
Source: Forbes, Gartner, Schulte Research.

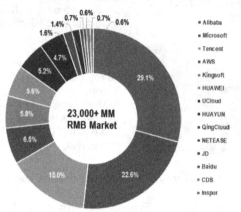

China Public Cloud Market Share 2017

- China government views cloud computing as a strategic priority and included it in the nation's 12th Five-Year Plan.
- China's cloud technology industry is expected to grow to US$103 billion by 2020.

Figure 12.5:   China cloud market share and trends: Intense competing and emphasis.

Source: IDC, CGTN, Gartner, Schulte Research.

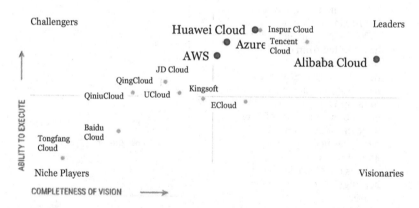

Figure 12.6:   China cloud market: Infrastructure as a service performance breakdown.

Source: IDC, Gartner, Schulte Research.

12.2   Cloud Services Comparisons

	amazon web services	Azure	Alibaba Cloud	HUAWEI
Market Share	33%	13%	4% 29.1% (47.6% Iaas) in China	5.6% in China
Revenue(2017)	$17.5 billion	$27.4 billion	$2.1 billion	$500 million
YoY growth	48%	98%	104%	N/A
Market Share Gain	+1/2%	+3%	+1%	N/A

Figure 12.7:   Cloud services comparison: Market share, revenue and growth.
Source: Business Insider, Statista, Zdnet, Techcrunch, Gartner, Schulte Research.

	amazon web services	Azure	Alibaba Cloud	HUAWEI
Launch Time	March 2006	February 2010	September 2009	March 2017
Available Countries	190+	140+	150+	China and Hong Kong
Regions	18	42(plan for 54)	18(8 in China)	4

Figure 12.8:   Cloud services comparison: Available regions and countries.
Source: Business Insider, Zdnet, Techcrunch, Gartner, Business Line, Data Economy, Investopedia, Schulte Research.

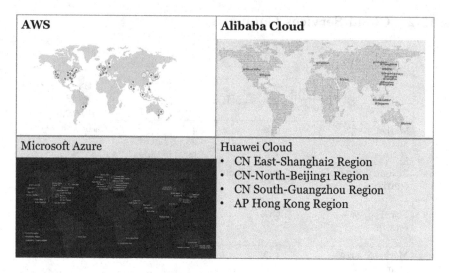

Figure 12.9:   Cloud services comparison: Locations.

Source: Amazon, Schulte Research.

Figure 12.10:   Cloud services comparison: Customers (1).

Source: Business Insider, Computer World, Schulte Research.

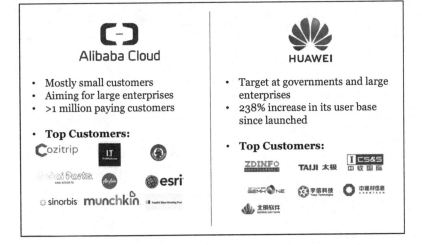

Figure 12.11:   Cloud services comparison: Customers (2).

Source: Sdx Central, ZD Net, Huawei, Schulte Research.

Chapter 13

Insurtech: Digitizing Head, Health, Hands and Heart

13.1 Introduction

Financial technology is the way it is because millennials are the way they are — not the other way around. Financial institutions, especially insurance firms, are coping with two problems as they navigate new technology with social change. The fundamental shift in attitudes and habits of millennials is more abrupt than the change in the technology itself, so there are two moving targets. This paper compares aging boomers and a richer millennial in four areas: work, personal, finance, and worldview. Their new and profoundly shifting needs as well as rapid technological prowess — both in the US and in the emerging world — present great challenges for insurance companies. We provide a framework for the following: (1) restructuring processes within an insurance firm to integrate data analytics to understand and process the needs of millennials; (2) retraining staff using age-old treatment methods for fear-based patients in substance abuse rehabs; (3) understanding how healthcare data and financial data come together; (4) how blockchain will challenge incumbent insurance companies; (5) how manpower can be reorganized by functionality; and (6) the changes in insurance products and processes that are necessary as millennials and technology interact, including the necessary creation of separate digital units for millennial insurance products. Fundamentally, the golden opportunity for insurance is to transform from paper-based creatures into data-centric powerhouses, transforming digital provenance into digital collateral to become wholly owned, asset-light, digital bancassurance entities aimed at millennials. This process is also entirely applicable to banks and brokers.

13.1.1   Introduction: The Problem is Losing Track of Customers as New Technology Explodes

Insurance has changed very little in the past several decades and is still manpower-intensive, paper-heavy, backward-looking, and often led by very conservative boards. They have only recently begun to take action to combat a new and rapidly growing pure digital foe which has no marginal cost of acquisition and which will be further revolutionized by blockchain. So, it is at greater risk of disruption than banks, media, or retail over the coming years. New pure online insurance providers which are based in the cloud — and whose policies and claims will be processed through the cloud — have no agents, no paper and no antiquated vertical structures. They have automated claims processes with no wasted time or human error. They have access to mountains of vast and fine-tuned data on the risks we pose to ourselves and others on a minute by minute basis for any human activity. This will logically lead to better pricing for risk and more manageable reserves. Those who cannot create secure yet porous data centers where large data sets can enter *en masse* for all insurable objects must act quickly to remain relevant.

Many surveys indicate that current customers of insurance are unhappy with their experience and the way it is being delivered.[1] There is a general dissatisfaction with the process. It is unintegrated, a paper-driven slog and it is confusing. Moreover, it seems to start with the premise that people are basically dishonest. Dan Ariely of *Lemonade* said it well: "If you tried to create a system to bring about the worst in humans, it would look a lot like the insurance

---

[1]There are many surveys which show dissasisfaction. One is "Health Insurance Survey Finds Consumer dissasisfaction" by Dan Gorenstein. This is from the Kaiser Family Foundation.

of today."[2] It is a business, many think, which begins with the supposition by the insurance company that the submission of a claim is fraudulent.

Insurance is not the only sector where there exists a certain "millennial fatigue" with institutions. Consulting firms have published many surveys which show that people, especially millennials, have trust issues with religious institutions, banks, government, and law enforcement. The Global Financial Crisis (GFC), in particular, has contributed to much of the distrust in financial institutions which continues today.[3] So, social networks are proliferating for these millennials to form new kinds of institutional relationships which can satisfy needs while they offer a sense of trust. (The recent episode of Facebook and its activities with Cambridge Analytica and Palantir may shake the confidence in social networks). In the meantime, millennials bypass corporate websites and seek out crowd-sourced advice to satisfy needs, wants, desires and the securing of a financial future.

This movement in millennials seeking out crowd-sourced solutions is happening at a time when a proliferation of new technologies puts into their hands data sources which were undreamt of only a few years ago. Countries like China are arguably ahead because it was starting from a technological tabula rasa and could create vast new digital infrastructure without the intellectual, physical, lobbying and regulatory baggage of the past. As millennials seek out an entirely new way to hedge against risk in their lives outside of traditional insurance products — lives which may not include having a long-term career in one firm or owning one car or buying one home — they leave behind a contrail of data. So far, they are allowing this to be vacuumed up by ubiquitous sensors. These sensors then offer a feedback loop of billions of data which allow enterprising

---

[2]Please find attached Schulte research chart which shows surveys indicating that financial services are significantly behind in terms of customer satisfaction and other issues like honesty: http://www.schulteinstitute.org/wp-admin/upload.php?item=1099

[3]KPMH International Guardians report (Feb 2018) notes that only 21% have a high level of trust in financial institutions in the US. Contrast this to 65% in India.

entrepreneurs the ability to gather, sort, and refine in detailed ways the risk of — and the ability to insure against — accident, default, sickness, and so forth. Quantifying these risks — and then creating financial platforms to fund these risks — has generated great fortunes for such entrepreneurs. Kabbage, Lemonade, Zhong An, Oscar, Slice, Trov are but a few.

Why is this taking so long for insurance companies to create a coherent response to these upstarts? Insurance companies are in the business of quantifying risk. When will we die? Will we get hurt at home? Will this plane crash? Will this oil rig blow up? Will my house get robbed, burn down or flood? The old way which worked for decades was to use backward-looking, paper-based and static data like previous insurance, arrests, education, FICO scores, zip code, ethnicity, marital status, or age. The new way to assess risk is by harnessing the power of smartphones ($\sim$20 sensors on 5 billion phones will offer a vast store of risk data on each individual 24/7) as well as the tens of billions of sensors (which offer risk data on homes, ships, rigs, cars, boats, bridges, children's cribs, pipes, and buildings). This lets firms which can handle billions of data sets create a whole new science of risk to: (1) calculate premiums and reserves more accurately than ever; (2) diversify portfolios of different types of integrated insurance products on one platform; and (3) get robust data sets which can, in fact, offer any kind of insurance anywhere, anytime, to anyone. Paper-based firms which cannot adapt to the technology — as well as the ways in which new technology users want their goods and services delivered — will die off. We will discuss this new species of users in the following section.

13.2 Why This Massive Shift is Happening: Millennials are a Different Species Than Boomers

Before we go into detail about 'how' this technological explosion is happening with all its myriad consequences, we need to discuss the 'why'? This is often overlooked, since a powerful narrative is "build a technology and they will come!" This is done without recognizing that there is a profound difference in what millennials want and it is not just because new forms of technologies have been created. Technology did not create the millennial — it is the other way around. So we ask: Why is there a sudden yearning for much different insurance products among these millennials. The chart below shows the differences between millennials and boomers.

Perhaps the best way to explore the vast changes in the demands of customers — and the way their tastes, preferences, sources of data, and networks are evolving — is to contrast the boomers with the millennials on four different levels. Figure 13.1 shows characteristics that we think encompass the boomers. We call them *Homo papyrus*. (The authors together have spent a total of more than 50 years in equity research understanding their habits!). These are the "over 48 crowd". They were born into a postwar world which we refer to as *Pax americana*.

Boomers	Homo Papyrus -1946-1966	Product
1. Financial	1 more work, less meaning	Full time work dominated the model
	2 pricing power	Stable premiums
	3 mortgage, broker account-agent	Few unique $ obligations
2. Professional	1 One office. One PC. One job.	Group plans dominate
	2 institutional trust	Traditional brands, products.
	3 Same home, job. Data is static.	Group plans with little data deviation
3. Personal	1 own home, car,	Traditional life, car, home products
	2 All paper	Little digital contrail
	3 Dependable holiday schedule	Little variety in travel
4. Worldview	1 YES / NO	Fill in forms. Get product
	2 Black/white	Why hybrid, unique products?
	3 Grew up in Pax Americana	America-centric patriotic view

Figure 13.1: *Homo papyrus.*

The generation of the millennials was a "yes/no" world where many of these people had parents who served in the military. There was a black-and-white approach to life. Segregation was commonplace and not questioned. Work was primary and vacation was at the same time each year. These boomers often had only one job, one office, one desk, one computer. There were two loans — one for a house and one for a car. Credit cards did not begin to proliferate until the mid- to late-1980s. Financial transactions were done with ONE mortgage broker and ONE stockbroker. ALL of it was paper.

The chart above shows what naturally flowed from this worldview in terms of insurance products. In addition, this group of Americans did not know the war on terror as it was "over there". Looking back, the US was a safe cocoon of regularity and certainty with work, life, and leisure. The result of these lifestyle choices was group plans which had little deviation. Life insurance products were plain vanilla and based on zip code, sex, marital status and a few checked boxes for health risks. It was plain, group, homogenous, static, paper and boring.

13.3   Millennials: Individual, Diversed, Toughened

Now, we go to the millennials. They have grown up in the shadow of 9/11, the GFC war on Islamic terror in the West, immense data thrown at them from a young age, the Iraq war and a spate of school shootings. (Many of the older generation did not have access to the newspaper at breakfast when they were growing up. Millennials have seen 8–10 news sites before they even leave for school!). In addition, these millennials enter the adult world with a much higher degree of financial uncertainty, no wage bargaining power, and this is often punctuated by a high student loan burden. Let's not forget, college for veterans after World War II was free. Moreover, an in-state student in California until the 1990s could go to a UC system university for virtually nothing.

As a result, the life and times of millennials — their tastes, sources of information, traits, and habits — are very different from boomers. Figure 13.2 below shows interesting traits of this new generation. On the financial end of things, these people are struggling to make ends meet with declining earnings power, fewer jobs in traditional firms and a higher debt burden. Does their preference for meaning over workflow from a resignation or realization that they are just not going to be able to accumulate wealth? Do they genuinely look to a higher sense of meaning as a rejection of — or in response to — the materialism of their parents? This is hard to know, but it is probably a mix of both.

Millennials	Homo Digitalis – 1980-2000	Product
1. Financial	1 less work more meaning	Mental health, holistic life
	2 little wage power	Higher Volume over lowerprice
	3 student loan burden	Debt relief insurance
2. Professional	1 pure mobility.No office, pc	Temporary plans
	2 institutional distrust	Digital subsidiaries are perfect
	3 job hopping "Gigs"	Partnership, collaboration
3. Personal	1 non-owners	Any time anywhere products
	2 quality of life	Lifestyle IS the product.
	3 pure digital in everything	ZERO paper
4. Worldview	1 Questions boundaries, structures	Seeking solutions in diverse apps
	2 No black/white – gray	LGBT, womens, internat'l products
	3 Grew up w/ Terror: wiser, stronger	Tougher, grittier advertising

Figure 13.2: *Homo digitalis.*

1. **Millennials and Finance:** So, in the area of their *financial lives*, it is no surprise that the outcome of these difficulties among millennials is the need for products which are radically different from Boomers. It is not just a matter of offering partial policies which focus on mental health among the young, ways to refinance student loans through insurance products or offering lower price products to young people who simply do not have steady or high income. It is a matter of reaching them where they go. They are NOT looking for any of these on the websites of either banks or insurance companies. Insurance companies must find them, draw them in, offer products with very different advertising campaigns and seal the deal. This requires a very different sense of advertising, platforms, processes, price points, and structures. It is root; it requires melding the product mix into a fluid arrangement on a platform which can reach people where they are. They are

not interested in seeking out the static website of the insurance company.

2. **Millennials and Work:** In the *professional lives* of these millennials, we can see that the above characteristics flow into their desires for work. These people between the ages of 18 and 36 are, in general, looking to create value and find satisfaction outside of the traditional work environment either because they are repelled by the values of the institutionalized work environment or because those opportunities are harder to come by and, therefore, highly competitive. Is it worth it, they ask.

In addition, is it worth the compromise in values that may be necessary to survive in a cut-throat environment run by older people who simply do not understand the needs of younger people, not to mention the types of technologies they use and respond to? As a result, millennials may be drawn to part-time situations with other, younger people who are creating new forms of institutions in a pure entrepreneurial situation. This is a combination of new values meeting new technologies, and neither involves the boomers.

These features are not only true but also magnified in emerging markets. This is because in markets like Indonesia, there are 48 million millennials who are not held back by aging and ossified systems. The banking system is still fairly primitive, as is the case in Bangladesh (50 million), Pakistan (56 million), India (360 million), Vietnam (30 million), Philippines (30 million) and Iran (29 million).[4]

3. **Millennials and Personal Lives:** Thirdly, in the *personal lives* of this younger generation, they are more comfortable with not owning much of anything. There is no great desire to own a car, which was the first thing a boomer wanted to own. Nor is there a need to own a house, even if it is because that goal is out of reach.

---

[4]This comes from AT Kearney, Where are the global Millennials, Global Business Policy Council.

In addition, they find it ridiculous to fill out a piece of paper for any type of loan — that is a non-starter! This attitude has created — or been created by — the world of Uber and Airbnb. These are easy to use, all digital apps which offer anywhere, anytime, anyplace action for movement, vacation, living, working. Where is the equivalent in insurance? Why is it taking so long for traditional insurance operators to adjust to this reality even as they see billion-dollar companies being born who can offer effective anywhere, anytime insurance?

4. **Millennials' World View:** The last part of this section on *worldview* is that this group is much tougher than many perceive. They were born in households traumatized by 9/11. Millennials have been exposed to more news than their parents had in their entire lives by the time they are 18 years old. They have lived in the world of domestic terror in both the US and Europe. They are aware of alternative lifestyles. They have parents of different cultures and ethnicities. They are not lazy or indifferent or checked out. This misses the point. They have access to a revolutionary new set of technologies which are absent in the established world and are trying to navigate a way to create meaning and value with no roadmap. Confused? Yes. Lazy or afraid? NO.

In summary, we can describe this group as a more nomadic, questioning diaspora of young people who see the current professional *status quo* as unexciting and lacking in technological prowess. They see professional progress in the context of starting from scratch and questioning the edges of the professional sandbox. They think they will increase their value and learn more by being around other younger — who live and breathe technology — people rather than established older professionals who lack technological skills. They are much more open to experimentation and live in the grey areas. A large portion of them are mixed race and are totally liberal in outlook in terms of race, language, religion, nationality, and gender. More of these are professional women. One implication for this is that it requires a fundamentally different or slightly edgy type of advertising which traditional insurance companies may find uncomfortable or to break away from tradition.[5]

13.4   Summary Comparison: *Homo Digitalis* (Millennial) vs *Homo Papyrus* (Boomer)

Different in Everything	Homo Papyrus: Boomer. 1946-1966	Homo Digitalis: Millennials 1980-2000
1. Financial	ONE mortgage, ONE broker, ONE agent. ONE DATA SOURCE	No wage power, Indebted, GIG apps. MULTIPLE DATA SOURCES
2. Professional	Institutional trust; One job. Locals. STATIC	Distrustful;Experimental;Tech driven. DYNAMIC
3. Personal	Steady schedule, finance.Same vacation, newspaper, splendid isolation. Suburbia. GROUP.	Anywhere, anytime diverse digital cultural diaspora w/out assets. INDIVIDUAL.
4. Worldview	Pax Americana. Patriotism. Black &white. INSULATED	9/11. Age of Terror. Multi-racial. TOUGHENED.

Figure 13.3: Comparison of Boomers and Millennials.

Looking at the contrast between the two: Night and Day!

In conclusion, Figure 13.3 summarizes *in general* the bottom line differences between Boomers and Millennials. We represent it in terms of products that each used or currently use. In the area of financial activity, Boomers and Millennials differ in the same way as the old Merrill Lynch and the new Robin Hood. It is the difference of boomers who looked to one agent and millennials who look to crowd-sourced confirmation from thousands of people to arrive at a conclusion. In the area of professional activity, it is the difference between working at Sears Tower in Chicago (or any tower in LA, NY, Atlanta, Charlotte, or Boston) for one firm for 20 years and working

---

[5]The link shows legacy value vs new digital values. It is interesting but looks at it from organizational structure rather than end users shifts btw boomers and millennial: https://www.constellationr.com/system/files/uploads/user-16818/new_digital_power_values_for_2018_ceo_cio_cdo_cmo_cco.png

in Wespace in a dynamic startup in gigs that may only last for 12 to 18 months.

In the personal area, it is the difference between a static world of Yellow Pages and a steady schedule vs a new world of anytime, anywhere, anything on any app. In their worldviews, millennials and boomers could not be more different. One grew up in Pax Americana that was a turbulent Cold War, but there was no domestic terror incident or frequent school shooting. Millennials grew during 9/11, the age of terror in US borders and school shootings.

13.5   Implications for Insurance Products

The intention is not to generalize but to highlight the characterizations. The point is that, even if there is only a degree of truth here, the approach of insurance companies to processes, services, differentiation, advertising, and products needs a radical rethink. The answer perhaps lies in simply creating a separate digital unit for a different customer. It takes a different platform, different products, and different advertising.

Not only is there a fundamental shift in technology but also an entirely new customer. We get lost in the weeds of technology, but it is really about a very different customer. The customer has a different worldview (they might call it 'evolved') and the approach to explosive new technologies must be adapted dramatically to hit the mark. We have two sets of rapid change occurring — the technology itself and the customer who is using this new technology. Getting the mix right is key.

The implications for insurance products are that insurance companies need to create a parallel ecosystem which can exist within the millennial world as a separate and distinct universe from the traditional forms, processes, and products which are a relic of the boomers. This speaks to a parallel digital company (a wholly owned subsidiary) for these people who will be wealth creators and will have families for the next 25 years. It speaks to new forms of partnerships and methods for collaboration for delivering products to this group. We will show how this phenomenon is creating a blurring of the lines among and between virtually every type of insurance product.

Standard group insurance plans will soon be a thing of the past. This is especially true of life insurance. Anywhere, anytime insurance is the only way forward. In the same way that banks are having a difficult time creating accounts and products for entrepreneurs, so, too, are insurance companies. Where is the "entrepreneur's insurance package"? There is no such thing. There are so many potential products to be created for a market starving for different kinds of insurance that do not exist.

13.6   How This Massive Shift is Happening: He Who is Flexible and Controls the Data Will Win

As banks and insurance companies try to get their collective heads around this group of millennials globally, they need to gather and understand the data from the internet of things and how this allows the tech-savvy to gain knowledge of a new customer base. There is a gusher of billions of pieces of data from cell phones, wearables, cars, pipe systems, oil rigs, locks, thermostats, window sensors to discover every potential risk before it happens. Risks can be monitored through merchandise receipts and sensors to anticipate and offer products to help people deal with, prevent, anticipate or plan for: the imminence of pregnancy, frozen pipes, burglary, depression, parental dementia, alcoholism, divorce, genetic predisposition for disease and much more.

Where are the short term insurance products which can insure people against catastrophe but also create a positive feedback loop which can offer them a reduced price in return for data which can prevent unnecessary payouts? Where is the possibility for a person to offer his or her own solution for potential risk and then agree on a price? Where is the flexibility that must correspond to frequent change in the professional and personal lives of millennials? Where are the policies that correspond to frequent job changes as the

GIG economy[6] of short term jobs is the one they inhabit given the disappearing white collar job market?

Every facet of life — from the cradle to the grave — can and will be anticipated with greater accuracy than ever. Algorithms anticipate divorce, bankruptcy, cancer, accidents BEFORE the person knows it.[7] Why don't product offerings from banks and insurance companies have the same degree of accuracy and anticipatory capability? This presents a host of ethical problems which are beyond the scope of this paper. These include whether any insurance company has a right to genetic information which can offer insight about the likelihood of chronic diseases in later life. This does, in fact, make sense since having anticipatory information on future disease can incentivize a person to take remedial action to prevent chronic diseases from developing. As a result, they live longer. Both sides win.

13.7 Five Ways to Understand the Customer in the New World of Data

The proliferation of data and increasing transparency of the products' customers need and how these customers want their insurance products delivered will inevitably create disintermediation in the way they are (1) calculated; (2) distributed; (3) assessed; (4) reserved for; (5) bundled; and (6) processed. In Section 13.6, we discussed the process by which the delivery of insurance products must change. In this section, we show this evolution in sensor data and the way that data is analyzed forces change in the way insurance firms must change on a manpower and organizational level. The organization (Figure 13.4) basically separates the world of financial institutions into people, things, money, and ideas.

---

[6]A gig economy is an environment in which temporary positions are common and organizations contract with independent workers for short-term engagements.

[7]See our research on this which looks at quantifying risks through algorithms. Link: http://www.schulteinstitute.org/wp-content/uploads/2018/04/understand-risk.pdf

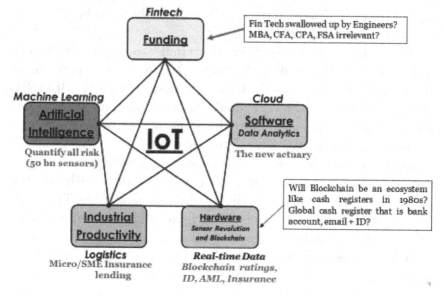

Figure 13.4: The whole ecosystem: How the old system is being reinvented in insurance: Will actuaries, CFAs fade away?

Source: Schulte Research Estimates.

People are the data that offers insight into their wants and needs. The data lets any company create a bespoke understanding of the person and offer them products that benefit them the most as an individual. The 'money' is how to get products to people which they want and need. The 'things' are how Artificial Intelligence (AI) connects to the Internet of Things (IOT) to turn things into digital markers. Moreover, 'ideas' are how the risk of default and accident can be quantified. Of course, this causes a complete rethink of what an actuary is. Because of vast data points on individuals, a new science of actuarial analysis becomes a science of clustering and pattern matching to estimate life expectancy, among many other health and physical issues. Let's break these down into five constituents.

1. **A radical rethink of FinTech as a small subset of the picture:** Retraining the finance people.

   First, we look at the area of finance. The area of FinTech (top of chart in Figure 13.4) will no longer be its own sphere but will be a subset of a data pool management and analysis business which feeds into all parts of the system. So, actuaries will need to return to school and learn Python. Actuarial science will go through a very exciting renaissance as we discover dozens of new ways to use vast amounts of data to measure life spans, catastrophes, weather problems, floods, drug addiction, depression, disease, suicidality and other mental disorders which can be life-threatening. Advances are being made to use facial expressions, searches, physical activity through the cell phone, among many other sensor activities, to gauge the current wellbeing of people.[8]

   For mid-level executives from banks and insurance companies, the suggested educational tools are R and Python. The underlying premises are simple and do not need to be overwhelming. We are generally looking for (1) clustering of events to find out how many people do a certain thing at a certain time; (2) pattern matching along a line of best fit to see what X factor corresponds to what Y factor; (3) rate of change along a curve (differential analysis) to see how people's preferences are changing. The only challenge is that these various subsets can have millions of data points. Furthermore, they need to be cross-matched in meta-analysis in order to find robust and meaningful trends in preferences, activity, product types, locations, website activity, etc., which are occurring simultaneously and may be related activity.

2. **The cloud-based insurance platform.** The cloud is beyond the scope of this paper, but we have included research on this in the footnotes.[9]

---

[8] In discussions with students at University of Science and Technology, the actuarial sciences course now has the study of data analytics for the first time in 2018. This is a good start.

[9] Please find attached a comparison of Alibaba and Amazon with a special section on the cloud businesses: Link: http://www.schulteinstitute.org/wp-content/uploads/2018/04/BABA-AMZN-CLOUD.pdf

3. **Advanced hardware and quantum computing are coming soon.** Get ready.

   Second, hardware is being developed which can offer quantum change in individual credit ratings. This is the ideas part. Traditional credit scoring methods need to be altered. Firms like student loan mega startup SoFi have more than 100 unique data sets for every person to anticipate credit-worthiness. As a result, their Non Performing Loan (NPL) ratio for student loans is a fraction of the market rate. Companies like sub-prime lender Aliya have discovered dozens of ways in which the FICO score is inaccurate for millions of Americans whose scores are below 660. Using data analytics, Aliya goes to the core of a person's monthly cash flow with multiple independent checks in order to weed out those who have been unjustly cut out of the credit market. This is a political issue in which politicians will be supporting these smaller technologically-savvy entities to help create jobs. Ping An's Lufax has detailed measures of creditworthiness for hundreds of millions of Chinese.

   It is interesting to see that FICO has invested in Quadmetrics. FICO knows the traditional static game of using historical data to analyze personal credit ratings and other forms of risk is up. So, it invested in Quadmetrics. This company was created by University of Michigan engineers and does extensive cyber security on the personal level. The program assesses active threats, latent threats, and mismanagement indicators. This is a quantum leap from the sleepy use of insurance, college education, prison, and zip code to gauge the risk of individuals and board members. We are looking at the need for hardware to manage billions of pieces of data and to manage these data sets in ways which were undreamt of before.

4. **Logistics and insurance:** Specificity of insurance will spread to all human motion.

   The left side of the illustration in Figure 13.4 includes logistics and machine learning. This is turning things into digital markers — from two-pound chickens to two-ton cars. The growing level of detail about a person's character or responsible behavior offers accuracy into small pockets of action when it comes to leisure, for instance. Insurance products can be offered for an afternoon of surfing. They can be offered for a certain train ride or to guarantee return of purchased online goods. All of this is possible for those who can secure the technology to process, translate, and control the vast ocean of information.

Any item on a ship or truck bound for another country can be counted and made accountable from its primary location to the destination. This includes pharmaceutical pills for infections in children in Cambodia, chickens for Long Beach, cars for Hamburg, or high-quality handbags for London. They can become tradable assets with proven provenance. This has huge implications for insurance not only to cover loss, theft, damage or loss of life. It also allows insurance companies to transform assets into tradable goods. Ping An understands this and is morphing into a universal bancassurance company because it knows that it can and insure and trade these assets. In this sense, it can buy, sell, lend, and create credit products against these products because it knows more about these products than anyone else. This is the first of the cyber conglomerate — the cyber bancassurance model.

5. **The AI function tries to make sense of all the data.**
   This is a good jumping off point to the following section which is to see how the analysis of all of this data on people, places, things and the analysis of risk translates into business lines. This must be the starting point for insurance or any industry for that matter. Insurance must learn to offer anything, anytime, anywhere insurance to people where they are at. Offering an array of services from a remote location — a corner store of sorts — no longer works. It is offering millennials what they need, on their terms, in their locations, in their way. This is because technology has created this circumstance. AI is a response to this diaspora of

hundreds of millions of millennials globally. It is also a golden opportunity for insurance to create integrated products whose value comes precisely from offering unique data on the needs and desires of customers as these data points offer independent and precise confirmation. Rings upon rings of data sets allow insurance firms to offer specific products for any period of time to anyone or anything in any country. This is a fact. Ping An is currently doing this better than any insurance company globally.

13.8 How Insurance Companies Can Restructure Their Process by Analysis of Business Lines

We present below the model for the way in which data needs to be collected, digested, processed, and sorted by business lines (see Figure 13.5). First of all, we need to think about the past data which is still sitting on paper that is outside of the new digital order. Many older financial institutions have immense amounts of paper sitting around which have yet to be turned into digital data. Some take the view that the old data may as well be from the Bronze Age and, therefore, be thrown out in favor of a torrent of current and more relevant live digital data. The conclusion makes sense for one simple reason: the conversion of paper to digital data is fundamentally flawed because the new digital data is precisely based on paper. The distinction here is that sensor data is live and dynamic and sorted in a digital format. There is no historical one-dimensional "paper trail." Focusing on current and future data sets is one option. In any event, we will discuss later that the model of a vertically integrated castle where there is air-tight cybersecurity (nothing gets in or out) defeats the purpose. Smarter cybersecurity is needed to allow a wide array of data into the machine and process it through six different areas.

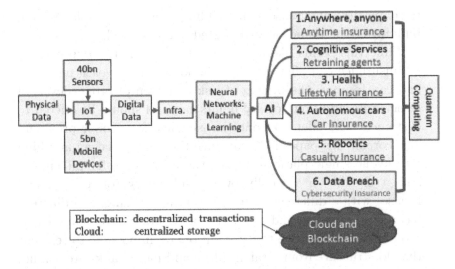

Figure 13.5: Artificial intelligence described on a single chart and how blockchain acts as foundation.

Source: Schulte Research Estimates.

1. The model for creating a data-driven anywhere, anytime insurance.

The platform ideal must be the starting point as it aims at offering on one platform a multi-pronged set of solutions for customers: anytime, anywhere, anything, anyone insurance. We place it as "primes inter pares" — first among equals and on top — because this is the center of all analysis. It should be the top of the waterfall and flow down to other businesses. This is where the lion's share of data coming in as a gusher can feed through to an increasingly 'on the go' group of people where everything is temporary: space, home, car, relationships, gyms, doctor, everything. It is the massive use of data which is incorporated into human empathy and compassion which can offer new and diverse products through a unified platform as well as better speed, lower cost and with greater choice. Data will be pouring in from wearables, sensor objects, location, and movement sensors, as well as geography sensors. It can come through phones, third parties, and sensors in buildings, cars, ships, planes, and buildings. It must be cleaned, sorted and analyzed.

While this seems like a huge challenge, a digital approach to life insurance can offer very specific knowledge of each customer and can improve the pricing of risk and even of proper levels of reserves. In addition, loyalty points can and should be offered to customers who offer over data in exchange for a lower premium. The core of this paper is to convince the reader that an anything for anyone single platform makes the most sense for millennials and it should be a wholly owned subsidiary (on the original license) with a different name, different advertising, and different technology. The old and new must be separated. We submit it is prohibitively expensive and institutionally impossible to reinvent the old structure from scratch. More and more banks are coming to this conclusion.[10]

2. Retraining staff and the addiction rehab model.

The second area for organizational change is the retraining of agents. Now that is a chore! Much of the insurance business is based on the agent. This must change. They must be upgraded. The best model for this retraining is slow gentle nudges at first and then a carrot and stick later. In this area, leveraging on the approach to changing mindsets in drug and alcohol counseling may be interesting.

When there is an obstinate and closed-minded individual who has a problem recognized by most and yet does not want to change, there is a need for intervention. The six steps in the change process for drug addicts shown in Figure 13.6 are a classically accepted method for creating change within the medical community. At its base, addiction is a disease of fear. Older people are in a kind of paralysis of fear. They have a sense of feeling overwhelmed by the realities of new technology in the same way a

---

[10]Please see out ppt on this which shows that more banks are creating separate digital entities. Link: http://www.schulteinstitute.org/wp-content/uploads/2018/04/DIGITAL-SUB.pdf

person with a substance abuse problem is overwhelmed by how to start a new life. It is just too much, so why bother? They are very similar, since in moments of elevated fear, the frontal cortex which drives future planning shuts down and the reptilian brain kicks in. Not surprisingly, this reptile brain offers poor outcomes (really bad judgment) because it has two outcomes: fight or flight. This lizard brain does not offer options that include compromise, negotiation, reflection on mistakes or capitulation. Some organizations are in this mode now. Something must give for them to survive. This is why fear-based decisions are usually wrong, especially ones made late at night when people are tired. They are based on the lizard brain amygdala. This part of the brain tends to form a cycle of self-sabotage rather than forward progress for the reasons mentioned above.

However, it is a fact that the human mind does, indeed, evolve and come around even in the scariest of circumstances, especially in a group setting where there is an established bond of trust, calm, and order. The ability of this group to intervene in an environment of trust through an interventionist (outside consultant) is a vital starting point. External advisors are a key part of this. That is why, for instance, an independent, technologically savvy and aggressive advisory board is advisable.

Part A: Precontemplation: Most insurance firms are here

The first part of this process is offering information frequently (often the same information) in a calm way to coax recalcitrant people into reflecting on change. These messages should be repeated often — in the shared pantry, in morning meetings, in clever and amusing ways on the walls, in a room dedicated to "thinking". Humor works best. But, serious conversations about change from different angles (frequently repeated) work best. Mind mapping exercise for senior partners and executives to see the importance for urgent change is useful. Offsites can be helpful here and an experienced external "interventionist" (outside consultant or advisory board member) is a key part of this.[11] Mind mapping experts are expensive but can be very useful.

1. **Precontemplation**: "I'm Fine." Get information; point out need for change from many sources

2. **Contemplation**: "What if." think about change; what if; get an ideal and agree on it.

3. **Preparation**: "How." what kind, how, when, timeline, resources, support, regimen. Agree and move.

4. **Action**: "Let's go." Proposals, budgets, partnerships, affiliations, external help. Agree and move.

5. **Maintenance:** 'Milestones.', Disciplined quarterly assessment. Discussion of progress. Self assessment.

6. **Relapse**: "Failure". Must be without anger or bitterness. Move on. Gently point out need for change. Agree to move on.

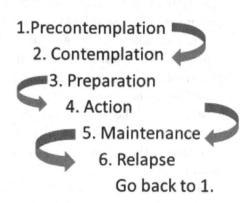

Figure 13.6:   Addiction relapse model.

Part B. Contemplating a roadmap — with total buy-in

The second part of this is to arrive at what an ideal is. This is the beginning of building a roadmap. What if this company was a pure digital company? What would it look like? How would it function? What new skills would people need? What would the technology look like? How would space change? What would the regulator want? How has the firm lost touch with the customer? Where is the moral compass? What are ethical considerations? Then the group agrees on the plan and sticks with it.

---

[11]Please find attached the result of this which is called "How to turn cities into hubs". This applies to both universities and firms. The ingredients are the same. Link: http://www.schulteinstitute.org/wp-content/uploads/2018/04/HUB-FINAL-v2.pdf

Part C: Allocating the resources for change — with total buy-in

Once this ideal is agreed upon, the next part is agreeing on the resources, regimen, support, the kind of organizational changes. What are the requirements for space? How does recruitment for new employees change? What does a retraining program look like? What are definitions of old behaviors and new behaviors? What kind of collaboration is required? This is a broad brush approach to change. The group agrees on the plan and moves forward.

Part D: A detailed plan of action — with total buy-in

This leads to Step 4 which is action. This is the gritty detail? Proposals, training manuals, budgets, specific partnerships, internal working groups, external consultants, university collaboration, HR considerations, internal prizes to motivate employees for creative solutions, regulatory interaction, government affairs. Again, the group agrees on the action plan.[12]

Steps 3 and 4 will blend a bit but should focus on a few vital questions: (1) Who are the important stakeholders? (2) What is the best case scenario? (3) What is a realistic delivery time? (4) What is the ideal size of data to be captured? (5) What will naysayers assert? (6) What is the experience of people in other firms who have gone through change? (7) What does a retraining program for employees look like? (8) What are the consequences for those who do not want to get on the bus? (9) What kind of loyalty points can be offered to customers who offer their data? (10) What is the basic framework for an integrated platform? (11) How will the C suite of people change as data becomes the center of the business — not the boardroom? (12) What does a leapfrog strategy look like if a separate subsidiary is built or bought? (13) In a cost analysis, how much will it cost to upgrade the entire organization? Is this organizationally possible? Or should separate

---

[12] Again, the PPT entitled Turning cities into hubs applies equally to turning a firm into a digital city. Link: http://www.schulteinstitute.org/wp-content/uploads/2018/04/HUB-FINAL-v2.pdf

units be created? We strongly support the latter. Use vision, plan well, and agree. Then move.

Part E: Maintenance, monitoring, self-assessment

Step 5 is maintenance and occurs when the action plan is implemented. Discipline is needed on a quarterly basis for achieving milestones. Accurate self-appraisal is vital. Where are shortfalls? Who is ahead of schedule? What are the roadblocks to progress? Where is institutional inertia? Which consultants need to be replaced? Are the young people being used to greatest benefit (Often the young people have the best ideas, but their ideas are ignored)? What is the competition doing that your firm is not doing?

Part F: The most important part is how to deal with failure

This step may be the most important of all. Failure is part of life. A relapse into past thinking or behavior — or wanting to give up — is part of life. Failure does not mean doing an endless post mortem and breast-beating. In a world where technology is moving fast, failure will happen and it must be met without anger or bitterness. It is not blame but learning. The Chinese firms do this very well. They fail, move on quickly and integrate another new system. Coming with failure compassionately, honestly and in the context of learning is vital to get to the next level. What has been laid out above is a valid and time- tested way to get the most hardcore of delusional drug addicts to change. It works at the rehab facility, so it will certainly work on the firm level to allow a group of frightened people to move into the digital world quickly. It is a way to deal with the naysayers. Fear is the enemy of progress. The only antidote is a supportive group that makes a plan for change, agrees to it, sticks to it and supports each other in each step. It is so simple, yet so few use it.

3. **Health Technology should be its own center.**

The third part of this is in the area of health. Data and technology now exist which were not available a mere 5 years ago which allows an insurance firm to integrate almost all parts of the insurance process from the initial visit to a clinic to exiting surgery with antibiotics. There is no company doing this better than Ping An Insurance.[13] Ping An is arguably ahead of any insurance company globally in that it has already turned itself into a flat, horizontal, unified platform which has bashed down silos and created a unified system which connects the patient to the clinic, hospital, doctor, pharmacy, and payer through an integrated process.

Figure 13.7 shows how Ping An operates. This is likely the shape of things to come and is a solid ideal for which to aim. There is simply no insurance company globally which has done a better job of integrating all of its data sources into a cohesive whole. As we will see in a moment, it also shows how the company potentially spin-off companies for IPO listings on the Hong Kong Stock Exchange.

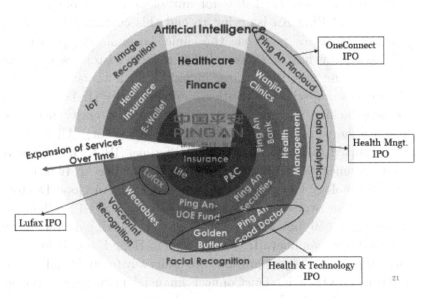

Figure 13.7: Ping An has made greater strides into becoming an autonomous financial ecosystem than any other insurance company globally.

---

[13] For a detailed analysis of this, see the link for a report on Pin An. Link: http://www.schulteinstitute.org/wp-content/uploads/2018/04/PING-AN.pdf

Ping An started out as a small insurance company in sleepy Shenzhen and then attached itself to a clunky state-owned bank. In the past few years, the leadership of Peter Ma has catapulted Ping An into a full-service universal financial institution. The illustration above shows how traditional insurance has morphed into a full-service banking institution by adding new business lines on one platform with even broader data sets which can translate into newer, better, cheaper, and more diversified offerings. These data sets offer new types of independent verification of what people want and need.

Furthermore, Ping An has created a connection between the patient on his first day to the clinic all the way to the operating room and then to the pharmacy in a one-stop shop. Consider how easy it is to get data on the stock of GE going back 75 years and then consider how hard it is for any of us to find our medical data. Why? Ping An is trying to end this confusion by linking the patient to his payer via the clinic, hospital, and pharmacy. The company has in excess of 200 million customers and has created some unique features, including an automatic patient–doctor meeting schedule (Butler), a credit entity (Lufax), a data storage facility (Health Management) and the entity which is the workhorse of all the connection (Good Doctor). All of these are reported to have Initial Public Offerings (IPOs) in 2018 and raise as much as $50 billion. In April 2018, Ping An Healthcare and Technology Company, also known as Ping An Good Doctor, attracted an oversubscription for its IPO of US$1.12 billion by more than 650 times by retail investors, making it the most sought-after mainboard IPO since 2009 in Hong Kong. It has also been reported that preparations are underway for an initial public offering (IPO) of its OneConnect financial management portal that provides cloud computing and other technology services to small and medium-sized financial institutions in China. The company's partners include 468 banks, as well as 1,890 other firms including insurers, brokerages, fund managers, and private-equity investors. The IPO could raise as much as US$3 billion.[14]

The Good Doctor IPO is the one company which is getting the most attention. It deals with internet healthcare. It currently has an annualized run rate of revenues of CNY 1.35 billion, but is still running net losses due to the heavy investment. This has attracted attention from Softbank. This IPO will be widely watched and is a legitimate harbinger of the coming change in healthcare and its ability to meld insurance with healthcare technology, appointments, imaging, diagnosis, payment, research, anti-counterfeiting for fake drugs, and many other elements of healthcare which should have come together many years ago. In addition, Ping An will raise foreign currency in Hong Kong in order to make acquisitions in the Far East and the West.

One of Ping An's favorite hunting grounds to develop newer and better technologies is Israel. It recently acquired a company called TytoCare. This company offers diagnostic kits (about $300) to mothers at home which can feed to a video seen by a doctor. The mother can send her child's temperature, video of ears and throat, and a live feed of breathing and heartbeat to the doctor. From this data, the doctor can offer a diagnosis without an appointment, and the video can remain a record of the patient's progress later. There are also four AI recognition technologies — facial recognition, voiceprint recognition, micro-expression recognition, and intelligent insurance claim handling. With AI recognitions, ratings of reliability and credit can be further improved for insurance pricing. With intelligence insurance claims, small claims can be processed and disbursed within hours with AI damage recognition technology without incurring high cost of human inspections and disbursements.[15]

---

[14]https://www.businesstimes.com.sg/banking-finance/chinas-ping-an-starting-work-on-up-to-us3b-oneconnect-ipo-sources
[15]https://www.prnewswire.com/news-releases/ping-an-technology-highlights-ai-applications-at-2018-gmic-300638526.html

We can only ask ourselves: Why did this take so long to develop? This aggressive hunt for technologies globally is a vital part of why Ping An is so dynamic.[16]

4. **Parts 4 and 5: Autonomous cars and robotics:** Beyond the scope of this paper.

The fourth and fifth parts are, respectively, autonomous cars and robotics. This is beyond the scope of this paper and will be discussed in later publications. Suffice it to say that the very recent fatal accidents with the Uber Technologies car in Arizona and Tesla car in California are problematic. Many questions arise. Would a human driver have been able to avoid the Arizona woman who seemed to move into fast moving traffic at night? Does a human driver in the driver's seat of an autonomously driving car have liability if the car warns him or her to grab the wheel and he fails to do so? What if the autonomous car has to make a decision between hitting one person who is jaywalking or slams into a crowded bus stop? The implications for insurance are huge.

The video of the Arizona fatality reveals that the woman walking a bike suddenly and dangerously veered into traffic. Even a very alert and skilled driver would not have avoided the person. These are the first two precedents to see who will have liability: the car company, the human driver who did not "grab the wheel" when directed to by the autonomous car or the person who was killed who was illegally veering into traffic.

5. **Cybersecurity: A vital link to the process.**

The sixth part of this is Cybersecurity. This is a vital part of the system. This is a thorny issue for the following reason. If an external consultant does a full audit on cyber security for a company, it then submits a big report to the board of directors called something like "Vulnerabilities and shortfalls in cybersecurity". This report will be very technical in nature and very likely beyond the understanding of any board member. If

---

[16]Ping An PPT link: http://www.schulteinstitute.org/wp-content/uploads/201 8/04/PING-AN.pdf

there is subsequently a cyber breach from blackmail malware and a large blackmail payment is paid to turn on systems, the board is liable. Therefore, it is unlikely that a Board of Directors is very interested in a candid account of such vulnerabilities.

Enter firms like Security Scorecard, Quadmetrics, and Envelop. These can detect cyberattacks through many other means besides an internal audit. For instance, they can look at the internal activity of the programmers. They can monitor the interaction between the internal systems and external ones. They can look at the frequency of changes internally as well as the frequency of attacks from the outside. They can detect internal sabotage. After these vulnerabilities are done, cyber insurance companies like Quadmetrics offers a kind of credit score for security. Envelop can offer a premium to insure a company against malware. Companies like these can remain agile technology companies and pass along their insurance policies to reinsurance companies. The two can work hand in glove.

This is a new and urgent reality after shipping company Maersk was hacked last summer and submitted an insurance claim for nearly $1 billion for interrupted global services for a few days.[17] We will discuss this in Section 13.9 in the context of blockchain, since blockchain could and should very well form the foundation of cybersecurity for most elements of insurance in general and financial institutions in general.

13.9   The New Foundation of Data: Blockchain and the Challenge to Insurance

Seventh and last, the foundation of this is six-fold separation of labor in the cloud and blockchain. The model for blockchain is the clock in Figure 13.8 with 12 parts. Chapters 3 and 4 of *Blockchain Revolution* by Dan Tapscott (Tapscott, 2016) are a must read, but we have taken the liberty to boil down much of the book by creating the

---

[17]One of the authors is an investor in Envelop through Qx Branch in Washington DC.

clock in Figure 13.8. We have essentially narrowed down blockchain to four key ingredients. This clock is also derived from hundreds of hours of discussion from leading thinkers in blockchain, much client feedback from founders of some of the largest hedge funds globally, participation in blockchain conferences with people like Vitalik Buterin, and discussions with clients who are boards of banks and insurance companies.

The main features of the blockchain connected to financial institutions are four-fold: it is (1) a living audit; (2) intrinsically innovative; (3) an open system; (4) highly efficient. One can also see that each of these groups corresponds to a threat to one of the cartels which was on the ascendant for a few decades but is surely feeling the pinch of competition and disintermediation. Of course, there is some overlap, but the categories suit our purposes. In his introduction

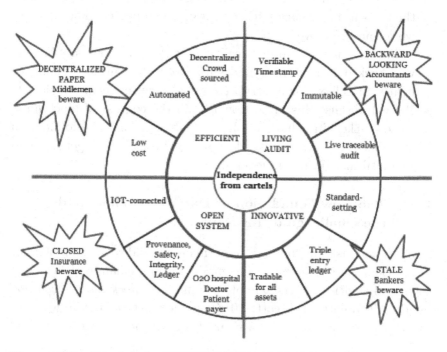

Figure 13.8: Blockchain SEEKS WAYS TO UPSET CARTELS: Turn your physical world into a digital reality ASAP.

Source: 'Blockchain Revolution', Schulte Research.

to the book, Tapscott makes the point that blockchain is "barreling down the entrenched, regulated and ossified infrastructure of modern finance. Their collision will shape the landscape of finance for decades to come."[18]

1. Blockchain as a living audit.

First, blockchain as a living audit is a mortal threat to accounting firms which produce annual audited reports by March or so for the previous year. There is also a history of human failure and fraud which has caused the collapse of a firm in the case of Enron. Blockchain is a life minute by minute audit which can be used to go back in time from now to see where there might be problems. It is not a historical record of activity from 3 months ago. It is live. Again, blockchain is to movies as traditional accounting is to a photograph.[19]

2. Blockchain as profoundly innovation: triple ledger book keeping.

Second, the clock hands 4–6 show that blockchain is a mortal threat to banks because of its innate capacity for innovation. The most profound innovation is the creation of an alternative financial universe because it is a triple ledger which can pull the balance sheet away from the bank and create a two-way financial transaction between a buyer and a seller (especially in trade finance) who do not know each other. In simple terms, the transaction is like a wedding ceremony. The blockchain "gathers these two together in holy matrimony" and confirms that these two people are who they say they are and that the community knows that these two people enter into this arrangement freely and without reservation. It is that simple. As long as a pre-determined consensus, such as agreement by a certain percentage, of the community is fulfilled, the agreement is live, permanent, indelible, and traceable. This

---

[18]Please see out PPT which is a summary of Tapscott's book. Link: http://www .schulteinstitute.org/wp-content/uploads/2018/04/Blockchain.pdf

[19]For a complete summary of Tapscott's book, please see our PPT on Blockchain. Link: http://www.schulteinstitute.org/wp-content/uploads/2018/04/Blockchain .pdf

has profound implications for all kinds of provenance for contracts, jewels, art, contracts, homes, mines, plantations, etc. We are only limited by our imagination on this score. One can easily see that this also has implications for insurance. It allows enterprising insurance companies to easily morph into a digital bancassurance model since insurance companies can adapt the triple ledger model by transforming provenance into collateral.

3. **Blockchain as an open source — the key for insurance to pay attention to.**
The third part of the clock is 7–9. This is the important area of insurance. This is a fundamental shift in our understanding of risk. Sensors can detect imminent explosions on rigs. They can be placed on a cow or a chicken and follow this animal from a hatchling to the market to ensure that no disease has spread. Sensors can reduce the millions of deaths of children each year by smart pills which are monitored from lab to store to guarantee their efficacy, especially for antibiotics. They can be used to allow machines in hospitals to talk to clinics and other doctors and to process claims quickly. The driving theme here is one of unity and openness. Various and separate parts of insurance are unified as the process of tracking and monitoring risks is unified through the monitoring of living beings and integrating these data sets to machines which can expedite, diagnose, treat, and pay out after treatment. There is a simple outcome: Life can be extended. Costs can be slashed. The system can be opened. Again, the intuitive conclusion here is that insurance products will have a strong tendency to blend into one platform.

4. **Blockchain as cheap and powerful disintermediary.**
Lastly, the clock hands 10–12 shows the imminent threat to the middleman in accounting, banking, and insurance. This is where the real mortal threat lies. Blockchain is a challenge to those in the middle of stock trades. It causes insurance middlemen to ask questions of how their business can be short-circuited. Moreover, it forced accountants to get out of the silent movie business of annual reports which come out 3 months after the end of the year

in question and get into the 'live and in color' movie business which is a universal and public ledger.

There is one important and poorly understood element of blockchain here. Some banks and accounting firms are talking about having their own blockchain product which is a living, indelible, and traceable product for their clients. This is fundamentally an illogical absurdity. Why? By its nature, blockchain is a public ceremony for all to see. It is a public and registered exchange of vows at City Hall. A bank cannot conduct private wedding ceremonies inside the headquarters in London or New York and then declare that two people are married for life. If they want to do this, just use a data base. But a data base is not confirmed by the mining community — it remains a private set of data in a closed system. So, this blockchain is false and unusable. Permissioned blockchains are one thing. An ersatz blockchain which is nothing other than an internal data base for one accountant or one shipping group to use is another. To remedy this and other issues, many insurance companies have formed BSI, the insurance equivalent to the blockchain consortium called R3 for banks. So far, the insurance firms that have joined are AIG, Aegon, Swiss Re, and Allianz. Of course, there are advancements in the computationally more expensive Zero-Knowledge cryptography that is used by Zcash, which is the first widespread application of zk-SNARKs (Zero-Knowledge Succinct Non-Interactive Argument of Knowledge). The strong privacy guarantee of Zcash is derived from the fact that shielded transactions in Zcash can be fully encrypted on the blockchain, yet still be verified as valid under the network's consensus rules by using zk-SNARK proofs.[20] This open decentralized blockchain technology provides a powerful privacy protection tool for regulated insurtech and at the same time, can "easily" satisfy the regulators' requirements of KYC, AML, and CTF by sending encrypted information to the central authority. It is the authors' view that without the token to incentivize the participants to

---

[20]https://z.cash/technology/zksnarks.html

ensure a good outcome, blockchain is just a cryptography ledger with a memory. Blockchain is more than production efficiency, it is a trust machine that improves relationship and aligns the interest of the participating economic agents for a favorable outcome. Without a token incentive to align interest so as to break down the silos, insurtech are not likely to enjoy all the benefits, especially the most important mechanism that improves the relationship among the untrusted parties. This is a good starting point for discussion but beyond the scope of this paper. (Interested readers can read Lee (2015) and Lee and Low (2018)).

13.10 A Significant Change in Products and Processes is Needed Now

The bottom line here is that insurance will be bought, sold, underwritten, and serviced differently than ever before. External data on physical movement as well as contextual information will become more important than historical internal data for risk pricing. This allows for a new type of just-in-time insurance through a person's mobile device. *This is the vital difference between internal data gathered on paper vs external data from the social network, cell phones, and sensors which offer a dynamic real-time flow of highly reliable data.* It is the difference between a photograph and a movie. The old is static, backward-looking, and on paper. The new is forward-looking, live, and digital.

There are two implications from this. This paper is about people and how the younger generation is a very different type of creature compared to older Boomers. People want and need help with education, child rearing, mental health, leisure, elderly care, self-diagnosis, music, entertainment, passion, mates, spirituality, and love. They need ways to insure against sickness, unemployment, crippling student loans, accident, death, bankruptcy, weather, and calamity. Change in these young people and the technology will very definitely cause products to bleed into each other.

All firms, not just insurance, must try on a daily basis to unify processes by gently removing silos, for silos are the best way to guarantee failure. Data is the center of a business — not the C suite. This must be the number one concept for all change in the digital area. Knowing one's customer as well as possible and treating the customer as a bespoke product is the key to success. It is an ecosystem — not a product line. The customers are a partnership — not a homogenous group. The product must be available on one interconnected platform — not a series of static web pages. The product must be better: better offerings, designs, and pricing. The process must be better: better distribution, engagement, speed, and advertising. Loyalty must be offered through incentive loyalty points.

13.11 Products Will Bleed into Each Other

As we showed in Section 13.2 on the difference between the static "yes/no" boomer and the "why not" millennial, products in many industries will bleed into each other. This is because the static array of products available to older customers is unacceptable to the younger customer. Since millennials are more of a diaspora (this is true in most developed countries as well as emerging markets) which are not tied down by cars and mortgages — and who just may prefer a 'gig' instead of a job — there is a radical need to reinvent the packaging of insurance.

For instance, pure medical insurance is not becoming supplemental medical. Group short term and long term care is becoming accident, critical illness and hospital indemnity. Group insurance for work is bleeding into life, dental and vision. Moreover, it is going the other way. An example of this is hospital insurance morphing into universal life. One can see that as the products bleed into each other due to shifts in generational preferences, the processes must change by eliminating silos. If the silos remain, the company will die as it will be unable to make the necessary shift to integrate data into a shifting array of centralized products which will become increasingly integrated. A life insurance company which wants to remain a life insurance is a goner.

The key concept here is that the world of blockchain transforms provenance into collateral. The need for proving that this product is insured can be transformed into a new business: this product can be traded or used as collateral. The guarantee of identity by something other than a bank or a government allows two parties to agree on the value of a thing, trade it, borrow against it, sell it, or offer it as an investment.

The implication of this chapter is that a company which offers only group plans is a goner. There needs to be an ability to offer plans along a spectrum of events, time, ages, locations, demographics, and needs. This will also extend into gene pools, personality types, and educational groups. The possibilities are endless, but so are the societal implications, ethical considerations, and legal complications.

Those who do not, for instance, create a separate digital unit for millennials may be in trouble. Those who do not create a unified platform with an array of services based on integrated data from people, places, and things are in trouble. Those who specialize in one form of insurance will be a thing of the past.

It will be increasingly difficult for an insurance company that makes billions to continue to increase premium without full transparency of risk pooling. The rise of mutual aid insurance among the millenniums with blockchain, Big Data, and AI is a very good example of where the industry may be heading. The customer is different, and the need can only be satisfied with a new approach. The best model globally is Ping An. We are only in the very early innings of this phenomenon. There is no time like the present to start the change.

PART 5

Singapore Back in the Game When the Titans Clash: Will it be SMART Enough?

Chapter 14

Singapore: An Outsider and Smart Nation Back in the Game!

14.1 Introduction

THE SWEET SPOT

We can remain in a sweet spot by playing our cards right, just as we have for the last five decades. We did this in the 1960s when we industrialised and caught the off-shoring wave.

In the 1970s, we bet on the right horse by doubling down on logistics management with our container port and airport. We then moved on to electronics in the 1980s and life science later. We can do it again with big data and smart technology. . . . Singaporeans need to be fully prepared to collectively face this age of disruption. This is why economic restructuring is a key priority for the Singapore Government.

Since independence, we have relied on our people's ingenuity and tenacity to survive. The times we live in require us to double down on these qualities. The Government will do its best to prepare Singaporeans. If we do not ride the technology wave, we will sink under it. But, if we succeed, we can navigate safely through our current digital Gilded Age into a new Golden Age.

— Foreign Minister Vivian Balakrishnan[1]

---

[1] https://www.straitstimes.com/opinion/foreign-policy-in-an-age-of-technological-disruption

In recent years, Singapore has hollowed out as a manufacturing base and upgraded its aggregate skill set to focus on pharma, logistics, fintech, and other biotechnology sectors. Monthly data for manufacturing has become more volatile, perhaps because it is an old measure that is more appropriate for the traditional economy (Figures 14.1 and 14.2). Vast changes in technology have caused a tectonic change in the overall economy. As with many other countries, the data fails to capture this digital activity. In the midst of these confusing numbers, Singapore is embarking on a "barbell" strategy of attracting not only tech giants like Google and Grab but also startups with global tech talents as well as homegrown companies.

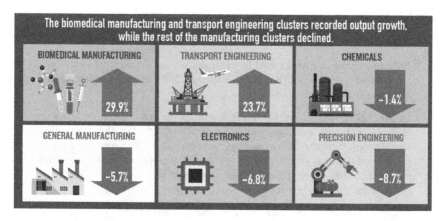

Figure 14.1: Singapore's manufacturing output increased 2.7% year-on-year in December 2018.

After Creative Technology, Singapore has few homegrown successful technology stories to tell. Services have become the main driver of economic growth while manufacturing remains as an

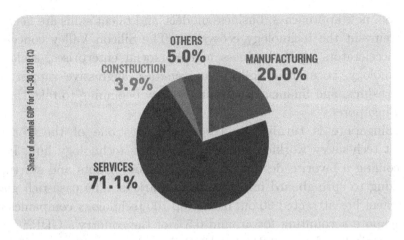

Figure 14.2: **The manufacturing sector continues to be an important pillar of the economy.**
Source: *DOS*.

important pillar. Traditional SMEs are struggling with the high cost, and new deep skill technology startups are small in numbers. Unlike, Silicon Valley or Shenzhen, it is not perceived to be a place for innovation. One reason is the lack of diversity in the manufacturing base to form an ecosystem for emerging technology. That may all be changing very soon. The other reason is that the mindset of "scaling, going global, and doing good" is not part of the business DNA. Clustering, industry support and receiving the grant are ingrained in the culture of Small and Medium Enterprises (SMEs). What is even more prominent is that some companies are surviving on grants, relying on government contracts, profiting from regulatory protection and forming unhealthy monopolistic dominance. Customers are seldom the focus.

So, new approaches, business models, and talent skills are needed to reinvent the technology ecosystem. The Silicon Valley concepts of accelerators, open business models, social enterprises, scalable technology, vocational skills in universities, aggressive support of internships, and financial inclusion are now beginning to take shape in Singapore.

Singapore is turning its labor pool into one of the world's most tech-savvy workforce and a world-class technology hub. It is becoming a favorite destination for technology giants and startups looking to springboard into the Asian markets. The cash-rich government has attracted 80 out of the top 100 technology companies to Singapore accounting for around 6.5% of the country's GDP.[2] The Economic Development Board (EDB) has always been successful in attracting the multinationals, and it is repeating what it does best for the technology giants. The Singapore Economic Development Board (EDB), a government agency under the Ministry of Trade and Industry, is responsible for strategies that enhance Singapore's position as a global center for business, innovation, and talent. Initially shocked by the GFC, Singapore is beginning to achieve its mission to create sustainable economic growth, with vibrant business and good job opportunities for Singapore. Figures 14.3–14.5 show the location of business parks and industries.

The Seletar Aerospace Park provides an integrated infrastructure dedicated to aerospace activities including the maintenance, repair and overhaul (MRO) of aircraft and components; manufacturing and assembly; as well as training and research & development. Rolls Royce, Bell Helicopter, Cessna Aircraft, and Hawker Pacific are some of the companies that maintain facilities here.

Figure 14.3:   Seletar Aerospace Park.

---

[2]https://www.edb.gov.sg/en/news-and-resources/insights/innovation/for-global-innovators-all-roads-lead-to-singapore.html

There are four specialised wafer fabrication parks located in Singapore – Woodlands, Tampines, Pasir Ris, and North Coast. Collectively, the parks house some of the world's top wafer foundries, such as Micron, GlobalFoundries, UMC, NXP Semiconductors. Our dedicated electronics manufacturing clusters also includes the Advanced Display Park in Tampines.

Industry: Electronics

Figure 14.4: Wafer Fab and Advanced Display Parks.
Photo Credit: JTC Corporation.

Aerospace ›

Consumer Businesses ›

Creative Industries ›

Electronics ›

Energy & Chemicals ›

Information & Communications Technology ›

Logistics & Supply Chain Management ›

Oil & Gas Equipment and Services ›

Medical Technology ›

Natural Resources ›

Pharmaceuticals & Biotechnology ›

Precision Engineering ›

Professional Services ›

Urban Solutions & Sustainability ›

Figure 14.5: Our industries.

The EDB does more than strategize. It invests in globally competitive businesses to create successful, sustainable industries with long-term profitability prospects such as Information and Communication Technology, Emerging Technology, and Healthcare. EDBI assists companies seeking to grow in Asia and globally through Singapore by leveraging on their network and expertise.[3] Among its emerging portfolio are Bitmain, Ambio Micro, Bright Machines, Greenwave systems, Joby Aviation, Magic Leap, RetailNext, and Savioke. Bitmain is a global leader in supercomputing hardware for Distributed Ledger Technology (DLT). EDBI is into the game of cryptocurrency mining equipment/pools, among other capabilities of Bitmain in AI deep learning chip under the Sophon brand, and a second-generation AI chip with five times the performance. These AI chip products extend capabilities in applications such as machine vision, data centers, supercomputing, and robotics. Savioke is creating autonomous robot helpers for the services industry and aims to improve the quality of life by developing and deploying robotic technology in the places people live and work. EDBI works with Alibaba, Baidu, and Tencent. Similarly, Google, Amazon, and Facebook all have well-established regional operations in Singapore, taking advantage of Singapore's government incentives for technology companies.

There are a few main reasons why many choose Singapore as their base.[4] First, it is the geographical location that is in the heart of Asia. Second, businesses treasure Singapore's transparency and consistency of law because this allows for long-term planning in a world that is volatile not only regarding financial variables but regulation and politics. The term open-sourced governance is peculiar to Singapore as it implies a government that (1) is pro-business, (2) encourages public–private partnership, (3) searches for

---

[3]https://www.edb.gov.sg/en/about-edb/who-we-are.html
[4]https://www.forbes.com/sites/alexcapri/2018/09/21/5-reasons-why-the-world s-tech-firms-are-moving-to-singapore/

win–win mutually beneficial deals, (4) invites ideas from partners and (5) continuously expands business capabilities.

Third, Singapore has implemented 22 bi-lateral and regional FTAs[5] with EUSFTA[6] and CPTPP[7] that have high standards for promoting digital trade and the networked economy. Fourth, Singapore open data mindset is friendly to digital trade and data storage. Singapore has been working on the ASEAN-Australia Digital Trade Frameworks Initiative,[8] which aims to put in place formal legal frameworks and standards for e-commerce, digital money, IP protection, and data management. Fifth, there are well-funded structure programs to transform its research and workforce into a leading technology hub. The Research, Innovation and Enterprise 2020 Plan (RIE[9,10]) under the National Research Foundation has SG $19 billion to promote a focus on advanced manufacturing and engineering, the digital economy and services, health and biomedical sciences and urban, sustainable solutions. Figures 14.6 and 14.7 show the ecosystem and plan of RIE. Singapore already has an ecosystem of world-class universities, multinationals, startups, accelerators, incubators, venture capitalists, financial institutions, logistics, commerce, tourism, real estate, advisory, legal, accountancy, and IT firms.

---

[5]https://www.enterprisesg.gov.sg/non-financial-assistance/for-singapore-comp anies/free-trade-agreements/ftas/overview

[6]The European Union-Singapore Free Trade Agreement, https://www.mti.gov.s g/Improving-Trade/Free-Trade-Agreements/EUSFTA

[7]The Comprehensive and Progressive Agreement for Trans-Pacific Partnership, https://www.mti.gov.sg/Improving-Trade/Free-Trade-Agreements/CPTPP

[8]https://www.opengovasia.com/initiatives-on-cybersecurity-digital-trade-and-s mart-cities-announced-at-inaugural-asean-australia-summit/

[9]https://www-nrf-gov-sg-admin.cwp.sg/docs/default-source/press-releases/201 601082039441690-20160108_rie2020-press-release-(final).pdf

[10]https://www.nrf.gov.sg/rie2020

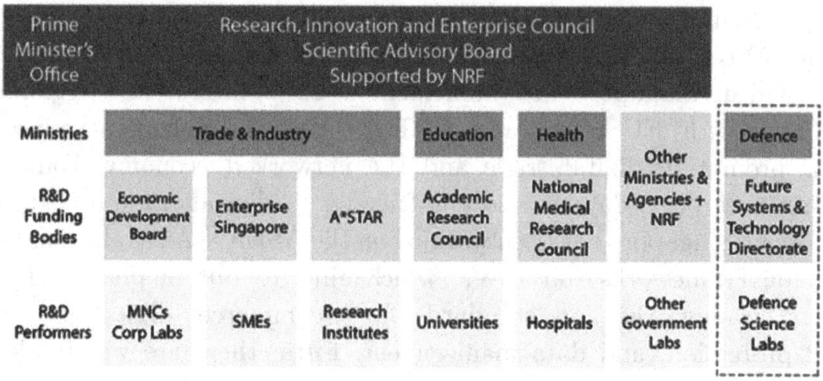

Figure 14.6: RIE Ecosystem.

RIE2020 Plan	Advanced Manufacturing and Engineering
Health and Biomedical Sciences	Urban Solutions and Sustainability
Services and Digital Economy	Academic Research
Research Manpower	Innovation and Enterprise

Figure 14.7: RIE2020 Plan.

Traditionally, startups and technology funding are from the stock market, venture capitalists, and financial institutions. The finance sector has been driving innovation in Singapore hand in hand with subsidies and incentives from the government. That strategy attracted international firms to set up the base of relocating the headquarters to Singapore. These strategies to attract large companies with capabilities to employ and train the Singaporeans have created jobs with technology transfer. However, in the vision of Services 4.0 in the digital economy, Singapore still lacks scalable technology to serve global customers.

In the case of financial services, the regulator was focusing on traditional financial innovation. The focus was on markets development such as cross-border capabilities but limiting the damage caused by complex financial products.[11] The other focus was on expanding the wealth management and insurance capacity,[12] and at the same time eyeing the opportunities in the US\$ 8 trillion hard infrastructure financing linked to the Belt Road Initiative[13] in the region.

The direction has changed somewhat since the set up of the FinTech and Innovation Group, but the underlying principles have not. The underlying thoughts have always been capitalizing on certain advantages that Singapore has, such as its credibility of the financial regulatory framework, the competencies of its workforce, the connectivity it has with the rest of Asia, and the collaboration MAS (Monetary Authority of Singapore) with the financial industry.[14]

In particular, the monetary authority always works closely with the industry, be it in formulating regulatory policies or growth strategies. Compliance cost is always on the mind of a smart regulator, and Singapore has not faltered to imposing high prudential standards defined to be above international norms. Given the rapid change in China and Silicon Valley, Singapore is in need to develop a new financial ecosystem, a deeper and broader market with critical infrastructure, and a competent financial sector workforce. There is little doubt that Singapore's financial sector remains the most innovative sector of the economy, each time reinventing itself as innovations emerge such as hedge funds, REITs, structured products and others. Figure 14.8 shows the three divisions of FTIG at MAS.

---

[11] http://www.mas.gov.sg/news-and-publications/media-releases/2014/mas-proposes-stronger-safeguards-for-investors.aspx

[12] http://www.mas.gov.sg/Singapore-Financial-Centre/Overview/Wealth-Management-and-Insurance.aspx

[13] http://www.mas.gov.sg/News-and-Publications/Speeches-and-Monetary-Policy-Statements/Speeches/2016/Shanghai-and-Singapore-Synergies-amidst-a-Dynamic-Asian-Financial-Landscape.aspx

[14] http://www.mas.gov.sg/News-and-Publications/Speeches-and-Monetary-Policy-Statements/Speeches/2014/Remarks-at-JP-Morgan-Singapore-50th-Anniversary-Dinner.aspx

MONETARY POLICY & INVESTMENT / DEVELOPMENT & INTERNATIONAL / FINTECH & INNOVATION
FINTECH & INNOVATION GROUP

Fintech and Innovation Group (FTIG) is responsible for regulatory policies and development strategies to facilitate the use of technology and innovation, to better manage risks, enhance efficiency, and strengthen competitiveness in the financial sector.

FTIG comprises three divisions:

Payments & Technology Solutions Office formulates regulatory policies and develops strategies for simple, swift and secure payments and other technology solutions for financial services.

Technology Infrastructure Office is responsible for regulatory policies and strategies for developing safe and efficient technology enabled infrastructure for the financial sector, in areas such as cloud computing, big data, and distributed ledgers.

Technology Innovation Lab scans the horizon for cutting-edge technologies with potential application to the financial industry and works with the industry and relevant parties to test-bed innovative new solutions.

Figure 14.8:   FTIG of MAS.

The challenges of technological innovation are always met with skepticism and naysayers. For a while, there was pessimism that large elements of the Singaporean economy — far too much dependent on the cozy real estate and banking arrangements — were not ready for digital disruption. As new plans were announced, the doubts subsided in the financial sector and direction becomes much clearer for other sectors too. Singapore has begun to take a different road, and the focus has switched to financial inclusion, networked finance, and green financing[15,16] in the financial sector and emerging scalable technology in manufacturing, services, commerce, tourism, logistics, and trade.

The National Trade Platform (Figure 14.9) was renamed Networked Trade Platform at the launch event attended by close to 700 industry representatives from the trade, logistics, and the public sector in September 2018. This change in name is very significant and to signal that Singapore is determined to build the network that is going to retransform its economy by adding another layer of the

---

[15]https://www.straitstimes.com/business/finance-sector-key-in-battling-climate-change-says-masagos

[16]https://www.mof.gov.sg/Newsroom/Speeches/keynote-speech-by-ms-indrane-rajah-minister-in-the-prime-minister%27s-office-second-minister-for-finance-and-education-at-the-launch-of-wwf-asia-sustainable-finance-initiative-(asfi)-on-21-january-2019-at-parkroyal-on-pickering

open virtual economy to its existing open physical economy. Here is how it works:

> "The Networked Trade Platform (NTP), launched on Wednesday (September 26), brings together four government certification services required for trading in and out of Singapore, as well as another 25 value-added services by third-party firms geared towards trade. The new platform will include more government services.

SINGAPORE CUSTOMS

Networked Trade Platform

The Networked Trade Platform (NTP) is a national trade information management platform that provides the foundation for Singapore to be the world's leading trade, supply chain and trade financing hub. At its core, it represents a concerted effort to drive an industry-wide digital transformation to build a trade and logistics IT ecosystem which connects businesses, community systems and platforms and government systems.

Replacing TradeNet for trade-related applications and TradeXchange for connecting the trade and logistics community, the NTP is designed to provide beyond the service offerings of the incumbent systems. Specifically, it aims to be a:

- One-stop trade information management system linked to other platforms
- Next-generation platform offering a wide range of trade-related services
- Open innovation platform allowing development of insights & new services with cross-industry data
- Document hub for digitisation at source that enables reuse of data to cut costs and streamline processes

Figure 14.9: Networked Trade Platform.

The NTP was developed by the Singapore Customs and the Government Technology Agency of Singapore (GovTech) over four years, and supported by more than 20 ministries, government agencies and working groups."

The comments by the Minister for Finance spell the direction:

"If we can stitch the disparate standalone systems or digital islands together, and bridge the government agencies and business community, the potential value to the economy is significant and transformational."

— Minister Heng Swee Keat[17]

The word *Networked* should be emphasized as we see the same idea of building the Hinternet where businesses can scale via a network of customers. PSA International Pte Limited (PSA), an independent commercial company owned by the Government investment arm Temasek after the restructuring of the Port of Singapore Authority in 1997, has embarked on a blockchain journey for trade. In February 2018, PSA increased its shareholding in CrimsonLogic Pte Ltd from 15% to 45% with the purchase of shares of Singapore Telecommunications Limited (Singtel) and Civil Aviation Authority of Singapore (CAAS) with Enterprise Singapore (ESG) owning the balance 55%.

"CrimsonLogic, a world leader in trade facilitation and eTrade products & services, developed the world's first Single Window "TradeNet" for Singapore in 1989. Since then, the company has implemented close to 20 other similar large-scale projects globally, including eGovernment portals. Its subsidiary Global eTrade Services (GeTS), launched in 2017, offers a comprehensive suite of services to help traders meet regulatory and compliance requirements from Government agencies and trade associations around the world. GeTS is currently connected to 23 customs nodes globally and has connected more than 174,200 parties, facilitated more than 13.5 million trade transactions annually and moved more than 9 billion tonnes of cargo."[18]

The GeTS launched a cross-border permissioned Open Trade Blockchain (OTB) for trade communities in anticipation of the Belt Road Initiative and the Southern Transport Corridor in 2018. Partners include China-ASEAN Information Harbor Co., Suzhou Cross-E-commerce Co. Ltd (operator of the Suzhou E-commerce

---

[17]https://www.straitstimes.com/business/new-trading-platform-to-digitalise-pa per-based-trade-processes-grow-trade-opportunities
[18]https://www.singaporepsa.com/-/media/Feature/PDF/News-And-Media/20 18/nr180222.pdf

Single Window), and Commodities Intelligence Centre (CIC, a joint-venture company between ZALL SMARTCOM, GeTs and SGX). Entities from Korea, Indonesia, Taiwan, and others are coming onboard to take advantage of the Digital Silk Road.

Shipping company PIL, PSA International, IBM Singapore has also been working on a blockchain-based supply chain platform to track and trace cargo movement from Chongqing to Singapore via the Southern Transport Corridor.[19] In November 2017, The MAS and the Hong Kong Monetary Authority (HKMA) started developing the Global Trade Connectivity Network (GTCN), a cross-border infrastructure based on Distributed Ledger Technology (DLT), to digitalize trade and trade finance between the two cities and potentially with an aim to expanding the network in the region and globally.[20] Association of Banks in Singapore[21] have jumped onboard, and several banks are ready to take advantage of DLT. MAS has signed an agreement with other central banks, launched an SG $27 million AI and Data Analytics initiative to promote adoption and integration of AI within financial Institutions and joined the MIT Media Lab to strengthen Singapore's talent pool in the FinTech Sector.

The building of the Hinternet was not solely a private sector initiative but led by a government who understands the importance of providing a conducive ecosystem and actively building it. These ideas of Hinternet and related ideas of LASIC[22] in ASEAN and BRI are explored in detail in a previous work of one of the authors.[23] Some of the earlier work was written in Chinese.[24]

---

[19]https://www.joc.com/technology/singapore%E2%80%99s-pil-psa-internation al-ibm-say-blockchain-trial-success-sign-mou_20180227.html
[20]http://www.mas.gov.sg/News-and-Publications/Media-Releases/2017/Singap ore-and-Hong-Kong-launch-a-joint-project-on-cross-border-trade-and-trade-fin ance-platform.aspx
[21]https://abs.org.sg/docs/library/gtcn-joint-release-(website).pdf
[22]https://skbi.smu.edu.sg/sites/default/files/skbife/research_papers/Emergence %20of%20FinTech%20and%20the%20LASIC%20Principles.pdf
[23]https://www.smu.edu.sg/news/2016/03/23/fintech-promise https://www.sm u.edu.sg/news/2016/03/23/fintech-promise-0
[24]https://www.zaobao.com.sg/byline/li-guo-quan http://beltandroad.zaobao.co m/beltandroad/analysis/story20150803-584721 https://www.zaobao.com.sg/for um/views/opinion/story20150515-480417

That works well in the face of deglobalization and the face of climate change. The mindset change has led to many technology companies moving to Singapore especially in the areas of ABCDE, namely, AI, Blockchain, Cloud and Cybersecurity, Devices, and Data Analytics, as well as Environment-Friendly Technology. There are thousands of blockchain companies and foundations being formed in Singapore and Dyson is moving their HQ to Singapore focusing on electric cars.[25] The move of Dyson shocked the British Public and established Singapore as an innovation island as Silicon Valley is finding it difficult to contain cost. However, the transformation of Singapore started in Silicon Valley.

14.2   It All Began in 2014: Silicon Valley and Singapore

In November 2014, Singapore's Prime Minister Lee Hsien Loong announced a wide range of initiatives (including aging, mobility, and data sharing) to transform Singapore into a Smart Nation. The aim is to harness technology to make lives more convenient; improving connectedness within the country. Smart technology can also expand Singapore business into the region, especially the ASEAN countries.

Soon, the Asia leaders were all looking towards west America for inspiration. In 2015, both Japanese[26] and Indian[27] Prime Ministers visited Silicon Valley in May and September. A conference in November 2015 organized by Sim Kee Boon Institute for Financial Economics and Business Families Institute in Stanford University was a forum for Singapore institutions to explore Silicon Valley technologies and their potential, and also offered insights into the latest technologies which could be relevant and beneficial to the Smart Nation program.[28]

---

[25] https://www.ft.com/content/02a636d8-1f2f-11e9-b2f7-97e4dbd3580d

[26] https://www.japantimes.co.jp/news/2015/05/01/national/abe-california-pitc h-shinkansen-tour-silicon-valley/#.XEblps1S82w

[27] https://www.scmp.com/business/global-economy/article/1861855/indian-pm -visits-silicon-valley-he-seeks-funds-and-skills

[28] https://skbi.smu.edu.sg/conference/131141

In January 2016, The Committee on the Future Economy (CFE) was convened to develop economic strategies for the next decade.[29] Over 9,000 stakeholders, including trade associations and chambers (TACs), public agencies, unions, companies, executives, workers, academics, educators, and students were consulted in this process.

The CFE's vision is for the Singapore people to have deep skills and be inspired to engage in lifelong learning, as well as for businesses to be nimble and innovate. The ideals were to have a vibrant city that connects to the world, and continually renewing itself, with a coordinated government that is inclusive and responsive.

Seven mutually-reinforcing strategies were identified to achieve this vision:

1. deepen and diversify international connections;
2. acquire and utilize deep skills;
3. strengthen enterprise capabilities to innovate and scale up;
4. build strong digital capabilities;
5. develop a vibrant and connected city of opportunity;
6. develop and implement Industry Transformation Maps (ITMs);
7. partner each other to enable innovation and growth.

As in all transformation, besides the resistance from the incumbents, it is difficult to change the mindset of the people and businesses. In particular, such a high level and ideals set by a committee that included many incumbents would be asking a lot from a population that was used to multinational driving innovation. It is difficult for anyone to understand and digest, let alone to know what actions are needed for the individual company and citizen to respond to disruption from a bottom-up perspective. Few startups and disruptors were involved in the initial CFE, and even if they were, their ideas were not prominently reflected in the report making concrete actions a lot more difficult. However, the direction was right and transparent, there had to be a change, or Singapore was not going to retain its competitiveness.

---

[29]https://www.gov.sg/microsites/future-economy/the-cfe-report/read-the-full-report

The government of Singapore was determined to create an environment for new businesses and innovation to thrive, but less so to take a bet at which industry or segment. To build an ecosystem requires more guidance on direction and technology mapping. One of the strongest attributes of Singapore is that its policymakers are innovative and nimble! Changes were quick to implement, and it was also clear that if the mindset does not change, then changing the approaches or people, or organizations may help to accelerate innovation.

The Info-communications Media Development Authority (IMDA) was officially formed on October 1, 2016, with the restructuring of the Media Development Authority (MDA) and Infocomm Development Authority (IDA). In September 2017, the government announced the merger of two agencies International Enterprise (IE) Singapore and SPRING. From mid-2018, the new agency Enterprise Singapore (ESG) would help companies grow and internationalize. The merged agency will work at developing more streamlined and comprehensive assistance programs for companies. ESG will also continue to be the lead agency for trade promotion, support the internationalization needs of large companies and retain its role as the national body for standardization, accreditation, and legal metrology.[30]

Earlier in February 2016, Singapore Prime Minister visited Silicon Valley and commented that Singapore must make engineering more attractive as a career.[31] While India was seeking fund and skills from Silicon Valley, Japan and Singapore are mapping out how to harness the technology for their next stage of development. Few in the world have realized that the commercialization of Chinese technology has already led to mass adoption. One main reason was that few could have imagined China has moved so rapidly and more interestingly, language was a barrier, and the Chinese except Alibaba was not even aware that their Internet Finance business models had already

---

[30]https://www.channelnewsasia.com/news/singapore/ie-singapore-spring-to-me rge-and-form-new-agency-9185630
[31]https://www.channelnewsasia.com/news/singapore/singapore-must-make-eng ineering-more-attractive-as-a-career-pm-l-8177098

surpassed the world in 2014.[32],[33] Alibaba was listed on NYSE in September 2014 with a proceed of US $25 billion, the highest in history for Initial Public Offering.

Before Alibaba, it was difficult for technology companies without profits to list in China and when China realized that many of its best technology companies were seeking for listing outside China, it changed the regulation to attract them back to Chinese stock exchanges. Interested readers can refer to articles in several publications by one of the authors.[34],[35]

Groups of senior policymakers and civil servants were sent to Silicon Valley to experience the innovation vibes and to China to study the commercialization of technology. The Global Innovation Alliance was formed to allow Singaporeans to be experienced and learn from other technology centers. In the budget of 2017 announcement, SG $100 million was set aside for new schemes so Singaporeans can boost skills to operate overseas.[36]

Minister for Finance Heng Swee Keat officially launched the Global Innovation Alliance (GIA) (Beijing) in Beijing, China in September 2017. Led by ESG (it was still IE Singapore at that time), the purpose of the GIA (Beijing) was to strengthen connections between Singapore technology companies, entrepreneurs and investors within China's established digital ecosystem of accelerators, incubators, tech giants, start-ups and venture capitalist funds. The intention was to help more Singapore tech companies access opportunities and find business partners in China.

The setup of the GIA was a key recommendation in the CFE report, targeted to strengthen linkages and partnerships with leading

---

[32]https://www.sciencedirect.com/science/article/pii/B9780128122822000012

[33]https://skbi.smu.edu.sg/sites/default/files/skbife/pdf/The%20Rise%20of%20Chinese%20Finance%20%E9%A2%A0%E8%A6%86.pdf

[34]David, L. K. C., & Robert, D. (2017). *Handbook of Blockchain, Digital Finance and Inclusion*, Volume 1 and 2, Elsevier, Academic Press, Massachusetts, US.

[35]David L. K. C., & Low, L. (2018). *Inclusive FinTech*, World Scientific, Singapore.

[36]https://www.businesstimes.com.sg/government-economy/singapore-budget-2017/sg-budget-2017-s100m-for-new-schemes-to-boost-skills-to

innovation hubs around the world, increasing access and opportunities for Singapore students and enterprises. The Singapore Economic Development Board (EDB) set up a GIA Programme Office (GIA PO) to coordinate the overall initiative with other government agencies.[37] Singapore was on full steam to create an ecosystem for linking Singapore and Chinese Technology firms.

Meantime, the MAS and IMDA are working actively, independently and subsequently jointly in getting the technology ecosystem going in the FinTech and Blockchain Space. Many conferences were organized by the Singapore Management University, Singapore University of Social Sciences in conjunction with ACCESS and Singapore FinTech Association. The MAS decided in 2016 that they should start the Singapore FinTech Festival. That propelled the FinTech and Blockchain industry even further.

There were national efforts to upgrade the skills at every level and every sector. While MAS was working to improve the skillset of the financial sectors, many other government organizations were doing the same with different emphasis.

Skillsfuture Singapore (SSG) is a statutory board under the Ministry of Education (MOE). "Skillsfuture is a national movement to provide Singaporeans with the opportunities to develop their fullest potential throughout life, regardless of their starting points. Through this movement, the skills, passion, and contributions of every individual will drive Singapore's next phase of development towards an advanced economy and inclusive society."[38]

Workforce Singapore (WSG) is a statutory board under the Ministry of Manpower (MOM). "It will oversee the transformation of the local workforce and industry to meet ongoing economic challenges. WSG will promote the development, competitiveness, inclusiveness, and employability of all levels of the workforce. This will ensure that all sectors of the economy are supported by a strong, inclusive Singaporean core. While its key focus is to help workers

---

[37]https://ie.enterprisesg.gov.sg/Media-Centre/Media-Releases/2017/11/Minister-Heng-Swee-Keat-launches-Global-Innovation-Alliance-Beijing-to-connect-Singapore-companies-to-tech-and-start-up-players-in-China-for-partnerships
[38]http://www.skillsfuture.sg/AboutSkillsFuture

meet their career aspirations and secure quality jobs at different stages of life, WSG will also address the needs of business owners and companies by providing support to enable manpower-lean enterprises to remain competitive. It will help businesses in different economic sectors create quality jobs, develop a manpower pipeline to support industry growth, and match the right people to the right jobs."[39] Its mission is to enable individuals to adapt and employers to transform. Its visions are 'very individual in a fulfilling career with progressive employers'.

Besides all the efforts to transform the workers and the industry, the efforts did not stop there. There were a lot more initiatives such as the Industry Transformation Map (ITM), Project Ubin (PU), ASEAN Financial Innovation Network (AFIN), National Digital Identity (NDI), Services 4.0, and Technology Map among others. The ITM is to position Singapore as a key node for technology, innovation, and enterprise in Asia and around the world that will group 23 industry sectors into six clusters to maximize opportunities for collaboration.[40] The PU is at its second of sixth phases as a collaborative project with the industry to explore the use of Distributed Ledger Technology (DLT) for clearing and settlement of payments and securities. The AFIN is spearheaded by MAS, the ASEAN Bankers Association, and the World Bank's International Finance Corporation. In September 2018, it launched the API Exchange (APIX) that is an online Global FinTech Marketplace and Sandbox platform for financial institutions (FIs). APIX is the world's first cross-border, open-architecture platform which will enable[41]:

---

[39] http://www.ssg-wsg.gov.sg/about.html

[40] https://www.straitstimes.com/business/economy/singapores-23-key-industrie s-to-be-grouped-into-6-clusters-as-economy-begins-next

[41] https://afin.tech/index.php/2018/09/20/worlds-first-cross-border-open-archi tecture-platform-to-improve-financial-inclusion/

   i. FIs and FinTech firms to connect to one another through a globally curated marketplace;

  ii. collaborative experiments in a sandbox among financial industry participants; and

 iii. adoption of APIs to drive digital transformation and financial inclusion across Asia-Pacific.

Services account for 70% of Singapore's GDP. "Services 4.0[42] (Figures 14.10 and 14.11) was identified through the Services and Digital Economy Technology Roadmap (SDE TRM) as a potential engine of growth for Singapore's digital economy, as the services industry accounts for 72.2% of Singapore's GDP and 74.3% of national employment. The TRM, conducted in consultation with local and international business leaders and technology experts, provides a scan of the digital technology landscape in the next 3 to 5 years, identifying the impact of key shifts and technology trends."

SERVICE 4.0

- Singapore's response to socioeconomic shifts accelerated by emerging technology, to enable Singapore businesses to capture the new opportunities made possible by services to customers that are end-to-end, frictionless, anticipatory, and empathic.

- Delivery is automated for repetitive and mundane tasks, and workers are augmented with emerging technology for creativity, analytical thinking, emotional intelligence, and innovation.

- The vision to capture the opportunities of the Future of Services.

Figure 14.10:   SERVICE 4.0.

---

[42]https://www.imda.gov.sg/about/newsroom/media-releases/2018/services-40-new-digital-capabilities-to-prepare-businesses-for-the-future-of-services

SERVICES &
DIGITAL
ECONOMY
TECHNOLOGY
ROADMAP
AND SERVICES
4.0

A confluence of emerging technologies has disrupted entire industries, business models, and jobs.

At the same time, these shifts have opened up new possibilities, enabling Singapore to shake off its conventional economic constraints such as its small domestic market and geography.

A better understanding of major technology trends and its implications, will enable Singapore to respond decisively.

Figure 14.11: SDE technology roadmap.

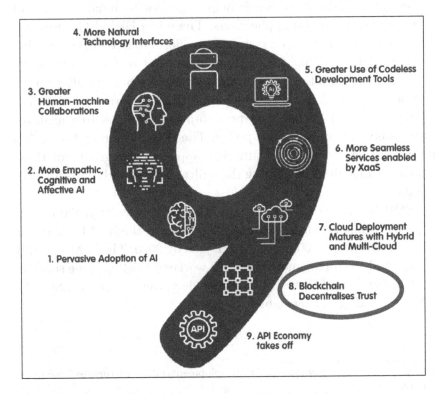

Figure 14.12: Services economy and the need for a response.

The Technology Roadmap[43] has identified nine key trends that will move the digital economy significantly over the next 3–5 years. They may be viewed either as challenges or opportunities. The most significant impact will be on the service sector as it forms the bulk of the global economy and Singapore's GDP. The eight trends will be one of the most interesting as privacy protection and distribution of trust become more valuable in a centralized digital economy that can produce digital dictatorship. Figure 14.12 shows that the focus on blockchain is the Distribution of Trust.

For personal KYC, the Singapore government has taken the first step with National Digital Identity (NDI) so that every resident of Singapore can use to establish his legal identity securely — with two-factor authentication and a public–private key pair — when making online transactions. A self-service platform, Singapore Government Technology Stack helps agencies to build better and more consistent digital applications. There is a Digital Government Blueprint which focuses on government efforts to maximize the value from its data and building interoperable and secure technologies. These are all crucial as privacy protection takes priority after online and offline leaks of medical records and data, investment in internet connectivity, IoT technology, peer-to-peer city apps to encourage more open and resilient data policy. The Business Grant Portal[44] is an interesting concept to ensure easy application of grants for those who have difficulties navigating the multi-agencies and cross-industry grants.

EDB is undergoing restructuring to move away from the cluster approach while the main aim of ESG is to grow the ICM Industry by 6% and to create 13,000 new PMET jobs by 2020. The ICM industry is expected to employ more than 210 workers! ESG has nine schemes to help founders. Figures 14.13–14.16 show the website of ESG, the schemes, and grants available.

---

[43]https://www.imda.gov.sg/industry-development/infrastructure/technology/technology-roadmap

[44]https://www.businessgrants.gov.sg/

Figure 14.13: Enterprise Singapore government website.

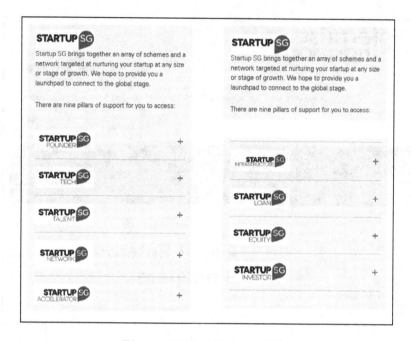

Figure 14.14:   Startup SG.

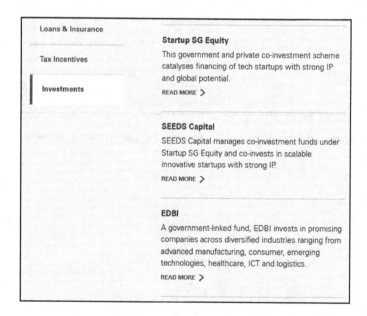

Figure 14.15:   Startup SG equity, SEEDS capital, and EDBI.

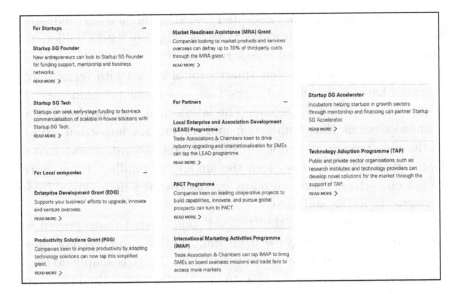

Figure 14.16: More schemes for startups.

The Singapore Digital Government Transformation is key to the entire exercise and where many initiatives were started in the Smart Nation. Government Technology Agency or GovTech advocates innovative technology to shape the way business is done in the government, from delivering digital services to developing Singapore into a Smart Nation.[45]

Others can learn from Singapore while Singapore is learning from Silicon Valley and China. Singapore has become a regional powerhouse for technology backed by a cash-rich and supportive government. It learns from the past lesson of being too top-down and grant-giving cluster culture, and reinvents itself to become an ecosystem builder. It knows that geographical location, well-developed IT infrastructure, strong investment opportunities, and a robust regulatory regime alone is not enough. What it needs is a mindset change to have an open and inclusive physical and virtual economy.

---

[45]https://www.tech.gov.sg/digital-government-transformation/?utm_source=top_nav

The government has allocated SG$150 million in industry research for the AI hub. The MAS is dangling US$5 billion to attract private equity and infrastructure fund managers to anchor in Singapore. It believes that smart capital comes with technology, business know-how, and networks that are useful to companies to grow and scale from Singapore. The latest change in mindset is the acknowledgment that translation and scaling are important. Most of the research and development have slanted towards research and translation is far and few given the amount of funding. If one were to search further, there are many grants covering many areas from startups, to SMEs and MNCs.

Allocating of funds to the "trusted" and "tested" have proven to be a suboptimal strategy with virtually nothing to show not only for Singapore but globally for private equity. There is simply no easy formula to ensure success for investment. So deliverables and targeted job creation have become the more pragmatic approach. Impact measurement that includes metrics such as financial inclusion, stability, job creation is a lot more "smart" investing than return measurement in terms of money terms. These impact measurements will show up GDP and general welfare growth for the population. The same can be said of the funding for academic research, and the future direction will surely be towards deep translation rather than deep fundamental research.

What is in stall for Singapore is the potential of expanding an open, real, and physical economy to one that has an external wing and Hinternet built on an open, secure, inclusive, neutral, distributed, and virtual economy. Singapore is getting ready for the fourth industrial revolution and leading in governance in areas such as AI[46] and eventually DLT and blockchain.[47]

Government leaders are known to be outspoken digital champions. Few governments have had their leaders showing so much

---

[46]https://www.pdpc.gov.sg/-/media/Files/PDPC/PDF-Files/Resource-for-Org anisation/AI/A-Proposed-Model-AI-Governance-Framework-January-2019.pdf
[47]https://www.imda.gov.sg/-/media/imda/files/industry-development/infrastr ucture/technology/technology-roadmap/wg4-executive-summary-for-artificial-i ntelligence-and-data-and-blockchain.pdf

alignment and support for Singapore digital efforts. The visibility of both the leaders and agencies has certainly made Singapore the most desirable place to do a startup. Now, it is up to the networked economy to demonstrate that the small red dot is not only an attractive and easy place to do business, but a place that the business can scale! More importantly, the research talents that Singapore attracts are not just publication academics, but serious researchers that can excel in translation research. A policy innovation such as GIA is already in place for Singapore to partner the tech giants from China and Silicon Valley, as well as translation research institutions such as MIT, Stanford and those in China.

While Singapore can leverage on its strengths of a pro-business environment, trusted legal system with global professionals, political stability and a pro-open source environment in driving innovation, it is facing many challenges. The fact that the domestic market and population are small means that there are few big datasets for commercial use. Record of top-tier ranking local universities and research institutions has not been impressive at all in producing scalable technology. The future lies in translation research and not just pure fundamental research that seeks university ranking. It remains to see if the nation is stuck in the old mindset. Even if the mindset is changed successfully, the legacy issues from the tenure system and other historical baggage in the research fraternity will take time to resolve. This may prove to be the biggest hindrance to commercialization that Singapore is facing. The shift to AI and Big Data will require large datasets that are lacking in Singapore and whether the country can build a Hinternet with sticky customers around the region remains to be seen.

Semiconductors have become the essential factor of production in the digital age. Without integrated circuits, it is difficult to have new digital products. These integrated circuits go into and power everything from smartphones to quantum computers and driverless vehicles. Semiconductors are at the center of the fourth industrial revolution, a gap that China wants to close as it consumes more than half the world's semiconductor chips. However, only 33% is supplied domestically with much of the higher end of IC design located in the

US and South Korean, while Taiwanese foundry accounts for more 70% of the world revenue at the lower end. Singapore's ambition in being a smart nation is handicapped by the lack of expertise in prefab, embedded software, and blockchain. It is therefore not surprising that Singapore will focus on these areas.

Singapore is no stranger to the semiconductor sector with STAT ChipPAC offering chip assembly and test services together with Chartered's six fabrication facilities in the 2000s. Since the divestment of Chartered Semiconductor to GlobalFoundries in 2009, Singapore is no longer a competitor to the Taiwanese. Singapore distanced itself from high-end IC design and mass production of semiconductor chips for several years. In 2017, a small scale new specialty commercial factory specializing in silicon photonics technology named the Advanced Micro Foundry (AMF) was launched as a spin-off from A*Star, a national research institute. The foundry has since broken even and rekindled the interest in this area.

In the unlikely event that the tension between US and China escalates with even more trade wars, technology licensing, data protection acts, sanctions, and blockages, Singapore's importance will be even more prominent. Business strategists will see the value of using Singapore as a base to have a smoother supply chain flow and ring-fence technology ecosystem in order to lessen the impact of any sanctions and security considerations. Even if the tension diminishes, privacy protection and distribution of trust will become important as we enter into an era of the open virtual and augmented economy. Singapore is well poised to take advantage of these developments as it continues to hedge in all known aspects. No one knows how Singapore will do in the future, but at least it has given its best shot at the services and digital technology.

Further References

https://www.mti.gov.sg/Newsroom/Speeches/2018/05/Speech-by-SMS-Sim-at-BRI-Forum.

https://www.mti.gov.sg/Newsroom/Speeches/2018/04/Speech-by-SMS-Sim-at-the-Discussion-on-Alternative-Sources-of-Talent-for-SMEs.

https://www.smeportal.sg/content/smeportal/en/moneymatters/assistance-for-startups/startup-sg-tech.html.
https://www.smeportal.sg/content/smeportal/en/moneymatters/assistance-for-startups.html.
https://www.edb.gov.sg/en/about-edb/who-we-are.html.
http://www.skillsfuture.sg/AboutSkillsFuture.

Acknowledgment

Figures from these chapters are captured from the official websites, reproduced from official documents or sourced from official statistics.

Figures 14.1 and 14.2: EDB Infographic and sourced from DOS.

Figures 14.3 and 14.4: Website of JTC.

Figure 14.5: Website of the Economic Development Board.

Figures 14.6 and 14.7: Website of National Research Foundation.

Figure 14.8: Monetary Authority of Singapore website.

Figure 14.9: Singapore Customs website.

Figures 14.10–14.12: sourced from Infocomm Media Development Authority and documents.

Figures 14.13–14.16: Enterprise Singapore and EDB websites.

Appendix 1

Alibaba

Alibaba Cloud – Platform for AI (PAI)

Schulte-Research

Alibaba Cloud
aliyun.com

I. Overview of Alibaba Cloud

Alibaba Cloud – Suite of Cloud Computing Services

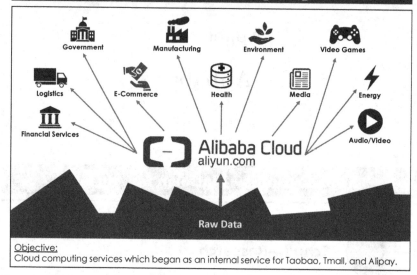

Objective:
Cloud computing services which began as an internal service for Taobao, Tmall, and Alipay.

Source: Alibaba press release, Alibaba Cloud Website, Schulte Research.

Alibaba Financials

		As of March 31st, 2016 (CNY mns)	As of March 31st, 2017 (CNY mns)
Assets	Cash/ Cash Equivalents	106,818	143,736
	Prepayments, receivables, other assets	16,993	29,060
	Short-term investments	4,700	3,011

		As of March 31st, 2016 (CNY mns)	As of March 31st, 2017 (CNY mns)
Liabilities	Accrued expenses, accounts payable and other liabilities	27,334	47,186
	Merchant Deposits	7,314	8,189
	Deferred Revenue, Customer Advances	10,297	15,052

Source: Alibaba press release, Alibaba Cloud Website, Schulte Research.

Alibaba: AI is the Future!

Artificial Intelligence/ PAI 2.0

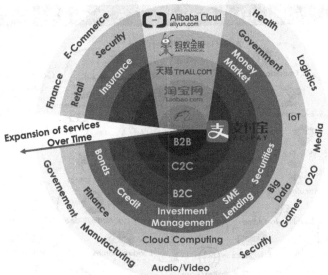

Source: Alibaba press release, Alibaba Cloud Website, Schulte Research.

II. Overview of PAI 2.0

PAI 2.0 – Comprehensive Suite of AI/Big Data Services

Platform for Artificial Intelligence (PAI)

Objective:

Give businesses of all sizes access to AI technology and the ability to incorporate AI into their business model. PAI provides an open-source suite of services, designed to be as user-friendly as possible, allowing any business with a basic technological expertise to use and adopt.

Services:

Banking
- A. Financial Services
- B. E-Commerce
- C. Cyber-Security

Lifestyle
- H. Media
- I. Video Games
- J. Audio/Video

Business
- D. Manufacturing
- E. Online-to-Offline
- F. Logistics
- G. Websites/Apps

Government
- K. Energy
- L. Health
- M. Environment
- N. Government

Source: Alibaba Website, Schulte Research.

III. Services Provided by PAI 2.0

Services Provided: Overview

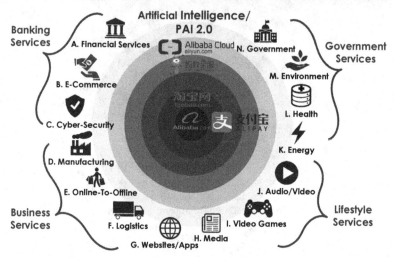

Source: Alibaba Cloud Website, Schulte Research.

Banking Services

Services Provided: Financial Services

a. Financial Services

 A1. Insurance
1. Cloud Document Storage
2. Cloud Service

 A4. Financial Big Data
7. Big Data Warehouse
8. Financial Risk Control

 A2. Banking
3. Hybrid Cloud Deployment

 A5. Exchange
9. Alibaba Cloud Exchange Core

 A3. Mobile Banking
4. Internet Finance (loans, securities, etc.)
5. Internet Financial Security
6. Financial Live Solution

Source: Alibaba Cloud Website, Schulte Research.

Services Provided: Financial Services

 A1. Insurance

1. Cloud Document Storage

Documents which require review and signature by clients can be transferred bilaterally through the new cloud platform. Large-scale query service as well.

Allow for users to quickly build new e-policy distribution platforms. Platforms include secure communication capabilities (via SMS, e-mail, etc.).

End-to-end solution provided within one package. Having met all compliance requirements, this platform can be quickly deployed, and quickly integrated into business practice.

Entry-Level Pricing:	**Promotion:**
CNY ¥ 968 /year	6 month free trial

Source: Alibaba Cloud Website, Schulte Research.

Services Provided: Financial Services

 A1. Insurance

2. Cloud Service

For new insurance firms which do not yet have the man-power, skill, or infrastructure to handle large scale business.

Solution: Complete technical structure to support insurance business services via hardware and software components.

Functions include insurance applications management, document storage, communication services, asset management, etc.

Entry-Level Pricing:	**Promotion:**
CNY ¥ 968 /year	6 month free trial

Source: Alibaba Cloud Website, Schulte Research.

Services Provided: Financial Services

 A2. Banking

3. Hybrid Cloud Deployment

A "dual-engine, hybrid cloud deployment" is provided through Alibaba Cloud to upgrade the banking industry's technology.

A. Internet Banking Application with real-time risk control and process automation.

B. A direct marketing core for mobile internet, real-time big data analysis, and management to support business. Use of cloud computing to drive innovative development.

Snapshot:

 Zhejiang Bank of China is completely run on this cloud service. From financial research, to operations, to management. All on the cloud!

Source: Alibaba Cloud Website, Schulte Research.

Services Provided: Financial Services

 A3. Mobile Banking

4. Internet Finance

Providing cloud computing services to P2P lending platforms, small loan businesses, pawn shops, etc. Suitable for start-up internet finance companies to quickly build a convenient platform that provides loans or matches lenders with borrowers.

5. Financial Security

Guaranteed network security, host security, and mobile security. Combines security and big data analysis to discover potential threats, deduce potential intrusion paths, and develop counter-measures.

6. Financial Real-Time Updates

Live technology SDK which allows users to choose real-time update of information to be broadcasted (market information, news, etc.)

Snapshot:

 HONGLING CAPITAL

A P2P industry leader which adopted Alibaba Cloud's platform to adapt to increasing platform users and user needs.

Source: Alibaba Cloud Website, Schulte Research.

Services Provided: Financial Services

 A4. Financial Big Data

7. Big Data Warehouse

Data storage architecture design. Integration of internal and external sources of data for financial institutions to enhance data mining business value of.

8. Financial Risk Control data

All-round risk control system which provides services such as off-line processioning, online analytical processing (OLAP) analysis, and real-time monitoring to produce real-time credit scores and early risk warnings.

Applications:
1. Identity authentication (facial recognition/ image analysis)
2. Verification of user identity
3. Information authentication
4. Real-time credit score analysis/ risk warning

Source: Alibaba Cloud Website, Schulte Research.

Services Provided: Financial Services

 A5. Exchange

9. Alibaba Cloud Exchange Core

Ali financial cloud for securities industry to provide a complete and integrated solution. Strong network services to provide stable and reliable transaction services, quotes/information services, and disaster recover services.

Snapshot:

 中泰國際
ZHONGTAI INTERNATIONAL

ZhongTai Securities is one of the pioneers in applying Alibaba Cloud's exchange platform to the mobile securities brokerage industry.

Source: Alibaba Cloud Website, Schulte Research.

Services Provided: Financial Services

b. Online Retail Services

B1. Infrastructure
10. Consumer Goods: Full Channel Integration
11. Big Data Analytics: Operations Refinement
12. Cloud POS Solution

B2. Lifestyle/ Search
13. Live Video Broadcasting
14. Personalized Search Recommendations

B3. Voice/Image
15. Image Recognition
16. Intelligent Voice
17. Customer Service AI

Source: Alibaba Cloud Website, Schulte Research.

Services Provided: Financial Services

B1. Infrastructure:

10. Consumer Goods: Full Channel Integration

Advanced internet architecture with dynamic scheduling of resources, used to deal with the rapid growth of business volume in young companies.

Functions include resource management, precision marketing services, data operations, and on-demand cloud capabilities.

Snapshot:

Xtep is a major fashion sportswear brand enterprise based in China. Developing and manufacturing branded sports footwear, apparel, and accessories, Xtep utilizes PAI 2.0's full channel integration service.

Source: Alibaba Cloud Website, Xtep corporate site, Schulte Research.

Services Provided: Financial Services

 B1. Infrastructure:

11. Big Data Analytics: Operations Refinement

Multi-dimensional factor analysis which explores user behavior, and purchasing preferences, to develop operational enhancements. Develops a personalized marketing strategy for each user based off of customer data. Data can be better visualized in their "360 degree data center".

Functions:
1. New media delivery program
2. Personalized marketing strategy
3. Marketing Strategy – Effectiveness Testing
4. Future Purchase Recommendations

Price Range:
CNY ¥ 20,000 - CNY ¥ 480,000
per year (negotiable)

Snapshot:

Durex, one of the world's most recognized intimacy product companies, utilizes Ali Cloud's Big Data Analytics within their business practices. German auto-manufacturer, Mercedes-Benz, is another customer of this Ali Cloud service.

Source: Alibaba Cloud Website, Schulte Research.

Services Provided: Financial Services

 B1. Infrastructure:

12. Cloud POS Solution

 A flexible and dynamic billing system which provides storage, computing, and network services. Significant reductions in price compared to market price, and fully integrable with other AliCloud services (e.g. Big Data Analytics, etc.)

Based on Ali Cloud's cloud computing and big data technologies, Cloud POS brings a new and innovative twist to conventional point of sales methods.

Snapshot:

Bosideng International Holdings Limited, a renowned down apparel company in China, utilizes Ali Cloud's Cloud POS solution.

Source: Alibaba Cloud Website, Schulte Research.

Services Provided: Financial Services

B2. Search/Lifestyle:

13. Live Video Broadcasting

Technical solution to allow for live video broadcasting that adapts to poor network connectivity, millions of concurrent users, and multi-channel transcoding.

14. Personalized Search Recommendations

Historical search data of each user is stored, and analyzed to customize the search recommendations for each user. Easy to use development kit allows for non-professionals to assemble as well.

Snapshot:

DaMai, a small business, uses the personalized search recommendation service in its business. By doing so, it has been able to increase its daily customer conversion rate by more than 10%!

Source: Alibaba Cloud Website, Schulte Research.

Services Provided: Financial Services

B3. Voice/Image:

15. Image Recognition

Image recognition engine which can analyze pictures entered into the search bar to find similar products or entries.

16. Intelligent Voice

Derived from machine learning principles, intelligent voice converts recorded phone calls into a very large set of valuable, and usable data.

17. Customer Service AI

Also derived from machine learning principles, the customer service AI moves towards providing automatic Q&A with customers about a wide range of topics related to the business.

Snapshot:

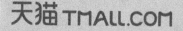

Currently, Tmall uses a wide range of tools offered by PAI 2.0. Of the services it uses, are Image Recognition (17), Intelligent Voice (18), and Customer Service AI (19).

Source: Alibaba Cloud Website, Schulte Research.

Services Provided: Financial Services

c. Cyber-Security Services

C1. Cyber Security, Defense
18. Secure Server Hosting
19. DDoS Attack Prevention/ Countermeasure

C2. Attack Detection/ Prevention
20. Potential Intrusion Warning System/ Prevention
21. Discovering Business/Network Vulnerabilities

Source: Alibaba Cloud Website, Schulte Research.

Services Provided: Financial Services

C1. Cyber-Security, Defense

18. Secure Server Hosting
Personalized cloud server hosting security. Includes manual safety check, removing Trojan viruses, system consolidation, AI technical support, and service reports.
Security experts from Alibaba Group provide 24x7 online monitoring of any potential security threats.

> **Price Range:**
> ~CNY ¥ 3,000 per year

19. DDoS Attack Prevention/ Countermeasure
DDoS Attacks (Distributed Denial of Service) can be very dangerous for companies with an online presence. Drainage of attacking traffic to high anti-IP ensures secure and stable connections. Industry leading defense algorithms + Ali Cloud's cloud computing expertise provide a strong defense and prevention of potential attacks.

> **Price Range:**
> ~CNY ¥ 16,800 per month

Source: Alibaba Cloud Website, Schulte Research.

Services Provided: Financial Services

C2. Attack Detection/ Prevention

20. Potential Intrusion Warning System/ Prevention
    Big data security analysis platform which alerts alerts all assets on the cloud of any potential intrusion and uses machine learning to prevent future intrusions. Visualization of all security issues is presented in a real-time UI.

UI Functions:
- Business operations monitoring
- Emergency response data
- Security awareness monitor
- Defense system monitor
- Host connection topology
- Etc.

Price Range:
~CNY ¥ 200 per month

21. Discovering Business/ Network Vulnerabilities
    Private security services which help companies to discover any potential network vulnerabilities. By utilizing Ali Cloud's emergency center security experts' resources and abilities, network vulnerabilities are quickly located and fixed.

Source: Alibaba Cloud Website, Schulte Research.

Business Services

Services Provided: Business Services

d. Manufacturing Services

D1. Equipment Operations/Services
22. Industrial SCADA System
23. Equipment Operation/Maintenance
24. Intelligent Device Interconnection

D2. Smart Factory
25. Intelligent Factory/ State-Owned Enterprises
26. Big Data Solutions

Source: Alibaba Cloud Website, Schulte Research.

Services Provided: Business Services

D1. Equipment Operations/ Services

Price Range:
~CNY ¥ 460 per month

22. Industrial SCADA System

Supervisory Control and Data Acquisition (SCADA) System which leverages on cloud computing technology to provide efficient control of very large operations.

23. Equipment Operation/Maintenance

Price Range:
~CNY ¥ 460 per month
+ CNY ¥ 0.02 per hour

Leveraging on cloud computing and IoT, this platform provides remote status monitoring of industrial equipment, life cycle management, and data storage solutions.

24. Intelligent Device Interconnection

Platform which allows for a variety of network protocol management, intelligent equipment interconnection, and machine-to-machine agreement. Combine real-time data analysis with current system to improve efficiency.

Snapshot:

XCMG Group, a Chinese, government-owned, heavy machinery manufacturer, uses Ali Cloud's enterprise-class architecture, EDAS as well as the platform for intelligent device interconnection.

Source: Alibaba Cloud Website, Schulte Research.

Services Provided: Business Services

D2. Smart Factory

25. Intelligent Factory/ State-Owned Enterprises

Provide factory workers with transparent production and data management solutions. Combine historical data and specific data within algorithm to produce future predictions and course of action.

26. Big Data Solutions

Price Range:
Pay by volume
CNY ¥ 500 minimum
per month

Combining Ali Cloud's big data analytics and AI capabilities to provide customers with real-time data on their manufacturing production, sales, and R&D processes to discover any potential inefficiencies.

Snapshot:

GCL–Poly Energy Holdings is one of the world's largest solar photovoltaic enterprises. They utilize Ali Cloud's intelligent photovoltaic factory platform within their own manufacturing plants.

Source: Alibaba Cloud Website, GCL-Poly corporate site, Schulte Research.

Services Provided: Business Services

e. O2O Solution

E1. OTA (Online Travel Agency) Industry
27. Business Process Support
28. Mass Information Collection/Storage/Analysis

E2. Wi-Fi Industry
29. Wi-Fi Certified Access
30. User Internet Behavior Records
31. Data Marketing

E3. Business Support System
32. Hotel Business Process Support
33. Online Shopping Support

Source: Alibaba Cloud Website, Schulte Research.

Services Provided: Business Services

E1. OTA (Online Travel Agency) Industry

27. Business Process Support

Technical infrastructure support using cloud services, servers, databases, and intelligent algorithms. Strong storage and network capabilities allow for business processes to run smoothly and efficiently, integral to the OTA Industry.

Price Range:
Pay by volume (GB)
CNY ¥ 1,000 minimum
per month

28. Mass Information Collection/Storage/Analysis

Storage and processing of image files, log data, and user behavior. Data is collected, analyzed, and used for rapid troubleshooting, error detection, and user experience enhancement.

Snapshot:

Tongcheng Travel is one of China's largest online travel agencies, with services offered ranging from hotels and airline tickets, to travel insurance, and financing. They utilize Ali Cloud's OTA Industry Service within their business processes.

Source: Alibaba Cloud Website, Schulte Research.

Services Provided: Business Services

 E2. Wi-Fi Industry

29. Wi-Fi Certified Access

To ensure certified access of Wi-Fi users, businesses can leverage on Ali Cloud's load balancing, authentication, and cache services.

30. User Internet Behavior Records

In order to be able to analyze user behavior records, Wi-Fi access data needs to be saved onto the cloud database. Technical services are provided to ensure the efficient process of collecting and storing Wi-Fi users' behavior records.

31. Data Marketing

Internet behavior data analysis to allow for personalized marketing schemes. Services include personalized search recommendations, and purchasing recommendations.

Snapshot:

Ai Xian Xia is a major Wi-Fi services provider. They utilize Ali Cloud's Wi-Fi services to provide storage capabilities, security services, and big data analytics.

Source: Alibaba Cloud Website, Schulte Research.

Services Provided: Business Services

 E3. Business Support System

32. Hotel Business Process Support

For hotels or hotel management system developers, this platform provides networking, storage, and big data analytics capabilities. These capabilities are utilized to enhance hotel operations, and management capabilities.

33. Online Shopping Support

For retail industry businesses, this platform provides operational enhancements, and big data analytics of its customers. Within online shopping, bandwidth allocated to each site is varied according to traffic, which ultimately reduces costs.

Snapshot:

Atour Hotel, a hotel chain in China, utilizes Ali Cloud's cloud computing capabilities, as well as big data analytics in their day-to-day business processes.

Source: Alibaba Cloud Website, Schulte Research.

Services Provided: Business Services

f. Logistics Solution

F1. Transportation Management
34. Commercial vehicle network
35. Logistics path optimization
36. Transport management system

F2. Warehouse Management
37. Warehouse management system

Source: Alibaba Cloud Website, Schulte Research.

Services Provided: Business Services

F1. Transportation Management
34. Commercial Vehicle Network

Safe and stable cloud services with big data analytics to provide large-scale vehicle monitoring, management, and enhancement. Functions include reduction in transport time, supply chain management, and efficient distribution management.

35. Logistics Path Optimization

Through machine learning and analyzing orders, vehicle paths, and traffic, this platform provides path optimization services, intelligent distribution systems, and path algorithm customization.

36. Transport Management System

Business process management system, managing billing, planning, sales forecasts, cost controls, and etc.

Price Range:
CNY¥2,500–CNY¥10,000 per month

Snapshot:

Yun Man Man is a logistics/shipping company based in China. Having to handle a high level of orders in many cities spread across China, Yun Man Man utilizes Ali Cloud's Logistics services.

Source: Alibaba Cloud Website, Schulte Research.

Services Provided: Business Services

F2. Warehouse Management

37. Warehouse Management System

The Warehouse Management System serves as an all-in-one platform for warehouse/storage management. Its functions include warehouse allocation, batch/inventory management, cost optimization, sorting optimization, global distribution, and sales forecasting.

This system is used to improve order processing efficiency, cope with peak traffic periods, and building a global logistics network.

Big data analytics is used for continual process optimization. By analyzing the streams of data that are constantly gathered, continuous optimization of business processes can be streamlined.

Snapshot:

vTradEx, a smart logistics software provider, provides solutions to many supply chain management issues, such as warehouse management, distribution services, and etc. They utilize Ali Cloud's platforms within their business.

Source: Alibaba Cloud Website, Schulte Research.

Services Provided: Business Services

g. Website/App Solutions

G1. Website Solution

38. Website Data Analysis Visualization UI
39. Real-time Risk Control
40. Real-time SMS Authentication
41. Customizable, Personalized, High-end Site Structures
42. Live Broadcast Function

G2. Mobile App Solution

43. Stable and Easy-to-use Platform, Mobile Access Acceleration
44. Quick, On-demand, Mobile Build
45. App Custom Development

Source: Alibaba Cloud Website, Schulte Research.

Services Provided: Business Services

 G1. Website Solution

38. Website Data Analysis Visualization UI

Customizable User Interface dashboard which visualizes website data through real-time data acquisition. Dashboard is easily customizable, and can display whatever relevant data the user requires.

39. Real-time Risk Control

Prevents any suspicious user activity, spam registration, hacking, or other malicious behavior. Through the use of data control services, such situations can be effectively prevented and controlled.

40. Real-time SMS Authentication

For secure sites which require dual authentication via SMS verification, this platform provides a solution. During popular holiday Double 11, this service can support the more than 200 million users (600 million SMSs').

Snapshot:

 Pipaw.com, an online vendor of video games, utilizes Ali Cloud's website developer tools on their own site.

Source: Alibaba Cloud Website, Schulte Research.

Services Provided: Business Services

 G1. Website Solution

41. Customizable, Personalized, High-end Site Structures

To meet the requirements of enterprises where marketing sites are of high importance. Building customizable, easy to use, fast, and technically-solid page layouts.

42. Live Broadcast Function

Protected, and smooth live broadcast function embedded in existing sites. Live authentication, URL encryption, and other security measures to protect user content.

Source: Alibaba Cloud Website, Schulte Research.

Services Provided: Business Services

 G2. Mobile App Solution

43. Stable and Easy-to-use Platform, Mobile Access Acceleration

By applying Ali Cloud's cloud service and technical infrastructure, customers can develop smooth and quick movement within their App. Adopting Ali Cloud's platform provides anywhere between 10 – 70% acceleration in the App's browsing speed.

44. Quick, On-demand, Mobile Build

For quick access to mobile videos, Ali Cloud's Quick, Mobile Build service allows for fast and efficient video uploading, transcoding storage, and playback distribution.

Snapshot:

XiaoKaXiu, a popular video sharing app in China, was originally built, and is currently running on, Ali Cloud's server. As per a technical director at XiaoKaXiu, Ali Cloud supports *"horizontal and vertical expansion"*, and has been very efficient during its inception, with *"a small entrepreneurship team, small capital, and huge explosive growth of users"*.

Source: Alibaba Cloud Website, Schulte Research.

Services Provided: Business Services

 G2. Mobile App Solution

45. App Custom Development

Platform for developing fully customizable Apps for any business industry. Customers only need to provide an idea, and can build their ideas into reality through the platform. There is a heavy focus on UI/UX through this platform for both customers, and future users of the App.

Snapshot:

KongGe, an app in China dealing with business customer service, has relied heavily on Ali Cloud's mobile development platform during it's inception. Not only was the app built on the platform, but the majority of it's services still remain through Ali Cloud's platform as well.

Source: Alibaba Cloud Website, Schulte Research.

Lifestyle Services

Services Provided: Lifestyle Services

h. Media Solutions

H1. News Services
46. Live Media Solutions
47. Media Release Operations
48. News and Program Production

H2. Infrastructure
49. Media Asset Management
50. Data Storage

Source: Alibaba Cloud Website, Schulte Research.

Services Provided: Lifestyle Services

H1. News Services
46. Live Media Solutions

Price Range:
CNY ¥ 14,800 per year

Integrated platform for live media broadcasting. Allows for not only streaming of live video, such as sports games, but also recording and streaming live video, for example in news stations. Features include load balancing, low delay, and low latency.

47. Media Release Operations

Logistical solution to the distribution of video content. Using Big Data analytics, based on video streaming demand data, distribution data, and other collected data, this service can determine effective distribution methods of media.

48. News and Program Production

Price Range:
CNY ¥ 340,000 per year

Hybrid cloud for news and content production. Cloud technology services and tools to support large-scale activities, or rapidly deploy content in emergencies.

Snapshot:

China Central TV (CCTV), the largest television channel in China, utilizes Ali Cloud's live media solution platform, and news/program production service within their business services.

Source: Alibaba Cloud Website, Schulte Research.

Services Provided: Lifestyle Services

H2. Infrastructure

49. Media Asset Management

A complete solution which integrates media content, storage management, content processing, global distribution, and media data analysis. Based on powerful infrastructure and big data analytics, this platform provides mass storage, management, and video/audio content analysis capabilities.

50. Data Storage

Provide unlimited space, multiple back-ups, and hierarchal storage of media data. Based on the flexibility of capacity extension, customers can store virtually as much as required. This data can also be transcoded, to change the size of data required, in order to cut costs.

Snapshot: Zhejiang University of Media and Communications has adopted Ali Cloud's media platforms and integrated them into their education curriculum.

Source: Alibaba Cloud Website, Schulte Research.

Services Provided: Lifestyle Services

i. Video Game Solutions

I1. Game Server Deployment
51. MMO Game Solution
52. MOBA Game Solution
53. Global Server Deployment (Game Distribution)

I2. Game Data Operation
54. Big Data Operation

I3. Game Cyber-Security
55. Cyber-attack Prevention/ Counter-measures
56. Cheating and Cracking Prevention

Source: Alibaba Cloud Website, Schulte Research.

Services Provided: Lifestyle Services

 I1. Game Server Deployment

51. MMO Game Solution

MMO Games (massively multiplayer online games), require very strong network and servers to be able to host the sheer magnitude of players. To support this high performance, Ali Cloud offers high network, server, memory, and storage solutions.

52. MOBA Game Solution

To serve Multiplayer Online Battle Arena (MOBA) games, Ali Cloud offers BGP network services, cross-regional server deployment, and industry-leading security solutions for game stability.

53. Global Server Deployment (Game Distribution)

Global distribution of games is highly dependent on global networks, servers, and data centers. To reduce network delays, improve stability, and improve game synchronization, Ali Cloud offers strong server, database, network, and security solutions.

Snapshot:

 Age of Empires, one of the world's most popular gaming franchises, uses Ali Cloud's global network services to ensure smooth and stable global deployment.

Source: Alibaba Cloud Website, Schulte Research.

Services Provided: Lifestyle Services

 I2. Game Data Operation

54. Big Data Operation

A large amount of data is collected through video games. This data has a wealth of application scenarios, from business, to technical applications. To maximize the utility of this data, Ali Cloud offers Offline data analysis service, and real-time data analysis.

In order to better understand behavior preferences, Ali Cloud's offline data analysis service applies data mining and big data analysis techniques on game logs, and other data to produce OLAP reports.

Through strong network services, Ali Cloud also offers real-time data analysis. The application of this analysis ranges from information regarding user information and behavior, to technical applications such as data usage or network connectivity. The service can handle very large-scale data at a low cost.

Snapshot:

 Pokemon, another major global franchise, utilizes Ali Cloud's video game services in it's deployment for its games.

Source: Alibaba Cloud Website, Schulte Research.

Services Provided: Lifestyle Services

13. Game Cyber-Security

55. Cyber-attack Prevention/ Counter-measures

Depth analysis of game security issues. This service uses Ali Cloud's strong cyber-security experience with preventing and defending against DDoS attacks, and other large traffic attacks. The service also deploys cheap and affordable counter-measures to reduce the effects of any attack.

Price Range:
CNY ¥ 500 per month

56. Cheating and Cracking Prevention

As player incentives to unlock certain rewards or accumulate items within games increase, so do cases of hacking and cheating. To prevent such situations, Ali Cloud deploys automated mechanisms which prevent spam registration, cheating, and access to internal networks.

Secondly, to prevent players from "cracking" the game in order to bypass any payment, Ali Cloud also deploys automated counter-measures.

Snapshot:

Popular FPS franchise Crisis Action, deploys Ali Cloud's network, server and security services to ensure a safe and stable gaming experience.

Source: Alibaba Cloud Website, Schulte Research.

Services Provided: Lifestyle Services

j. Audio/Video Solutions

J1. Live Video Solutions
57. Interactive Global Entertainment Broadcast Control
58. Live, Real-time Viewer Analysis
59. Live Video Solution

J2. Video Applications
60. Short Videos Solutions
61. Video Surveillance
62. Video Quality Enhancement

Source: Alibaba Cloud Website, Schulte Research.

Services Provided: Lifestyle Services

 J1. Live Video Solutions

57. Interactive Global Entertainment Broadcast Control

For global entertainment businesses to provide their customers with low-cost, low latency, and high capacity broadcasting systems. Ali Cloud offers live transcoding, HD video streaming, and global deployment services.

58. Live, Real-time Viewer Analysis

Real time data analysis of the number of viewers, viewer behavior, and viewer preferences. This service both collects data, and conducts real-time analysis.

59. Live Video Solution

Quick deployment of end-to-end broadcasting, live video service. Ali Cloud provides easy access, high-definition, low latency, and high concurrent audio/video broadcasting service.

Snapshot:

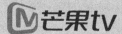 Mango TV, an online and satellite video platform, utilizes Ali Cloud's live video solution to provide high quality videos through their online site.

Source: Alibaba Cloud Website, Schulte Research.

Services Provided: Lifestyle Services

 J2. Video Applications

Price Range:
CNY¥50,000 per year

60. Short Videos Solutions

Personalized recommendations for what to watch next, video streaming and viewing services, and video editing.

61. Video Surveillance

Real time data analysis of the number of viewers, viewer behavior, and viewer preferences. This service both collects data, and conducts real-time analysis.

62. Video Quality Enhancement

By increasing the frame rate and processing speed of the servers, Ali Cloud is able to stream higher quality videos at a lower cost. Using advanced video algorithms, Ali Cloud is able to further increase video quality.

Snapshot:

 Danale, an IoT firm, utilizes Ali Cloud's video/audio services within their business.

Source: Alibaba Cloud Website, Schulte Research.

Government Services

Services Provided: Government Services

k. Energy Solutions

K1. Energy Production and Planning
63. Power Plant Construction
64. Energy Field Planning/ Investment Income Forecasts

K2. Electric Vehicles
65. Electric Car Sharing/Rental Cloud
66. Electric Vehicle Charging Operation/Service

Source: Alibaba Cloud Website, Schulte Research.

Services Provided: Government Services

 K1. Energy Production and Planning

63. Power Plant Construction

Cloud computing, big data analytics, and infrastructure solutions for enhancement of integrated power plants. Services include operations/ maintenance, monitoring production processes, power grid monitoring, and other supply chain operations. Ultimate goal to create an integrated platform which connects individual users, power plants, and businesses.

64. Energy Field Planning/ Investment Income Forecasts

Help to develop a big data driven, fully intelligent, Internet platform for investment and development. Services include energy planning management, analysis of energy needs, and provide project development.

Snapshot:

 Beijing East Environment Energy Technology, a grid connection and operating system provider, utilizes Ali Cloud's energy field planning platform.

Source: Alibaba Cloud Website, Schulte Research.

Services Provided: Government Services

 K2. Electric Vehicles

65. Electric Car Sharing/ Rental Cloud

Development of a real-time electric car sharing platform. Such a platform helps customers quickly find the nearest available car, manages data about each vehicle, and ensures the smooth running of the service. Big data analytics and cloud computing are utilized to ensure the smooth running of the platform, and of the business operations.

66. Electric Vehicle Charging Operation/ Service

Platform for connecting customers with nearby charging stations. Furthermore, this platform is also able to handle payments and administrative services. Big data analytics utilized to ensure smooth and efficient operations.

Source: Alibaba Cloud Website, Schulte Research.

Services Provided: Government Services

I. Health Services

 L1. Intelligent Hardware
67. Medical Intelligent Hardware

 L2. Medical Data Storage
68. Medical Images Storage
69. Medical Information
70. Medical Data Storage
71. Telemedicine Platform

L3. Cloud Services

72. Cloud HIS
73. Cloud PACS
74. Cloud Hospital
75. Cloud HRP

 L4. Diagnosis
76. Grading Diagnosis
77. Biological Gene Analysis

 L5. Hospital/Patient Interaction
78. Video Consultation
79. Hospital/ Patient Exchange Platform
80. Medicine Circulation

Source: Alibaba Cloud Website, Schulte Research.

Services Provided: Government Services

 L1. Intelligent Hardware

67. Medical Intelligent Hardware

Real-time collection of patients data via smart hardware, storage, analysis, and warnings systems. Leverages IoT application building platform, massive storage capabilities, and big data analytics to build intelligent devices which collect patient data, analyze it, and stream to Doctors.

Snapshot:

Mindray Medical International Limited, China's largest medical equipment developer/manufacturer, utlizes AliCloud's Intelligent Hardware platform to utilize AliCloud's IoT, storage, and big data analytics capabilities.

Source: Alibaba Cloud Website, Schulte Research.

Services Provided: Government Services

 L2. Medical Data Storage

68. Medical Images Storage

Dedicated cloud storage for medical image storage, 3-D image reconstruction, strong network security.

70. Medical Data Storage

Electronic medical data, payments, prescriptions, and other data storage. Online authenticity can be confirmed.

69. Medical Information

An integrated, comprehensive information system which manages all functions within a hospital.

71. Telemedicine Platform

A medical platform which shares medical information between hospitals to allow for remote diagnosis and education. Document sharing, secure communication, data management, etc.

Snapshot:

Shanghai Jingyi Technology Company is a leading provider of intelligent medical technology. Their services include telemedicine platforms, cloud technology, and etc.

Source: Alibaba Cloud Website, Schulte Research.

Services Provided: Government Services

 L3. Cloud Services

72. Cloud Hospital Information System (HIS)

An integrated, comprehensive information system which manages all functions within a hospital (administrative, finance, medical, etc.) sold by SaaS.

73. Cloud Picture Archiving and Communication System (PACS)

A cloud-based service which integrates image data from different devices to provide transfer and analysis services.

74. Cloud Hospital

Online portal for patients to access appointment registration, online payment, online medical report viewing, and other services.

75. Cloud Hospital Resource Planning (HRP)

To fully support a hospitals expanding management and operation requirements.

Snapshot:

 Neusoft is a China-based IT services and medical equipment provider. They are customers of AliCloud's Picture Archiving and Communication System (PACS).

Source: Alibaba Cloud Website, Schulte Research.

Services Provided: Government Services

 L4. Diagnosis

76. Grading Diagnosis

Centralized sharing of medical data, regional synergy and collaboration, mixed with great computational power/analytics, brings a central system for enhanced diagnosis and prognosis.

7. Biological Gene Analysis

Cloud storage and big data mining of gene data. Enhanced interpretation and efficiency of gene data analysis. Services include data storage, computational resources, and security.

Snapshot:

 The Chinese Academy of Agricultural Sciences utilizes AliCloud's Biological Gene Analysis platform on their sequencing project of 3,000 super rice genomes, the world's largest plant genome sequencing project.

Source: Alibaba Cloud Website, Schulte Research.

Services Provided: Government Services

 L5. Hospital/Patient Interaction

78. Video Consultation

Integration of cross-video terminal types to achieve remote video consultation. Integration with medical business to allow remote hospital/patient interaction.

79. Hospital/Patient Exchange Platform

Online communication platform which allows for quick communication as well as administrative tasks (e.g. scheduling, etc.).

80. Medicine Circulation

Pharmaceutical B2B, B2C, and O2O business. Leverage on Alibaba's heavy e-commerce presence, applied to pharmaceutical industry.

Snapshot:

 Pexip is a video conferencing cloud platform system. Used in a wide range of applications, and applying the AliCloud technology, they provide an affordable video conferencing system.

Source: Alibaba Cloud Website, Schulte Research.

Services Provided: Government Services

m. Environment Solutions

M1. Analysis/Protection

81. Ecology Cloud
82. Analysis and Judgement
83. Intelligent Dispatch

Source: Alibaba Cloud Website, Schulte Research.

Services Provided: Government Services

 M1. Analysis and Protection

81. Ecology Cloud
Environment data management platform which supports big data analysis, model development, and convenient environment data services. Specific environmental factors, such as pollutant levels, can be monitored through this cloud/ user interface. These factors are automatically monitored by the cloud and alert proper authorities of any potential issues.

82. Analysis and Judgement
Analysis of factors that are monitored by the cloud. Analyzed using big data analytics, and AI. Any behavior out of the ordinary is reported to the user.

83. Intelligent Dispatch
Big data and AI technology to provide comprehensive assessment of environmental issues, causes, and future course of action. Simulation of future prediction service which has an accuracy of 93% is also used for early detection and warning services.

Source: Alibaba Cloud Website, Schulte Research.

Services Provided: Government Services

n. Government Services

 N1. Smart Government Services
84. Mobile Police Services
85. Electronic Tax Services
86. Government Cloud Services

 N2. Traffic Management
87. Big Data Traffic Analysis
88. Illegal Traffic Reporting

Source: Alibaba Cloud Website, Schulte Research.

Services Provided: Government Services

 N1. Smart Government Services

84. Mobile Police Services

Smart and secure communication system for police officers. Also a secure and innovative platform for police business operations.

85. Electronic Tax Services

Electronic taxation platform which allows for tax bureaus to better cope with the sheer number of taxpayers and tax payments. Services include tax collection, and tax processing.

86. Government Cloud Services

Development of a unified government cloud which allows for document and data sharing between government departments, service coordination, data computing, and storage. Great for provincial governments who are not as technically developed, but adaptable for national governments as well.

Snapshot:

 Both the Chinese Ministry of Public Security and the Beijing State Administration of Taxation utilize Ali Cloud services.

Source: Alibaba Cloud Website, Schulte Research.

Services Provided: Government Services

 N2. Traffic Management

87. Big Data Traffic Analysis

Services ranging from collection of traffic data, storage of data, and ultimately big data analysis of traffic data. Traffic data analysis for smart route planning, future roadway construction, smart signal timing, and etc.

88. Illegal Traffic Reporting

Platform through which citizens can report traffic violations and upload video evidence. Platform used to automatically process and classify violations. Used to streamline operation.

Snapshot:

Zhejiang Provincial Department of Transportation uses Ali Cloud's Big Data Traffic Analysis service among many other services.

Source: Alibaba Cloud Website, Schulte Research.

IV. Top 6 – Exciting and Innovative Services

Top 6 – Exciting and Innovative Services

9. Alibaba Cloud Exchange Core: Universalizes banking to anyone

A strong network and cloud solution which allows users to deploy professional, and integrated exchange platforms for any financial product. Brings high-level technology to any broker/dealer, bank, NBFI, insurance company or SMEs'.

17. Customer Service AI: Universalizes all available data sets for anyone

Utilizing machine learning principles to automate labor intensive activity and customer services. Opens doors for future data analytics of customer requests.

25. Intelligent Factory: Enhancing production and business processes

Platform for managing production, business processes, and administrative processes within factories. Big data processing of historical data through algorithms to produce future predictions and suggested course of action.

Source: Alibaba Cloud Website, Schulte Research.

Top 6 – Exciting and Innovative Services

34. Commercial Vehicle Network: Universalizes all logistics

Machine learning and big data analytics processes to provide commercial vehicle monitoring, management, path optimization, and enhancement. Solves a large problem in supply chain management by offering an advanced and efficient solution.

64. Energy Field Planning: Smart Cities

Intelligent platform which integrates a city's electric grid network and drives "Smart City" development. Also can be used to monitor the financial viability of city-wide projects, ie hydro electricity, etc.

76. Grading Diagnosis: Universalizes hospital data

A platform which integrates patient data nationally, allowing for centralized sharing of medical data with hospitals across the country. Also has enhanced diagnosis/ prognosis tools. Big data and cloud computing leads to life-saving solutions.

Source: Alibaba Cloud Website, Schulte Research.

Appendix 2

Ping An

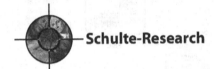 

I. Overview of Change in Insurtech

Digital insurance is 90% cheaper than traditional insurance.

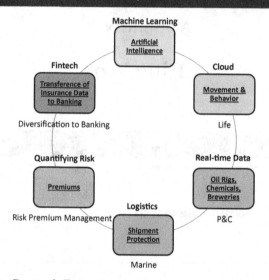

Source: Schulte Research Estimates.

Volume of Global Healthcare Data Growing at 48% CAGR. Companies that are not <u>fully</u> digitalized with full cloud services are <u>blind</u>! Ping An is ahead of everyone.

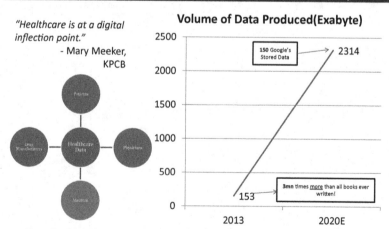

Source: Stanford Medical School, KPCB Internet Trend Report, Schulte Research Estimate.

US healthcare industry has highest spending with lowest performance. There is no insurance company in the US, or EU, that is undergoing the transformation that Ping An is undertaking.

• US spent **$3 trillion** in 2016 on national healthcare, **18%** of GDP.

• National Heathcare Expenditure should grow **1.2%** faster than GDP (**20%** in 2025).

• However, US healthcare performs **the worst** among the top 11 countries with the highest income per capita.

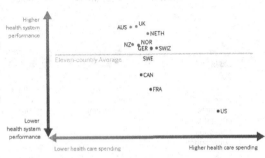

HealthTech is the way to solve this!

Source: Commonwealth Fund, Schulte Research Estimates, US CMS.

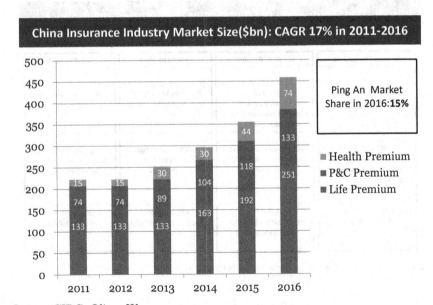

China Insurance Industry Market Size($bn): CAGR 17% in 2011-2016

Ping An Market Share in 2016:**15%**

■ Health Premium
■ P&C Premium
■ Life Premium

Source: CIRC, Oliver Wyman.

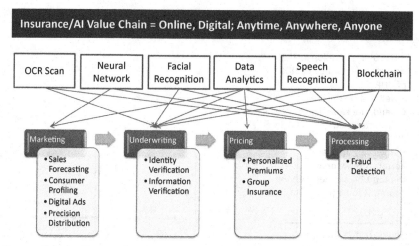

Source: Schulte Research Estimates.

Introduction: Four Levels of Complexities in AI Applied to Insurance

2012	2014	2016	2020
Reactive Machines	**Limited Memory**	**Theory of Mind**	**Self Aware AI**
Basic artificial intelligence.	Pre-programmed knowledge and observations carried out over time.	Requires knowledge that people & objects can alter feelings & behaviors.	Machines have self-awareness.
React to current situations, but cannot rely on past data to make present decisions.	Looks at an environment and adjusts if something changes.	Understands people's intentions and predict what they will do.	Computers can escape from common human failure to detect danger.
Example: Basic Premium Pricing Model	Example: Fraud Identification Based on Past Transactions	Example: Facial Recognition for Loan Approval	Example: Robotic Insurance Agent

Source: Betanews, Schulte Research Estimates.

II. Ping An Group

A. Overview: Global Financial Services Provider

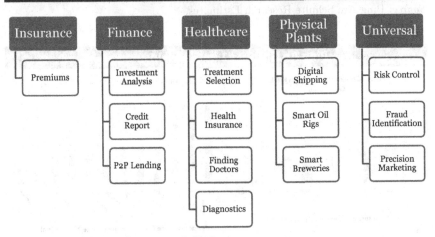

Source: Ping An, Schulte Research Estimates.

Monthly Active Users

- 62 mn total users, up 42% YOY.

- Within the 62 mn MAU, 20% of them are active users.

- 4 mn from Ping An Bank App.

- 18 mn customers link their car with Ping An Auto Owner app(No. 1 app in aftermarket)

- 1 mn DAU, ranking No. 4 in China, up 8 places from last year.

- Ping An Good Doctor has 450k daily inquiries.

- One Account has 30 mn MAU, up 63% YOY.

- E-Wallet has 7mn MAU.

Source: Ping An, Schulte Research Estimates.

Ping An Group: #1 in Global Insurance

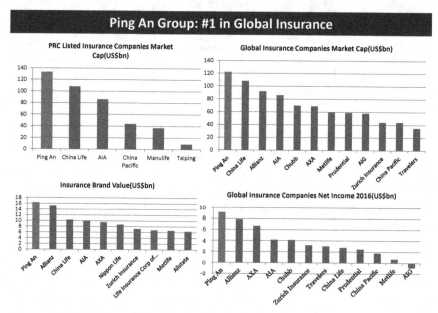

Source: Schulte Research Estimates, 4-traders.com, South China Morning Post.

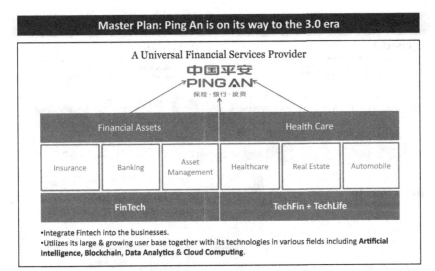

Source: Ping An, Schulte Research Estimate.

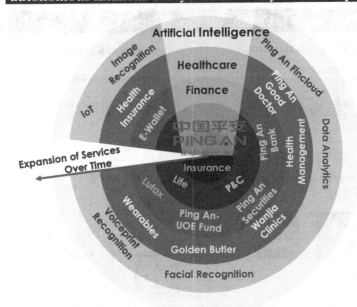

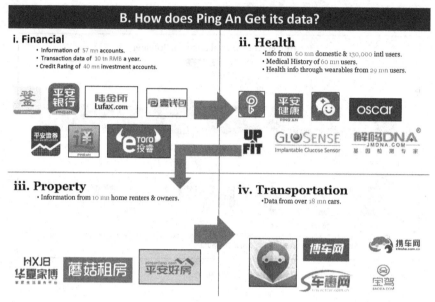

Source: Ping An, Schulte Research Estimate.

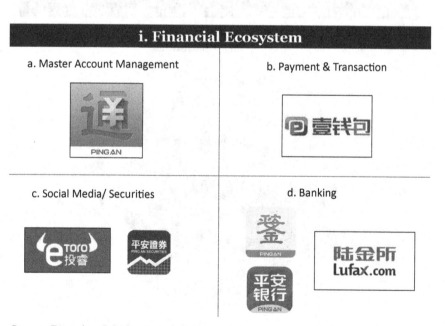

Source: Ping An, Schulte Research Estimate.

i.a. Snapshot: Finance One Account

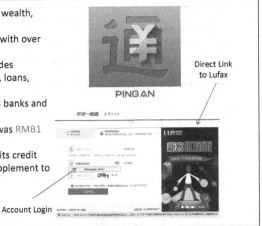

- Develops footprint of account, wealth, credit and life data.
- It has 185 mn registered users with over 30 mn MAU
- Through Ping An cloud, it provides e-banking services, credit rating, loans, and interbank trading.
- In 2016, it cooperated with 258 banks and 1,135 NBFIs.
- The interbank trading volume was RMB1 tn in 2016.
- With ~360 mn credit inquiries, its credit reference system is effective supplement to PBOC.

Direct Link to Lufax

Account Login

Source: Ping An, Schulte Research Estimate.

ii.b. Snapshot: E-Wallet

- E-wallet develops an O2O life and financial service platform integrating prepaid cards with E-wallet app.

- 76 mn registered users with 6 mn MAU

- 2016 transaction volume was 2tr RMB.

- Loyalty points worth ~ RMB 14 bn in 2016.

- It makes money through withdrawal, transaction fees; fees collected from shops which distribute loyalty points through platform.

Source: Ping An, Schulte Research Estimate.

i.c. eToro: discover smarter investment by automatically copying the leading traders in the community.(1)

- Can discuss their "alpha trades" while amateur investors can learn/copy the pros.
- Ping An Venture, together with Sberbank of Russia, invested $27 mn in total during D round financing.
- Linked to Ping An Securities through ONE ACCOUNT, generating 420 mn RMB in stock trading.
- Has a cryptocurrencies copy fund for users to invest in.

eToro   ONE ACCOUNT   Ping An Securities

6M   **Registered Users**

140   **Countries**

220mn **Trades**

370   **Employees in London, Moscow, Shanghai, Tel Aviv & Cyprus**

$62mn **Raised Capital**

Source: Ping An, Ventures, eToro, Rise Conference, Schulte Research Estimate.

i.c. eToro Cryptocurrency Copy Fund: The fund will be distributed among these 6 currencies(2)

Bitcoin is a cryptocurrency and a digital payment system. The system is peer-to-peer, and transactions take place between users without an intermediary. It is the first decentralized digital currency. Besides being created as a reward for mining, Bitcoin can be exchanged for other currencies, products, and services in legal or black markets.

Ethereum is an open-source, public, blockchain-based distributed computing platform featuring smart contract function. Ethereum also provides a cryptocurrency called "ether", which can be transferred between accounts and used to compensate participants for computations performed.

Ethereum Classic is a parallel version of Ethereum. It was started when the DAO project (Distributed Autonomous Organization), which raised $150 mn in Ether, was hacked with fund stolen. Investors who lost their fund were reimbursed with this new kind of currency.

Litecoin is a peer-to-peer cryptocurrency and open source software project. While in most regards identical to Bitcoin, Litecoin has some improvements over Bitcoin such as the adoption of Segregated Witness & the Lightning Network. Litecoin also has almost zero payment cost and facilitates payments approximately four times faster than Bitcoin.

Ripple is a real-time gross settlement system, currency exchange and remittance network. Ripple purports to enable "secure, instant and nearly free global financial transactions of any size with no chargebacks." It supports tokens representing fiat currency, cryptocurrency, commodity or any other unit of value such as frequent flier miles or mobile minutes. Currently, Ripple is the third-largest cryptocurrency by market capitalization, after Bitcoin and Ethereum..

Dash is an open source peer-to-peer cryptocurrency that offers all the same features as Bitcoin with advanced capabilities, including instant transactions, private transactions, and decentralized governance. Dash's decentralized governance and budgeting system makes it the first decentralized autonomous organization.

Source: Schulte Research Estimate, Ethereum, Litecoin, Ripple, Dash.

i.d. Snapshot: Lufax by the Numbers(1)

- 6 mn **investors**
- 2 mn **borrowers**
- RMB 300 bn **retail customer AUM**
- RMB 1.5 tr **retail transaction volume**
- 320 **third party asset providers**
- 600 **outlets (to verify borrower information)**
- 1,800 **employees**, 800 **engineers**
- NPL: 6% for unsecured loans; 1% for secured loans
- **The numbers should grow substantially as Lufax just received licenses from MAS to operate in Singapore.**

Source: Ping An, Schulte Research Estimate.

i.d. Snapshot: Lufax Models & Services(2)

1. **Financing**
   - P2P
   - Online consumer finance
   - Online SME lending
   - Online supply chain finance

2. **Services**
   - Credit scoring
   - Online credit asset trading(securitization)

3. **Investment Management**
   - Online Investment management (Equity + FX)

4. **Secondary Trading Market**
   - Providing liquidity for financial products

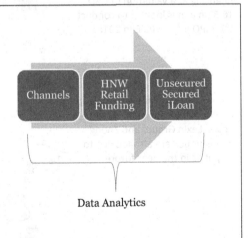

Source: Ping An, Schulte Research Estimate.

i.d. Snapshot: Lufax Advantage(3)

Borrowers Needs
- Fast & efficient application
- Affordable credit
- Risk-adjusted pricing
- 24/7 customer service

Investors Needs
- Attractive returns
- Breadth of products
- Easy-to-use
- Transparency
- Help HNW Chinese do overseas investments

Fintech Platform(Lufax)
- Scalable Platform
- Risk management
- Innovation
- Regulatory Compliance

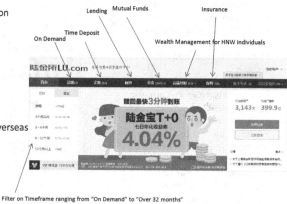

Source: Ping An, Schulte Research Estimate.

i.d. Snapshot: Lufax IPO(4)

- Lufax has secured two rounds of financing at a valuation of USD 18.5 bn and is looking to conduct their IPO in the HKSE in 2018.

- Other fintech giants also looking to IPO: Ant Financial, JD Finance, Ppdai, Lexin Group, and Zhong An Insurance are also looking to go public in the near future.

Source: Ping An, Schulte Research Estimate, WSJ.

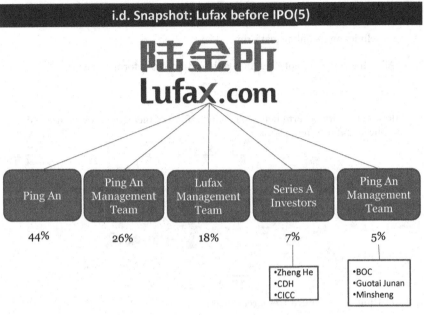

Source: crunchbase, wngdaizhijia, Schulte Research Estimate.

Source: Ping An, Schulte Research Estimate.

ii.a. Snapshot: Wanjia Clinics

- •10k clinics on its online platform

- •All of the clinics accept Ping An health insurance as a form of payment

- •Serve 110 mn users from Good Doctor

- •Resonates with government's upcoming reform of the nation's overburdened public hospital system.

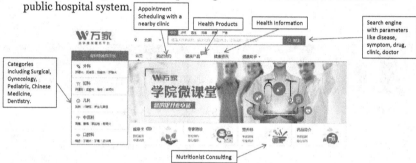

Source: Ping An, Schulte Research Estimate.

ii.b. Snapshot: Jin Guan Jia(Golden Butler) with 29mn MAU

- •Online service for Ping An Life Insurance.
- • 100 mn registered users with 29 mn MAU.
- • Users purchase insurance & investment products on the platform.
- •Supports online doctor appointments so users will process their health data through the App.
- •Makes money through insurance purchases, cash deposit and doctor references.

Ping An Jin Guan Jia
A Trustworthy Financial
Lifestyle Assistant

Policy Management
Manage your insurance policies with 24/7 customer support

Wealth Management
Online financial product market with Wang Cai cash account to compound your wealth.

Community Engagement
Social Impact investment Opportunities and discuss your opinion on Ping An with other policies holders

Health Management
Offers customers with 1-on-1 doctor appointments. Customers can ask for help online regarding disease, dosage, nutrition and mental consulting with the app.

Source: Ping An, Schulte Research Estimate.

ii.c. Snapshot: Ping An Good Doctor(1)

•Provided health management services for 130 mn users; MAU of 26 mn with daily inquiries of 440k (300 per minute).

•Online: Online consulting, appointment making, online medicine purchase, health Podcast, health info, health plans, etc.

•Offline: includes O2O services like health checkups, gene tests, glasses purchase, dental care, door-to-door medical care.

•In the first half of 2016, Ping An Good Doctor completed A-round financing of US $500 mn with valuation of US $3 bn.

(In thousand)	December 31, 2016	December 31, 2015
Registered users	131,500	30,260
Peak number of monthly active users	26,250	9,200
Peak number of daily inquiries	440	120

Source: Ping An, Schulte Research Estimate.

ii.c. Snapshot: Ping An Good Doctor(2)

•Ping An Good Doctor has 1k members and 60k contracted external doctors provide 24/7 online advisory.

•Doctor appointment scheduling at 2k hospitals, and 700 checkup institutions in 150 cities.

• 30k common drugs and dietary supplements.

•It makes money through health insurance purchases, brokerage fee between patients and hospitals and medical drugs purchases.

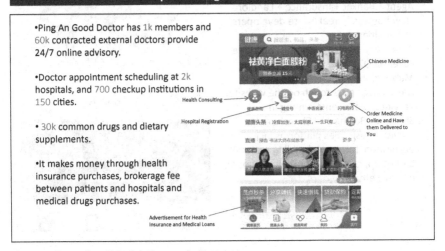

Source: Ping An, Schulte Research Estimate.

iii. Real Estate Ecosystem

a. Real Estate Purchasing

b. Real Estate Renting

c. Online Furniture Sales

Source: Ping An, Schulte Research Estimate.

iii.a. Snapshot: Pinganfang.com(Ping An House)with 10mn Users

- "AI + Finance + Real Estate"
- Life style Manager(Pinganfang.com) = Agent + Banker + Insurance + Landlord
- New homes: 70 real estate developers, in 125 cities
- Pre-owned homes: channels for 5 cities, covering O2O trading
- Makes money when real estate transactions go through on the website.
- Provides apartments renting in Shanghai, Beijing and Shenzhen.
- Provides flash sales platforms including decoration, furniture, building materials, bathroom, ceramics, flooring, cupboard, electric, etc.

New Homes   Preowned Homes

Rental   Overseas RE

Mortgage Calculator

Mortgage

Source: Ping An, Schulte Research Estimate.

iv. Automobile Ecosystem

a. Auto Insurance Management

b. Car Rental Service

宝驾
BAOJIA.COM

c. Post Sale Service(4S)

携车网
xieche.com.cn

d. Online Automobile Market

车惠网

博车网

汽车之家
autohome.com.cn

Source: Ping An, Schulte Research Estimate.

iv.a. Snapshot: Ping An Auto Owner 17mn Users(1)

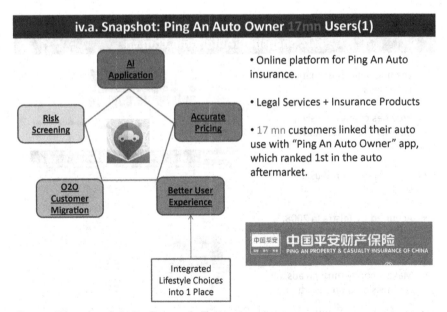

- AI Application
- Risk Screening
- Accurate Pricing
- O2O Customer Migration
- Better User Experience
- Integrated Lifestyle Choices into 1 Place

- Online platform for Ping An Auto insurance.

- Legal Services + Insurance Products

- 17 mn customers linked their auto use with "Ping An Auto Owner" app, which ranked 1st in the auto aftermarket.

中国平安 中国平安财产保险
PING AN PROPERTY & CASUALTY INSURANCE OF CHINA

Source: Ping An, Schulte Research Estimate.

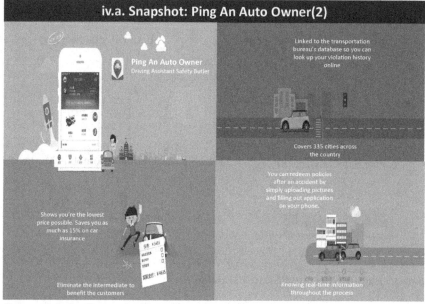

Source: Schulte Research, Ping An.

iv.b. Snapshot: Autohome.com.cn

- Leading online destination for car buyers.

- Provides comprehensive, independent, interactive content.

- 100 mn **registered users**, 10 mn DAU.

- Acquired by Telstra in 2008, went public in 2013 and was acquired by Ping An in 2016.

- Makes money through ads, and commission on transactions.

Source: Ping An, Schulte Research Estimate.

C. What does Ping An do with the Data?

i. Big Data Analytics

ii. Identity Recognition   iii. Blockchain

Source: Ping An, Schulte Research Estimate.

i.a. Customer Profiling

Source: Rise Conference, Schulte Research Estimate.

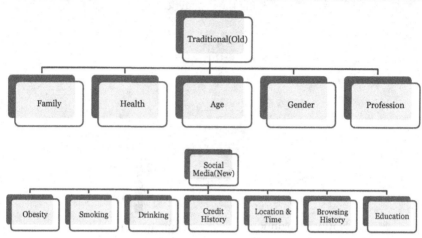

Source: Ping An, Schulte Research Estimate.

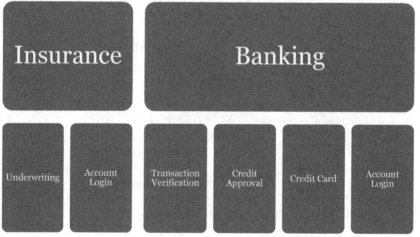

Source: Ping An, Schulte Research Estimate.

ii.a. Facial Recognition at Ping An(1)

• *"Facial recognition is the core of our financial business because we know if you are you and if you have doubts when you come apply for a loan."*

— Ericson Chan,
CEO
Ping An Technologies

•Ping An will incorporate the facial recognition technology into its wealth management and other services to create the fastest lending platform in China, only requiring 6 minutes to complete a loan application.

China PING AN Launches World's First 'Face Recognition Loan' Technology

NEWS PROVIDED BY
PING AN TianXiaTong →
18 Apr 2016, 06:41 ET

SHARE THIS ARTICLE
🅑 🅞 🅞 🅑 🅞 🅞 🅞

WASHINGTON, April 18, 2016 /PRNewswire/ – On April 15, app developer China PING AN TianXiaTong (PING AN) released its new face recognition technology during a launch event held at the 7-star Pangu Hotel in Beijing.

Source: Ping An, Technologies Rise Conference, Schulte Research Estimates.

ii.a. Facial Recognition at Ping An(2): It has data on 108mn PRC Citizens

•At present, human eyes are limited to 72% accuracy when it comes to comparing faces.

•In 2016, Ping An Technology's facial recognition technology attained an accuracy level of more than 99.8%.

•According to the latest result released by Labeled Face in the Wild(LFW), Ping An AI lab yielded the best result among the famous domestic and international competitors.

Rank	Name	Score
1.	PingAn AI Lab	0.998 +- 0.0016
2.	Youtu Lab, Tencent	0.998 +- 0.0023
3.	Dahua FaceImage	0.997 +- 0.0007
4.	THU CV-AI Lab	0.997 +- 0.0008
5.	SamTech Facequest, Siemens	0.997 +- 0.0018

Source: Ping An TianXia Tong, 10Tia.com, LFW, Schulte Research Estimates.

ii.b. Snapshot: Voiceprint Recognition for Account Login and ID Verification

Voiceprint Recognition used by Lufax, P&C	
95% Accuracy Rate	**30+ mn** Telephone Data Training Sample
18 seconds Registration	**10 seconds** Recognition

Source: Ping An, Rise Conference, Schulte Research Estimates.

iii. Blockchain at Ping An: Fraud Detection

Ping An is exploring blockchain opportunities in 10 areas:

1. Insurance Accidents
2. Mortgage
3. Credit
4. Supply Chain Finance
5. Cross-border payment
6. Bills
7. Inter-bank Transfer
8. P2P
9. Electronic Prescription
10. Loyalty Points

•Ping An became the first Chinese member of R3 in May 2016.
•Since then, Ping An's blockchain initiative has assembled a team of 15 people from Ping An Technology and finance division.

Source: Ping An, Schulte Research Estimates.

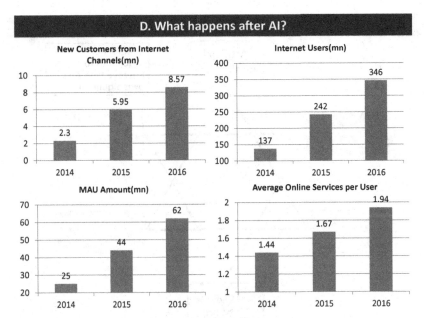

Source: Ping An, Schulte Research Estimates.

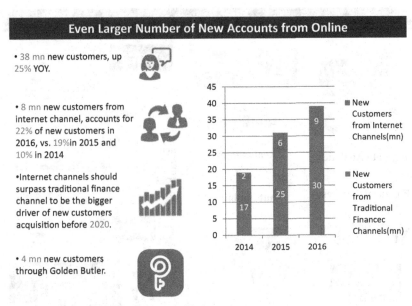

Source: Ping An, Schulte Research Estimates.

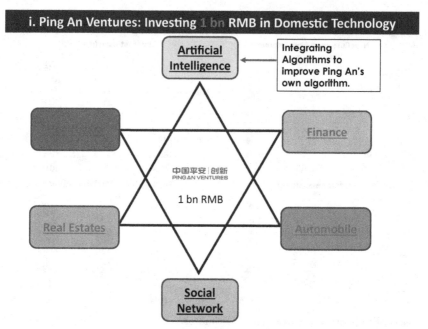

Source: Ping An, Ventura, Schulte Research Estimates.

Source: Ping An, Ventures, Schulte Research Estimates.

i.a. Oscar Insurance: a personal health management and insurance company

•Ping An Ventures participated in the $400 mn private equity deal to acquire equities of Oscar with Founders Fund, General Catalyst, Khosla Ventures, Thrive Capital, etc.

•No. 4 on KPMG's *Global Fintech 2016*

•Massive international users data can be crucial to Ping An's international expansion and its product-client matching algorithm will help Ping An's internal development.

•Jared Kushner's brother, Joshua Kushner is one of the founders

Source: Ping An, Ventures, Crunchbase, Schulte Research Estimates.

ii. Global Voyager Fund: $1bn Intl Fund

•In May 2017, Ping An set up a $1bn outbound technology acquisition fund led by the new CIO Jonathan Larsen, former Global Head of Retail Banking & Mortgages @ Citi.

•99% of the fund will be distributed outside of China.

•It will be a minority stake model.

•Portfolio Composition:

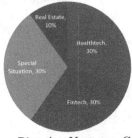

Source: Ping An, Ventures, Crunchbase, Schulte Research Estimates.

III. Appendix

Ownership(Notable Aggregation)

Name of Shareholders	Shareholding Percentages	Type of Shares
Shenzhen Government	5.27%	A Share
CP Group	9.59%	H Share

•The **Charoen Pokphand** Group (CP) is a Thai conglomerate company.

•It is the largest single shareholder in Ping An Insurance and a major shareholder in CITIC Group of China.

C.P.GROUP

Source: Ping An.

Line of Business(Strategic Investment)

Name	Principle Activities	Ownership
Ping An Venture	Venture Capital	99%
Ping An Voyager Fund	Venture Capital	100%
Ping An Real Estate	Asset Management	100%
Ping An Financial Leasing	Financial Leasing	100%
Ping An Commercial Property Investment	Real Estate Investment	77%
Shenzhen Ping An Real Estate	Real Estate Investment	100%

Source: Ping An.

Ping An Ventures Big Data Analytics Portfolio Companies

Celebrities' Wardrobe is a Pinterest-like App that provides structural fashion data to guide customer purchases. http://www.hichao.com/	Jike is a APP with content aggregation and personalized recommendations. Users can follow with interested people, news and events. Jike will track the update automatically. The users can receive their own information timely and efficiently. http://www.ruguoapp.com/	Cubee is a sport big data platform that provides a myriad of services for football fans and football-lottery customers. http://cubee.com/

Source: Ping An, Ventura.

Ping An Ventures Finance Portfolio Companies

WALLSTREETCN.COM provides financial information for market investors and professionals. https://wallstreetcn.com/	eToro is the world's largest social investment network. Users can join millions of traders who discovered smarter investing by automatically copying the leading traders in the community. http://events.pingan.com/etorobd/index_5.html?keywords=brand021

Source: Ping An, Ventura.

Ping An Ventures Automobile Portfolio Companies

SCAR is an Internet-based new car retailer. It can provide one-stop services such as group purchase, discounted booking, car insurance and loans. http://www.scar.com.cn/	Boche is a repairable accident car auction platform. The pricipals includes insurance companies, car rental companies and financial leasing companies. http://www.bochewang.com.cn	Baojia provides P2P car rental services based on internet. It will provide car rental insurance for users. http://www.baojia.com/

Source: Ping An, Ventura.

Ping An Ventures Healthcare Portfolio Companies

Oscar Health Insurance is a personal health management and insurance company, which provides health insurance plans mainly for personal account. The company improve transparency and effectiveness during the insurance matters. http://mp.weixin.qq.com/s?biz=MzA5MTYyNTgxOA==&mid=402267227&idx=1&sn=463d933a13ee6feca61fd67aa1921269#rd	Glusense is a company developing subcutaneously implanted glucose monitoring system, incubated by Israel incubator Rainbow Medical. http://www.glusensemedical.com/	Rani Therapeutics has developed a novel approach for the oral delivery of large drug molecules including peptides, proteins and antibodies. http://www.ranitherapeutics.com/

Source: Ping An, Ventura.

Ping An Ventures Healthcare Portfolio Companies

解码DNA®
——JMDNA.COM——
基 因 检 测 专 家

Ativa
M E D I C A L

Zhangshang Tangyi provides diabetes management services., helping diabetics to monitoring blood glucose and providing management services. http://www.91jkys.com/	JMDNA is a genetic testing company that aims to become the best brand of genetic testing and health management with such services as personalized drug prescription, disease screening and talent gene testing. http://www.jmdna.com/	Ativa Medical Corporation has not only recognized the void for comprehensive POC testing, but has also committed to delivering an affordable diagnostic solution to decentralized healthcare settings all over the world. http://www.ativamed.com/

Source: Ping An, Ventura.

Ping An Ventures Healthcare Portfolio Companies

Rogrand 融贯资讯

转化医学网
www.360zhyx.com

Medbanks
Medbanks Network Technology Co.,Ltd

Rogrand is a drug electricity supplier provides integrated services. Their goal is to"care life, guard health". http://www.rograndec.com	360zhyx, using molecular biotechnology to transform laboratory research results to clinical application of products and technologies. https://www.360zhyx.com	Medbanks, helping oncologists to improve efficiency, has built data structure of almost every tumor. http://www.sipaiwangluo.com

Source: Ping An, Ventura.

Ping An Ventures Healthcare Portfolio Companies

Nanopep is an early cancer screening and personalized diagnosis service provider. Its hige-tech: TumorFisher, can detect 1-5 mm tumors.
http://nanopep.com/

36 Kangzhi is a social Q&A site focus on healthcare. It has many columns including Reading, Q&A, Group, Interview, Public Welfare, Activity, etc.
http://www.kangzhi.com/search/康知网/

Purple Clinic provides multi-department follow-up consultation, providing online practitioners platform for doctors.
http://www.ziseyiliao.com/

Source: Ping An, Ventura.

Ping An Ventures Healthcare Portfolio Companies

UpFit is a platform connecting athletic directors and users online/offline, helping enterprises, personal and professional users to build exercise schemes.
http://www.upfitapp.com/

Jianmeng is an incubating and crowd funding platform for healthcare projects.
http://www.jianmeng.org.cn/

HealthcareCN is a news media platform that aims to become "China Healthcare Think Tank". It provides vertically-integrated websites for hospital administrators, community for doctors and other products like the mobile newspaper and mobile app.
http://www.cn-healthcare.com/

Source: Ping An, Ventura.

Appendix 3

Tencent

AI: The Big Picture

"Cloud computing will be the prime power of this industrial revolution. Cloud computing plus artificial intelligence is equal to electricity plus computers"

— Ma Huateng, CEO of Tencent

Tencent Development

Tencent Holdings Progression

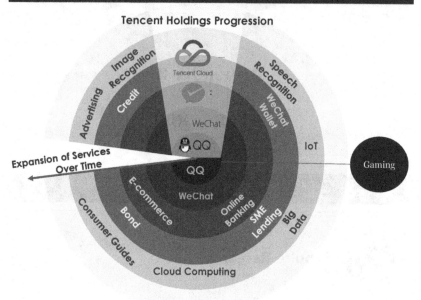

Source: WeChat, Schulte Research.

Brief Overview

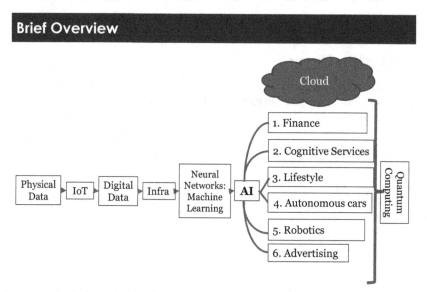

Source: Schulte Research Estimates.

Tencent Has Exceptional Foundation Which is World Class

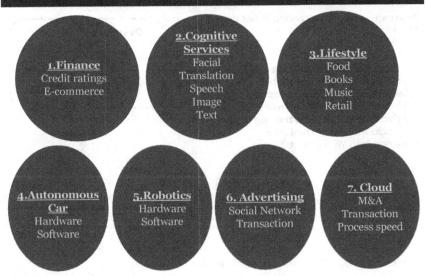

Why AI?

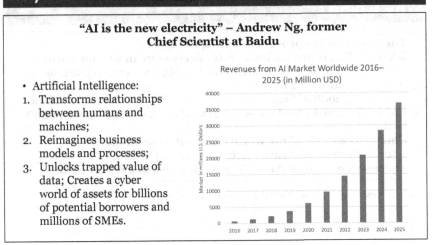

"AI is the new electricity" – Andrew Ng, former Chief Scientist at Baidu

- Artificial Intelligence:
1. Transforms relationships between humans and machines;
2. Reimagines business models and processes;
3. Unlocks trapped value of data; Creates a cyber world of assets for billions of potential borrowers and millions of SMEs.

Revenues from AI Market Worldwide 2016–2025 (in Million USD)

Source: Statista, Schulte Research Estimates.

Why Now?

Advances in technology and lower costs have laid the ground work for AI revolution.

- **Cost of connecting is decreasing**
- **Rapid growth of cloud/mobile Technology costs are collapsing**
- **Smartphone penetration rates up**
- **Advancement in analyzing Big Data**
- Broadband widely available
- Sensors and Wi-Fi capabilities

Average cost of one gigabyte, 1956–2010

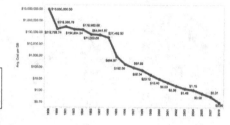

Source: Statista, Schulte Research Estimates.

Why China?

1. **The development process is getting cheaper.**
2. **R&D funding is increasing in PRC, decreasing in US and Europe**
   - China's 13th Five Year Plan: $1.2 tn into R&D; US decreasing R&D by 10%.
3. **China is betting big on IOT.**
   1. By 2020, 200B IOT connected devices globally; 95% produced in China.
4. **China's internet penetration rate is only at 52%, compared to 89% in the US.**
5. **China boasts 43% of the world's trained AI scientists**
   - Unstable visa policies in US force foreign talent to leave.
6. **There was no "there there" infrastructure was a fallow field**

Source: New York Times, Schulte Research Estimates.

Why China? Demographics Forces Automation

6. China's labor intensive economy has a problem: its population is aging and retiring
- China's dependency ratio, (those not working against those that who are working) could rise to 70% by 2050 versus the current 34% (Statista).
- With a falling labor force, China is seeking forms of automation to keep its economy growing.

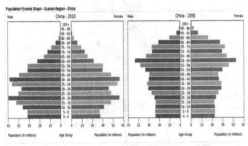

China's population pyramid (left, 2010) shows a large younger and middle aged population. By 2050, the population (and labor force) in China will have fallen.

Source: Statista, Forbes, Schulte Research Estimates.

Intentioned & Unintentioned Data will Blend into each other

Intentioned data: data collected based on online purchases

- Purchasing patterns allow companies to directly predict what products users will order next.

Many think that, intentioned data is more valuable because it is simple to discern, process and predict. But...

Unintentional data: data collected based on random communication about plans, fashion, tastes, ideas, entertainment, sexuality.

- Using text/photo recognition and sentiment analysis between two users of an app.

People may underestimate the power of unintentioned data to actually offer more insight than receipts. Underlying (and poorly understood) secret of AI.

Source: Schulte Research Estimates.

Tencent and AI

Tencent: Largest Unintentioned Platform Ever Created

- Through social platforms and digital contact services such as WeChat, Tencent has created China's largest internet community (938 mn users)
- With 50% of its employees in R&D, Tencent is continuously at the forefront of the internet industry.

> **55% of all time spent on mobile is spent within Tencent products**

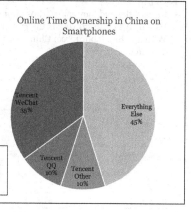

Online Time Ownership in China on Smartphones

Source: Forbes, Schulte Research Estimates.

Tencent: WeChat

- 938 mn users: 86% check it more than once a day
- Average user spends more than an hour per day on WeChat

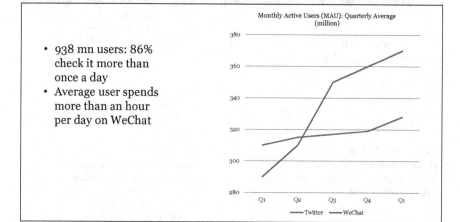

Source: Financial Times, Schulte Research Estimates.

WeChat's Key Functions: A Unified Ecosystem

Key Functions	Explanation	US Equivalent
Instant Messaging (微信)	Text, voice, photos, videos	WhatsApp
WeChat Moments (朋友圈)	Photo/text/article sharing with friends	Facebook
Wallet (微钱包)	$ transfer, red packet, in-store payments, bill/ticket payments, e-commerce, donation, etc.	Venmo
Wealth Management (腾讯理财通)	Investment fund options	Acorns
Official Accounts (公众号)	>10 mn accounts (blog idea sharing, media outlets, brand marketing, public service. New form of celebrity branding	Blogging

Source: Tencent, Schulte Research Estimates.

WeChat's mobile app is so well integrated, users are sucked into its ecosystem (page 1)

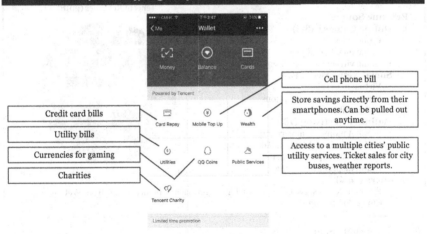

Source: Tencent, Schulte Research Estimates.

WeChat's mobile app is so well integrated, users are sucked into its ecosystem (page 1)

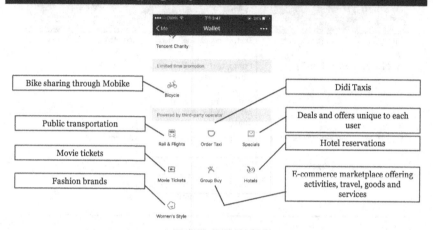

Source: Tencent, Schulte Research Estimates.

Tencent's Revenue Breakdown

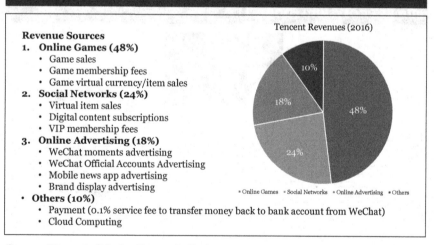

Source: Tencent, Schulte Research Estimates.

Tencent's Revenue Jump

- Tencent's Q1 2017 overall revenues surpassed expectations
  - **Tencent already has higher revenues and profits than Alibaba and Baidu**
- Revenue potential in cloud computing as an infrastructure for artificial intelligence

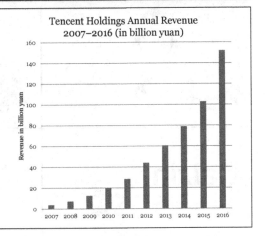

Source: Tencent, Schulte Research Estimates.

Q1 Revenues of Competitors

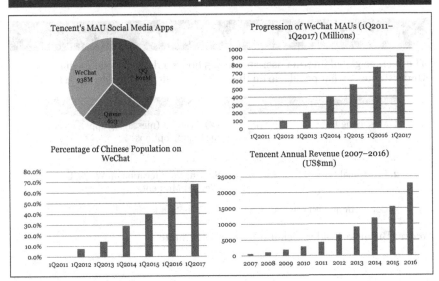

Source: China Internet Watch, Statista, Schulte Research Estimates.

Tencent's Revenue Jump

1. Social Network	2. Gaming 
3. History	4. E-commerce and lifestyle

Source: Tencent, Schulte Research Estimates.

a. Tencent: Social Network Data

The average global internet user spends 2.5 hours a day on social media. Understanding what people do on these sites reveals what makes them tick.

	Source
1. Messaging Data	QQ, WeChat (messaging apps), voice messaging options
2. Locational Data	Sharing your location on WeChat WeChat Moments
3. Connections with friends	"Add" option for friends

Source: Business Insider, Schulte Research Estimates.

b. Tencent's Most Popular Games

 League of
Legends: 100 mn
MAU

- Gaming data allows Tencent to
see what games are played the
most often and which are most
profitable.

 Honour of Kings:
200 mn MAU
- Tencent has recognized the
importance of socializing in
gaming through this.

- Honour of Kings: a game that
allows players to communicate
within the game.

 Clash of Clans:
100 mn MAU

Source: SCMP, Schulte Research Estimates.

c. Tencent: Historical Data Finding the Good and Bad in People

- Understanding past actions is key in developing credit scores based on social media.
- i.e., Photos posted on WeChat
  - Inappropriate behavior
  - Always drinking
  - Timing of photo posting after 10 PM? Trouble

- The collection and use of this data is important in understanding users' actions.

Good: Do you,
- Give to charity?
- Pay your parents' bills?
- Go to the gym at the same time daily?
- Act and purchase consistently?

Good: Do you,
- Post photos of hangovers
- 3 AM ATM withdrawals
- Frequent item returns?
- Contact previous girlfriends?

Source: SCMP, Schulte Research Estimates.

How does Tencent Collect Its Data?

- 938 mn users in Q12017 on WeChat **alone.**
- Data on Tencent Cloud allows it to collect real-time data, and respond immediately.
- Tencent's user generated content is a data gold mine for:
  - Marketing and advertising
  - Finance
  - Ecommerce (JD.com)
  - Healthcare
- More products are being introduced, thanks to the sticky nature of WeChat. Virtuous circle.

- **Data makes the business; not the other way around.**

Source: SCMP, Schulte Research Estimates.

Finance

AI & Finance. Question: How Will Tencent Transform Unintentioned Data into International Financial Data?

- **Computers can help us make decisions in this new world of immense amounts of data and many moving parts**

- AI: "augmented" intelligence to help make better financial decisions.

- Traditional financial institutions are becoming redundant. Bankers will need to learn new skills and join a new ecosystem.

- Traditional jobs of:
  - Assessing assets and funding them correctly
  - Digitalizing these assets so they can be counted correctly
  - Assessing asset flow so they can be valued correctly
  - Anticipating and controlling these assets so they can be traded correctly

- AI: ability to digitally collect data and analyze it better than an analyst.

Source: Schulte Research Estimates.

a. WeBank's Business Model

Business Model: Charges platform service fees (0.1%)

1. WeBank provides customer access, pricing through behavioral algorithms
2. Partners: banks with excess liquidity (i.e. Shanghai Bank)
3. Bank partners provide most funding and bear credit risk (not WeBank)

Benefit for bank partners:
- Access to mass customers through Tencent's social platforms
- Access to risk ratings for individuals off of AI & data analytics
- This is too often from the "too difficult" box for banks

Benefit for WeBank
- "Light" balance sheet business model. Credit risks with bank partners. Income from the platform service fees

WeChat is learning how to become a bank.

Source: WeChat, Schulte Research Estimates.

WeBank's Business Model

- WeBank provides the algorithm for risk analysis of users in its app, and connects those users to the banks.

- Banks provide the capital and receive the interest payments, but hold all of the risk.

- AI is used to create the credit rating. Banks are using these products not only because of the access to an immense part of the Chinese population, but also because the algorithm is very accurate.

Source: WeChat, Schulte Research Estimates.

WeBank's Features

- Weilidai is a "connect platform": customer insights, data, supply and demand matching, sourcing fund, risk control, value-added services.
- Excludes IOE (IBM, Oracle, ECM), built entire system internally
- Banks using IOE need to pay RMB ~60 per account p.a. just for maintenance
- Purely online. 50% of loans issued while traditional banks are closed
- Small amount for single loans, but tremendous with 932 mn users on WeChat

Big Data enabled:
- Behavioral data
- Payment data
- Financial data
- Location data
- Demographic data
- Interests and hobbies

Good User experience:
- Intelligent service
- Personalized service
- Predictive service
- Proactive service

Source: Company Data, Schulte Research Estimates.

b. WeBank: Weilidai

> **Key Features of Weilidai (微信)**
> - Total loans: RMB 100 bn
> - Whitelist — invitations only
> - Loans of RMB 500–300,000 to be repaid in 5, 10, and 20 installments
> - Borrow and pay back anytime
> - Entirely digital, 24/7 customer service available
> - Three-step application process, instant credit
> - Funds released in <1 minute
> - Risk-based pricing
>
> **Common Problems with Traditional Consumer Loans**
> - Not flexible (amount, repayment, interest method)
> - Minimum loan requirement may be too high
> - Complicated process
> - Credit conflicts

Source: WeChat, Schulte Research Estimates.

Tencent's Backing of Element AI

- In June 2017, Element AI (Montreal), a Tencent-backed machine learning company, raised a historic US $102 mn in its Series A funding round.
- This will likely be applied to WeChat's financial services product:

- **Bottom Line: Element AI will be far better at pattern matching and prediction than humans.**

- Except for a few leading hedge funds, many companies have not recognized the potential in machine learning and its potential to drive investment decisions.

- These AI Functions can: (1) Pick stocks, (2) do HFT, (3) create portfolios for a fraction of the price, (4) trade technical.

> **Machine learning will overtake traditional asset management. Humans are too risky. Tencent is getting ahead this way.**

Source: Element AI, Schulte Research Estimates.

Tencent is Establishing a FinTech Lab, should we assume Tencent will merge with BOC

- **The banks don't want to be left behind, and are partnering with technology companies to get ahead.**

- The Bank of China (BOC), one of China's big four state-owned lenders, is partnering up with Tencent to establish a joint financial technology laboratory.
- According to the Bank of China, the lab will work in:
  - Cloud computing
  - Big Data
  - Artificial intelligence
- The companies aim to create a unified platform of financial Big Data.

- **This will be through WeChat. With this joint laboratory, the BOC will have access to Tencent's user data.**

Source: Element AI, Schulte Research Estimates.

Big Banks are Blending with Tech Giants

- Will CCB be blended into Alibaba? Will BOC be blended into Tencent?

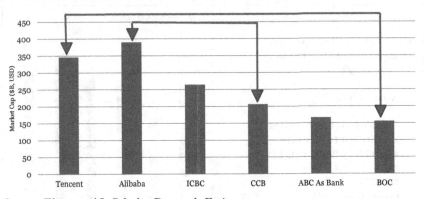

Source: Element AI, Schulte Research Estimates.

c. Tencent: Credit Ratings

- 33% people 18–29 yrs have a credit card. Of 1,000 Chinese millennials in a study by The Disruption House, **all of them used WeChat once a day** (Disruption House Study).
  - Sesame Credit, Tencent, Baidu, Xiaomi, NetEase, Qiho 360, and Didi Chuxing, proposed to jointly set up an individual credit scoring company in July 2017.
  - Credit will be based on non-traditional facets.
  - Tencent also recently ramped up operations with China Rapid Finance.

Company	Purpose
Sesame	Alibaba Everything
Tencent	WeChat, QQ, Qzone
Baidu	Search
Xiaom	Cell Phone
NetEase	E-Commerce
Qiho360	Antivirus
Didi	Taxi

Source: SINA, Business Insider, Forbes Schulte Research Estimates.

d. E-commerce

- China's retail e-commerce sales predicted CAGR to 2021 is 19%.

- A cashless society.

- China's e-commerce sales are forecast to be worth $840 bn by 2021. This is 2x the size of US e-commerce in the same period ($485 bn).

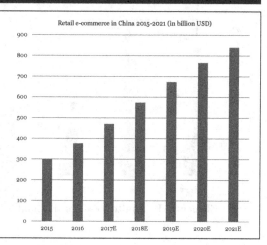

Retail e-commerce in China 2015-2021 (in billion USD)

Source: Business Insider, Schulte Research Estimates.

E-commerce

- China's mobile payments market is 50 times larger than the US.

- Alibaba and Tencent are fighting for control of the $5.5 tn market.

- By forging closer ties to other e-commerce platforms, Tencent and Alibaba will increasingly absorb the offline market.

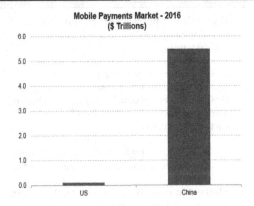

Source: iResearch China, Schulte Research Estimates.

E-commerce

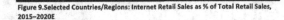

Figure 9.Selected Countries/Regions: Internet Retail Sales as % of Total Retail Sales, 2015–2020E

China's e-commerce is growing at a much more rapid rate than anywhere else in the world, and it is expected to continue to do so.

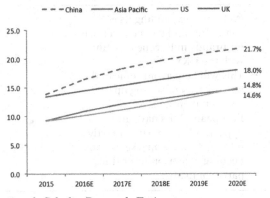

Source: Euromonitor International, Schulte Research Estimates.

Tencent: E-commerce

- Tencent's 15% purchase of JD.com (Jingdong) has fully completed the ecosystem within the WeChat app.
- It is now trying to expand outside of China.
- WeChat has a unique position as a top M-CRM* tool for global brands gives.
- WeChat Pay has promising potential to become a primary payment method for Chinese customers shopping abroad.

- While AliPay saw higher transaction amounts, WeChat Pay had almost double the number of active users.
- **There are roughly 1 mn transactions per minute within the app.**

*Mobile Customer Relationship Management

Source: Financial Times, Schulte Research Estimates.

Tencent: E-commerce

- WeChat is expanding its e-commerce and payment services for brands in Europe (starting with the UK).
- This will help avoid some of the bureaucracy of setting up its own retail options.
- European SMEs have an opportunity to expand directly into the Chinese market without needing a distributor or third-party company.

Source: Financial Times, Schulte Research Estimates.

Tencent: E-commerce

- WeChat has built an enormous ecosystem of public accounts of brands and independent publishers.
- Brands can now simply push notifications for sales and discounts, encouraging in-app purchases based on past purchases.
- WeChat will likely take lessons from social media platforms like Facebook on targeting, profiling, and bidding functionality.

Source: Financial Times, Schulte Research Estimates.

Tencent: E-commerce and Smart Homes

JD.com, Tencent's partner in e-commerce that is heavily integrated into the WeChat ecosystem, has begun to back iFlytek.co.

- The company focuses on smart home appliances and voice-control solutions.
- While JD.com works towards creating a channel for its sales and marketing, Pony Ma and the Tencent team are working towards connecting devices with WeChat.

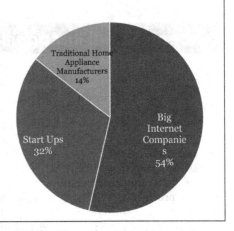

Source: Tencent, Schulte Research Estimates.

Cognitive Services

a. Tencent: Facial Recognition

- Tencent's Youtu Lab has set a new record of achieving an 83.29% recognition rate in the MegaFace Challenge.
- Test is based on 2 mn Western faces and 4 mn Asian faces.
- Youtu Lab has helped 10 families find missing members within 24 hours in Fujian Prince since March 2017.

Algorithm	Date Submitted	Set 1	Set 2	Set 3
Youtu Lab (Tencent)	April 2017	83.29%	83.27%	83.30%
DeepSense V2	January 2017	81.29%	81.30%	81.30%
Vocord	December 2016	80.25%	80.20%	80.20%

Source: PR Newswire, MegaFace, Schulte Research Estimates.

Tencent: Facial Recognition

- WeBank uses facial recognition to verify the identities of its users
- Allows users to bunch traditional bank accounts and WeChat bank accounts together.
  - Part of the loan taking process

Source: PR Newswire, MegaFace, Schulte Research Estimates.

Lifestyle

Lifestyle

- **Tencent has invested heavily in lifestyle to create a full ecosystem within WeChat. This includes:**
  - Transportation
  - Food
  - Music

- Major investments and third-party integrations into WeChat have created an ecosystem so that **the user can use WeChat for most lifestyle purchases.**

- In comparison to Google, Tencent excels in lifestyle.
- Google may have Google Hangout, but no one uses it. Tencent has WeChat Messenger, which is almost used by a billion people. **WeChat is fully integrative.**

- Though it has plenty of data to collect, consumer preference data and its use is the edge that they have over Google's ecosystem.

Source: Schulte Research Estimates.

a. Tencent: Food

- **Tencent also backed and led a $3.3 bnfunding round for Meituan Dianping, a ticket purchasing and restaurant reviewing platform.**
- Through this, Tencent has entered and dominated the O2O industry, as Meituan Dianping holds 80% of this $100 bn market.

- **Ele.me, a Shanghai based food delivery service, raised $350 mn in Series E funding from Tencent, JD.com, and Dianping.com, amongst others.**
- Because these three are so closely aligned, mutually invested in one another, and integrated into the WeChat ecosystem, the investment will help build their online-to-offline (O2O) strategies.

Source: Schulte Research Estimates.

b. Music: QQ Music & Joox

- QQ Music is Tencent's mainland music streaming and download service.
- Integrated into Tencent's ecosystem, the app provides data on preferences of its users.

- Joox is another of Tencent's apps that is focused on music streaming outside of China and specifically in Southeast Asia.
- As a significantly newer product, Joox aims to collect data on users in Hong Kong, Thailand, Malaysia, and Indonesia

50 M	Downloads in 2016
50 %	Of All Music Streaming in the countries listed above

Source: Analysys International, Tech Crunch, Schulte Research Estimates.

Infrastructure

a. Tencent M&A

Since 2012, Tencent has spent $62.5 bn in acquisitions, outspending both Alibaba and Baidu.

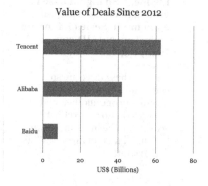

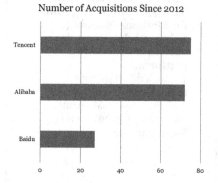

Source: Schulte Research Estimates.

b. Tencent: Processing and Transaction Speed

Cloud computing services provide the infrastructure for artificial intelligence. **Tencent Cloud recently sorted 100 terabytes of data in 99 seconds, beating the world record previously set by Alibaba Cloud in 2016.**

Sort Benchmark Competition	2015 World Records (Alibaba)	2016 World Records (Tencent Cloud)	2016 Improvement
Daytona Graysort*	15.9 TB/min	44.8 TB/min	2.8x
Indy Graysort*	18.2 TB/min	60.7 TB/min	3.3x
Daytona MinuteSort	7.7 TB/min	37 TB/min	4.8x
Indy MinuteSort	11 TB/min	55 TB/min	5.0x

*GraySort: TB/min achieved while sorting 100TB of data
MinuteSort: Amount of data that can be sorted in 60 seconds or less
Daytona (stock car): sort code must be general purpose
Indy (Formula 1): need only sort 100-byte records with 10 byte keys

Source: IBM, Sort Benchmark.

Marketing

Tencent: Advertising

- **97.2% of Facebook's $27.6B in revenue came from advertising in 2016 while this was only 17.7% of Tencent's.**

- Even though the company's online advertising revenue has increased by 485% since Q1 2014, it is a small revenue source.

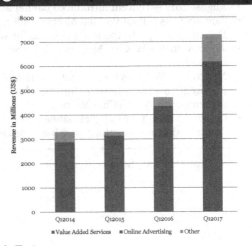

Source: Tencent, Schulte Research Estimates.

Tencent: Programmatic Advertising

- Programmatic Advertising automates the decision-making process of media buying by targeting specific audiences and demographics.
- Uses AI and real-time bidding (RTB) for advertising
  - 66% increase in click through rate (CTR)

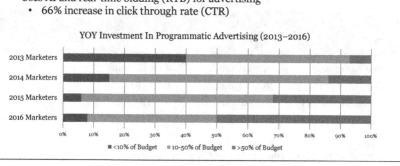

YOY Investment In Programmatic Advertising (2013–2016)

Source: Tencent, Schulte Research Estimates.

Tencent's Advertising Timeline

- August 2014 – Tencent introduced banner advertising at the end of articles
- January 2015 – First WeChat Moment Ads were introduced
- December 2015 – Introduced video ads for WeChat Moments
- January 2016 – WeChat Moment ads became manageable directly from backend
- May 2016 – Coupons for WeChat Moment ads are introduced
- August 2016 – Local-hosted Moment ads (原生广告) introduced
- September 2016 – WeChat Moments ads for local businesses are introduced
- October 2016 – Two-way pick banner advertising
  - Introduction of performance based pricing

Source: WalktheChat, Schulte Research Estimates.

Tencent: Programmatic Advertising

- Since WeChat has started to place ads in the "Moments" section.
- Users can be targeted for ads based on location, age, gender, interest, and device
- WeChat also has "Banner" advertising, (featured at the bottom of a message written by a WeChat official account)
- These banners also appear based on previous ads that the user has clicked on.

In 1Q2017, ad revenues at Tencent jumped 45% because of their recent push

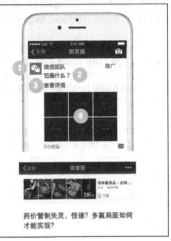

Source: WalktheChat, Schulte Research Estimates.

Tencent: Marketing and Advertising Revenue Potential

- Automated analyses of fashion trends in China.
- After processing billions of photos in Qzone and Moments, Tencent's research lab can identify popular fabrics/colors being worn by youth down to RGB* values.
- AI team: "light black," was the most popular color at the time (users prefer patterns).
- This information can be monetized quickly. This information will be worth $$ to clothing brands looking to expand into the Chinese market.

- Done through image recognition and the ability to sort real-time information quickly + accurately via the Tencent Cloud.

*Red Green Blue, color scale of over 16 mn possibilities.

Source: WalktheChat, Schulte Research Estimates.

Tencent's Data Collection: Remark Media

- In early 2017, Tencent began "Data Eco-System in Tencent Cloud" Campaign.

- Tencent and Remark Media, Inc., a global digital media company aimed to target the Millennial market, started collaborating in 1Q2017 in an effort to develop data-driven precision marketing solutions.
  - These will be powered by Remark's KanKan Data Intelligence Platform, and Tencent Cloud.

  - Help address sophisticated marketing needs and solve some of the issues in predictive analysis and precision marketing.

Source: Salesforce, PR News Wire, Schulte Research Estimates.

Tencent's Data Collection: AdMaster

- Tencent is also collaborating with AdMaster, China's leading data solution provider that specializes in data collection, analysis, and management.
- Together, they are launching the "People-based Measurement Solution (PMS)."
- Turning big data into smart data through AdMaster's SaaS platform, Tencent will be able to outguess humans.
- With more than 900 mn users and AdMaster's accumulated advertising measurement data in the past 10 years, they will use key metrics such as:
  - Gross rating points (GRPs)
  - Unique Audience
  - Reach
  - Frequency
- Will create a better advertising experience across computers, tablets, and smartphones.

Source: Salesforce, PR News Wire, Schulte Research Estimates.

Tencent's Future of Marketing and Advertising

- The use of AI in marketing and advertising still has immense potential in the marketing and advertising worlds.
- With increasingly complex algorithms, researchers will be able to access, combine, and filter multiple data sources. Gives an instant and deep understanding of individual preference.
  - AI will be used to:
    - Personalize offers;
    - Deliver connected and tailored consumers across channels;
    - Power branded chatbots that interact with consumers and provide relevant services via applications.

Programmatic advertising is due to account for **67%** of global display advertising by 2017.

Source: Martech Advisors, Schulte Research Estimates.

Tencent's Future of Marketing and Advertising

Though the US brought in $180 bn in global ad revenue, the Chinese market is closing the gap.

Factors that contribute to this are an increase in per capita GDP and improving online connectivity.

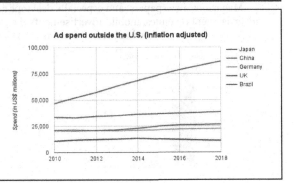

Ad spend outside the U.S. (inflation adjusted)

Source: Salesforce, PR News Wire, Schulte Research Estimates.

Tencent's Future of Marketing and Advertising

Mobile's share of global ad spend will increase by 11.7% in the next four years while other major forms of advertisement are decreasing. This is because of the personal appeal that mobile advertisements can display towards the user.

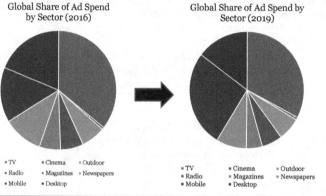

Global Share of Ad Spend by Sector (2016)

Global Share of Ad Spend by Sector (2019)

Source: Zenith, Schulte Research Estimates.

Tencent's Future of Marketing and Advertising

Magazines, newspapers, and especially desktops are expecting reduced advertisement revenues, mobile advertisements (i.e., in apps like WeChat) are expected to see an $8 bn increase by 2019.

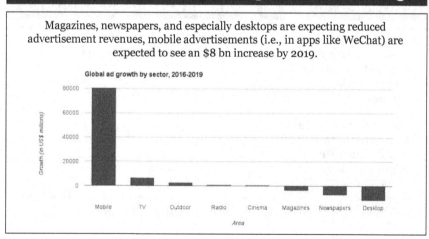

Global ad growth by sector, 2016-2019

Source: Zenith, Schulte Research Estimates.

Tencent Pros

1. Voice messaging allows for Natural Language Processing (NLP)/Speech Recognition.
2. Photos posted on "Moments" can be used for trends via image rec.
3. Strong platform for marketing. Companies are jumping at the opportunity to advertise on "Moments" and "Banners." Increase in ad rev.
4. Stepping away from companies like <u>Baidu</u> to advertise.
5. Fastest transaction speeds and cloud processing.
6. Fully integrative ecosystem allows them to integrate 3rd party providers and benefit from the data.

Source: Zenith, Schulte Research Estimates.

Tencent Cons

1. Companies like Alibaba have a further global reach.
2. Lacking much intentioned data:
   • Amazon and Alibaba have intentioned data that can be used simply.
3. Not involved in multiple AI sectors:
   • Autonomous driving
   • Robotics
4. Cloud business tiny compared to Alibaba.
5. Still very behind in financial services compared to Alibaba.

Source: Zenith, Schulte Research Estimates.

Cloud Comparison

• Amazon – Cloud is #1 globally with 33% market share
• Microsoft – Largest growth rate, 100% YOY
• Google – 45 terraflops, 4% market share in the US
• Tencent – 7.34% market share
• Alibaba – 40.67% market share in China, leader in Asia
• Baidu – Small cloud service
• Facebook – No cloud
• Apple – Relatively small but good at collecting private data, only for personal uses

Source: Zenith, Schulte Research Estimates.

Data Source Comparison

- Amazon – For information about consumers and 2 mn SMEs. IoT data from Whole Foods, Echo, and Kindle. Missing social network data
- Microsoft – Strong in business use. Office. Weak in collecting mobile data. Lacking consumer data.
- Google – #1 in mobile operating systems and map data. 2 bn users a month on cloud and Gmail. 2.2 bn on Android. 1B on Maps. Collect consumer data. Chrome, Snapchat 166 mn MAU.
- Tencent – Data from WeChat, QQ, Qzone, WeChat Pay, Didi, entire WeChat ecosystem. 938 mn users on WeChat, 700 mn QQ. Unintentioned data. Receives intentioned data from third party providers integrated into ecosystem, i.e. JD.com
- Alibaba –
- Baidu – Largest search engine in China. ~700 mn users. APIs, ba
- Facebook – No cloud
- Apple – Relatively small but good at collecting private data

Source: Zenith, Schulte Research Estimates.

Data Source Comparison

- Amazon – Best intentioned data
1. Intentioned Data: e-commerce, Amazon.com.
2. Unintentioned: Consumption search
3. IoT: Echo, Kindle, Whole Foods, Amazon Books, logistics
- Microsoft – Most corporate baased
1. Intentioned Data: Xbox
2. Unintentioned Data: Strong in business use. Office. Skype, Bing, IE, Edge, Weak in collecting mobile data. Lacking consumer data.
3. IoT: Kinekt, Microsoft Surface, Windows Phone Google – #1 in mobile operating systems and map data. 2B users a month on cloud and Gmail. 2.2B on Android. 1B on Maps. Collect consumer data. Chrome, Snapchat 166M MAU.
- Tencent – best social network data
1. Intentioned: WeChat Pay, third party providers integrated into ecosystem, i.e. JD.com, Didi
2. Uninentioned: Data from WeChat (938M), QQ (~700M), Qzone, entire WeChat ecosystem
3. IoT: Didi, Dianping review site
- Baidu – #1 search engine in PRC – strong IoT, best search in China
1. Intentional: search, Baidu Wallet weak 100 mn users
2. Unintentioned: search
3. IoT: food delivery service, strong autonomous driving, Little Fish

Source: Zenith, Schulte Research Estimates.

Data Source Comparison

- Alibaba
1. Intentioned Data: e-commerce, AliPay, Taobao, Tmall, AntFinancial, AliExpress
2. Unintentioned: Weibo second largest social network, Youku, food delivery ele.me
3. **IoT: Physical retail store (Hema, smartcar SAIC)**
- Comments: 450M users, strong transaction data, best intentioned in China
- Google –
1. Intentioned Data: Android Pay (not that big), Google Play (weaker than App Store)
2. Unintentioned Data: Search, email, Snapchat, maps, Youtube, Calendar, drive
3. IoT: Google Glass, Waymo (Autonomous car), nexus, android wear
4. Comments: Best in creating a consumer profile
- Apple
1. Intentioned: App Store (2.2M apps) Apple Pay
2. Uninentional: iTunes (800M users), iOS
3. IoT: Macbook, iPhone (1B), TV, Apple Watch, iPad
4. Comments: Strongest IoT
- Facebook
1. Intentioned: Messenger Pay
2. Unintentioned: Messaging, photos, posts, blogging
3. IoT: Oculus Riff VR
4. Comments: Best in unintentioned/social network

Source: Zenith, Schulte Research Estimates.

Finance Comparison

- Alibaba – 1.2 tn MMF, AntFinancial, Insurance, Payments, Money Market Fund (MMF), SME lending, personal loan, credit rating, investment management
1. Strongest FinTech company in the world. Amazon too small
- Tencent – WeChat Pay, third-party providers integrated into app (JD.com, Didi), credit ratings, insurance (Zhong An), investment management, personal loans, FILL IN DATA
2. Catching up with Alipay
- Baidu – 100 mn in Baidu Wallet, credit rating
3. Behind Tencent and Alibaba

Source: Zenith, Schulte Research Estimates.

Finance Comparison

- Amazon – 3 bn SME banking, electronic damage insurance UK ONLY
  - Slow development, considering the large data source
- Microsoft – Microsoft Wallet, very small
  - Lacking fintech
- Google – Google Wallet (small), Android Pay
  - Lacking
- Facebook – Messenger Pay but not used much, third parties host transactions. No lending, credit ratings.
  - Lacking in lending
- Apple – FIND A NUMBER Apple Pay (450% YOY), third party apps

Source: Zenith, Schulte Research Estimates.

Cognitive Services Comparison

- Alibaba — Customer service AI, image rec, Voice to text conversion, real-time translation services, gaming
  - COMMENT MISSING
- Tencent — Messaging and voice messaging data can be used for a natural language processing unit. Image rec to recognize fashion trends in Moments, Gaming
  - Unintentional data is an advantage
- Baidu — Image rec, text to speech, translation, opened up five AI labs. PRC support
  - Strong promise for AI, not much use has come of it yet

Source: Zenith, Schulte Research Estimates.

Cognitive Services Comparison

- Amazon — Image, face, voice rec, NLU, video understanding/analysis
  - Pretty good
- Microsoft — Computer vision, language, knowledge, pose understanding
  - Main investment is in this area. (Strong)
- Google — Tech analytics, find jobs, video, vision, speech, translation
  - MAIN INVESTMENT
- Facebook — Image rec, translation, text, no speech, image to text to speech (blind)
- Apple — Translation, Speech, facial rec, image API, NLP API, gaming

Source: Zenith, Schulte Research Estimates.

Lifestyle Comparison

- Alibaba — Internet financing, personalized search, online shopping support, media solutions, data marketing, video game solutions, audio–video solutions
- Tencent — Ele.me, Meituan Dianping, QQ Music, Joox, and the entirety of the social media platforms that Tecent provide contribute to lifestyle. Gaming
  - Strong in messaging data, photos
- Baidu — Food delivery startup via Baidu, locational data/food preference. Book recommendations are sold to third party providers
  - Strong in search data

Source: Zenith, Schulte Research Estimates.

Lifestyle Comparison

- Amazon – lifestyle books, food, music, TV, video stream, search recs, direct marketing, Whole Foods, TV and movies
  - Best in food in US
- Microsoft – xbox/gaming, skype (communication)
  - Not that strong
- Google – Snapchat, youtube
  - Best in Video and pictures in US and localtional data
- Facebook – Whatsapp, Blogging, posts
- Apple – health apps, iMessage, Apple Music
  - Best Music

Source: Zenith, Schulte Research Estimates.

Autonomous Driving Comparison

- Alibaba — "connected vehicle," software services connected to alibaba cloud inside a car.
- Tencent — N/A
- Baidu — Project Apollo, fully autonomous car by 2020, partners like Ford, Continental AG, Robert Bosch, Microsoft, Intel
- Google — Waymo — leading in data miles. Producing hardware and software. 3.2 mn miles
- Facebook — N/A
- Amazon — Focus on drone delivery, 12-person team on autonomous car logistics
- Microsoft — connected vehicle software
- Apple — Project titan — very behind, not making hardware but only focusing on software

Source: Zenith, Schulte Research Estimates.

Robotics Comparison

- Alibaba – Algo for robotics, path optimization, logistics
- Tencent – N/A
- Baidu – N/A
- Google – BostonDynamics
- Facebook - N/A
- Amazon – Warehouse robotics
- Microsoft – Researching robotic systems
- Apple – N/A

Source: Zenith, Schulte Research Estimates.

Tencent Holdings Financials

Tencent Holdings Assets (US millions)				
	2016	2015	2014	2013
Cash	$10,392	$15,021	$8,315	$4,000
Short-Term Investments	$7,433	$5,896	$1,728	$3,222
Accounts Receivable	$2,416	$2,061	$1,109	$707

Tencent Holdings Liabilities (US millions)				
	2016	2015	2014	2013
Stort-Term Debt	$17,565	$18,980	$6,311	$3,316
Accounts Payable	$30,583	$18,738	$10,854	$8,556
Income Tax Payable	$6,242	$3,059	$1,681	$2,478

Source: WSJ, Schulte Research Estimates.

Appendix 4

Baidu

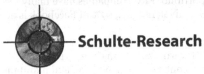

Baidu's New Strategy

Project Conclusions

1. Cloud computing is the infrastructure for all of AI.

2. US firms do not commercialize their AI technology as well as their Chinese equivalents, though they both invest largely into the sector.

3. All companies invest most into the cognitive services sector.

4. In the financial sector, US firms have data but do not include that their business model. Have potential for a credit profile, but they don't. They do not do SME lending, private lending, or insurance.

5. Chinese firms have a more diverse portfolio. PRC companies have many more revenue sources than their US equivalents (i.e., 97% of Facebook is ad revenue).

6. More support in PRC than from US Government. AI in China will be used to promote the country's technology, economy, social welfare, and maintain national security.

Source: Baidu Website, Gov.cn, Schulte Research.

China taking the international lead in AI

- China issued the "Next Generation Artificial Intelligence Development Plan.
- The PRC is embracing AI and have set targets for its development over the next decade.
- "Three in one" agenda will tackle:
    1. Key problems in R&D
    2. Pursuing a range of products and applications
    3. Cultivating an AI technology
- Upgrade of manufacturing sector and overall economic transformation.
- China has a lot to gain from AI because of its high proportion of output derived from manufacturing.
- Est. annual growth rate from 1.6% to 7.9% by 2035. More than $7 tn.
- China intends to be the "premier global AI innovation center".

Source: PWC, Bloomberg, The Diplomat, Schulte Research.

Baidu Development

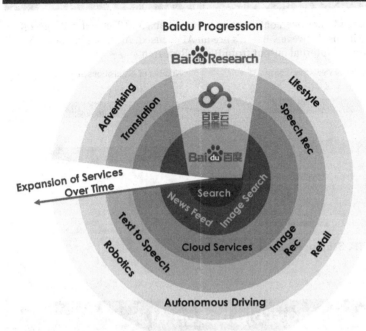

Source: Baidu Website, Schulte Research.

Baidu Finances

Baidu, Inc. Assets ($000)				
	2016	2015	2014	2013
Cash	$1,613,926	$1,548,590	$2,299,441	$1,643,788
Short-Term Investments	$11,359,532	$8,927,135	$6,882,468	$4,746,488
Net Receivable	$900,110	$938,563	$590,667	$414,245

Baidu, Inc. Liabilities ($000)				
	2016	2015	2014	2013
Accounts Payable	$5,057,120	$3,703,087	$2,783,643	$1,708,057
Sort-Term Debt	$1,409,458	$172,617	$373,590	$64,179
Other	$167,287	$144,069	$110,147	$50,197

Source: Baidu Website, Schulte Research.

Baidu's M&A Strategy is to buy – don't build

Since 2012, Baidu has been outspent by both Alibaba and Tencent, with $62.5 bn in acquisitions. However, Baidu's are all AI-focused. Also, Baidu does not have meaningful internal output from Baidu Research.

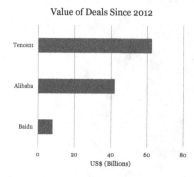

Value of Deals Since 2012

Number of Acquisitions Since 2012

Source: SCMP, Schulte Research.

Baidu is making a big bet on AI

- Baidu is pivoting away from the search engine and developing better algorithms and news feeds on its mobile.
- R&D spending on AI has increased 35% YOY ($412 mn).
- $3 bn in R&D over the past 2.5 years.

Source: Forbes India, Schulte Research Estimates.

Baidu is making a big bet on AI

- Baidu has a 1700-member AI research team, involved in four labs in Silicon Valley and China. Fifth lap opening March 2018.
- Also helping Chinese government build the National Engineering Laboratory of Deep Learning Technology on Biadu's Beijing campus.

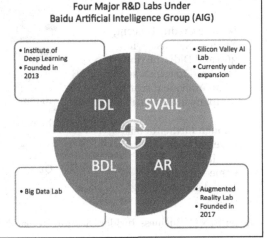

Four Major R&D Labs Under
Baidu Artificial Intelligence Group (AIG)

- Institute of Deep Learning
- Founded in 2013

IDL

- Silicon Valley AI Lab
- Currently under expansion

SVAIL

- Big Data Lab

BDL

- Augmented Reality Lab
- Founded in 2017

AR

Source: Medium, Schulte Research.

Finance

a. Baidu diving into credit ratings and ZestFinance

- **About 160 mn people in China received $180 bn in online loans in 2016. 50% CAGR over the next 3 years.**
- Baidu's invested in ZestFinance, an LA-based fintech company using Big Data for credit scoring purposes in two separate funding rounds. Steps are:

ZestFinance's Data Assimilation Discover and acquire data sources on a massive scale	ZestFinance's Modeling Tools Train, assemble and produce machine learning models in one workflow	ZestFinance's Modeling Clearly communicate economic value and support compliance

- Zest's goal is to help companies make smarter credit decisions to expand the availability of fair and transparent credit.

Source: iResearch, ZestFinance, Schulte Research Estimates.

Baidu diving into credit ratings with Pan-China consortium

- Baidu, along with Sesame Credit, Tencent, Xiaomi, NetEase, Qiho 360 and Didi Chuxing proposed to jointly set up an individual scoring company in July 2017.

- Though further details have not been released and it has not been named, it will be a "social credit scoring" system.

Company	Purpose
Sesame	Alibaba Everything
Tencent	WeChat, QQ, Qzone
Baidu	Search
Xiaomi	Cell Phone
NetEase	E-Commerce
Qiho360	Antivirus
Didi	Taxi

Source: SINA, Business Insider, Schulte Research Estimates.

b. Baidu: Biadu Wallet (Baifubao) and E-commerce

- 100 mn users by end 2016, up 88% from the previous year.
- Offers interbank transfers, lets users pay for online purchases and utility bills.
- Linked to Baidu's wealth management product, Bai Fa.

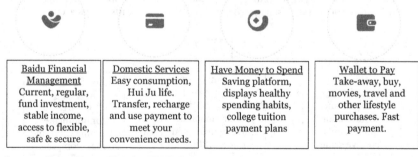

Baidu Financial Management	Domestic Services	Have Money to Spend	Wallet to Pay
Current, regular, fund investment, stable income, access to flexible, safe & secure	Easy consumption, Hui Ju life. Transfer, recharge and use payment to meet your convenience needs.	Saving platform, displays healthy spending habits, college tuition payment plans	Take-away, buy, movies, travel and other lifestyle purchases. Fast payment.

Source: Crunchbase, Baidu, Wall Street Journal, Schulte Research Estimates.

Baidu: Biadu Wallet (Baifubao)'s partnership with PayPay could mean expansion

- Baidu has entered into a strategic partnership with Paypal. **This is Baidu Wallet's first expansion beyond China.**
- Baidu Wallet will be accepted by about 17 mn Paypal merchants globally.

- Target cross-border payments between Chinese consumers and online businesses outside of China.
- Both Tencent and Alibaba have been expanding since 2015, and have begun infiltrating Europe and the US. Will have to make strong progress to get to an equal level.

Cognitive Services

a. Facial Recognition

- Baidu has achieved 99.7% accuracy with facial recognition.
- Chinese internet companies can take photos from their online apps to teach computers to analyze features.
- Currently being used to verify customer identities by insurance firm Taikang.
  - Sometimes required to tune in through live video, and can reference national ID photos to verify identities.

Source: Forbes, Schulte Research Estimates.

Facial Recognition Better than a Human's

Facial Recognition Errors (Smaller is better)

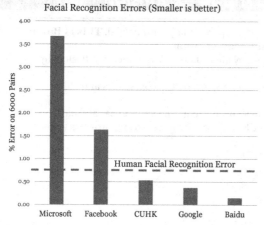

- Out of 6,000 test examples, Baidu's facial recognition technology made nine mistakes.
- Human facial recognition error is about 0.8%, meaning that Baidu's technology is over five times less likely to make a mistake.

Source: Baidu Research, Schulte Research Estimates.

Baobeihuijia and Facial Rec Bringing Families back together

- Baidu is helping Chinese families find missing children.
- Adult Fu Gui found it strange that he only had vague memories of his childhood, so he uploaded a photo of himself at 10 years old to a website called Baobeihuijia ("Baby Come Home").
- Parents uploaded childhood photos months later, and Baidu's cross-age facial recognition tech proved that it was able to match photos taken up to 6 years apart.

Source: Engadget, Schulte Research Estimates.

b. Baidu: Translation

- Baidu receives ~100 mn translation requests everyday.
- 6% error rate in translating Mandarin to other languages — this makes Baidu Translate the best at transcribing Mandarin voice queries in the world.

- Translation used to be developed based on decades of speech research and real human voices. With system now, input a .WAV file. There is a deep neural network that translates directly into characters.
  - Able to learn what's relevant from the input to directly translate the output with very little human intervention.

Source: Baidu, Medium, Schulte Research Estimates.

c. Baidu: Image Recognition and xPerception

- Baidu acquired xPerception (Mountain View, CA), perception software and hardware that have applications in VR and robotics.
- Perfecting its Simultaneous Localization and Mapping (SLAM) technology. Uses visual and inertial sensors
- Cameras, accelerators and gyroscopes for object detention and path planning.

 Visual Inertial Navigation

 Object Recognition

 3-D Sensing

Source: xPerception, Tech In Asia, Schulte Research Estimates.

Image Recognition: Two New APIs

- Pixlab API
  - Allows users to process, transform, and filter any images from any programming language with machine vision and deep learning APIs.
  - Includes:
    - facedetect
    - tagimg
    - facelookup

- WICG Shape Detection API
  - Shape detection acceleration platform for both still images and real-time image feeds.
  - The API is capable of detecting:
    - Barcodes
    - Faces
    - Shapes
    - Text

Source: Baidu Research, Schulte Research Estimates.

d. Baidu: Text to Speech

"Give it the right data and it can learn on their own what sort of features are important" — Andrew Gibianski, Research Scientist @ Baidu AI Lab

- In February 2017, Baidu announced the DeepVoice project. Can produce an actual human voice response, near real-time.
- Three months later, Baidu introduced DeepVoice 2.
- DeepVoice 2 can learn the nuances of a person's voice within 30 minutes & can learn to imitate hundreds of different speakers.
- It took Apple years to release Siri's regional accents because a human being was needed. Deep Voice 2 takes commonalities across all accents, tweaking the model via algorithms.
- Applications in voice assistants, eBooks.

Source: Baidu Research, The Verge, Schulte Research Estimates.

e. Baidu: Speech Recognition & Relationship with HTC

- 97% accuracy
- Baidu's DeurOS platform is its new voice-based digital assistant
  - typing speed three times as fast as humans'
  - Voice-interaction platform is used by Harman International, Lenovo, Xaiomi, Vivo
- Manages queries about calendars, weather, restaurants, etc.
- HTC's new smartphone, HTC U11, will be powered by DeurOS

Source: Forbes India, Medium, Schulte Research Estimates.

Baidu: Speech Recognition and Kitt.AI

- Baidu purchased Kitt.AI (Seattle) to strengthen DeurOS' natural language processing.
- Kitt.AI builds and powers chatbots and voice-based applications across multiple platforms.
  - Works in smartphones, speakers, appliances, web chats, cars, homes, conference rooms.
- Kitt.AI's three main products remain operational as before.

Source: Kitt.AI, Medium, Schulte Research Estimates.

Kitt.AI's Snowboy

- The Snowboy product is a customizable hotword detection engine for the user to create their own hotword, like "OK Google" or "Hey Siri".
- Powered by deep neural networks and is:
  - Highly customizable: users can freely pick their hotword (i.e., Hi, Jarvis)
  - Always listening but protects privacy: Snowboy does not require internet, nor does it stream the user's voice to the cloud.

- Currently supported by all versions of Raspberry Pi, Mac OS X, iOS, Android, Intel, Samsung, Arm-64.

- Comes in form of C++, made by Kitt.AI.

Source: GitHub, Schulte Research Estimates.

Kitt.AI's NLU

- NLU (Natural Language Understanding) is an interactive engine that provides "Understanding as a Service."
- In the engine, the user:
1. Creates a new application and begins to train it by entering phrases and key words to elicit a response in the NLU Editor.
2. After deploying the NLU editor, the Model Trainer is used to make corrections to the use of keywords and train the computer to understand further.
3. Made by Kitt.AI.

- A natural language processer can be created very simply through Kitt.AI's engine. The more time spent in the Editor and Model Trainer, the better it is able to understand.

Source: Kitt.AI, Schulte Research Estimates.

Kitt.AI's ChatFlow

- The ChatFlow framework makes it simpler for users to build more responsive and accurate chatbots that work on more than just one platform.
- For example, ChatFlow can allow multiple companies to be integrated into the same conversation.

- Using Yelp to find a restaurant
- OpenTable to make the reservation
- Uber to get a ride to the restaurant
- Venmo to split the bill

All in one conversation, without having to switch applications

Source: Kitt.AI, Schulte Research Estimates.

Kitt.AI's Summary

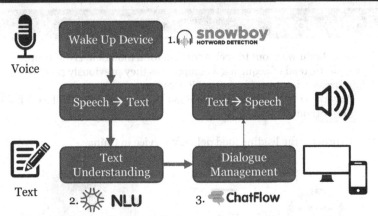

Source: Kitt.AI, Schulte Research Estimates.

Baidu's Speech Recognition and RavenTech

- Baidu's purchased RavenTech, a Chinese voice assistant platform, in February 2017.
- The company focuses on artificial intelligence, big data and the next generation OS.
- RavenTech's CEO, Cheng Lu, will lead Baidu's smart home device businesses and "work with the DeurOS team on new product development".

Source: TechCrunch, Schulte Research Estimates.

Lifestyle

a. Baidu: Food

- In 2016, Baidu went out to seek $500 mn for a food delivery startup.
- However, instead of acquiring a company as they previously planned to, Baidu began a joint venture with S.F. Holding.
  - According to the report, Baidu Food Delivery, and S.F. will each hold 50% of the JV.

- S.F. Express is the leading food delivery service in China.

Source: Tencent Tech, Schulte Research Estimates.

b. Baidu: Books

- Baidu uses searches to collect data on its users.
- The company looks at past searches and recommends books based on those searches.

Searching for symptoms of diseases? Baidu might recommend a book on health.	Based on movie streaming searches or movie ticket purchases, Baidu may recommend books of the same genre.	Based on searches of life situations (i.e. divorce), Baidu may recommend books on lifestyle change and coping with conflict.

- Baidu's search engine, along with its other services, can collect data and sell it to third-party providers.

Source: Tencent Tech, Schulte Research Estimates.

Autonomous Cars

a. Baidu's Autonomous Driving Software

- Baidu is hoping to have a self driving car on the market by 2018 and provide tech for fully autonomous by 2020.
- Baidu's "Project Apollo" is an open-source software from self-driving cars.
  - Lu calls it the "Android for autonomous vehicles"
  - Partners are:

China	United States	Other
• Chery Auto	• Ford	• Continental AG
• Great Wall Motors	• Intel	• Robert Bosch GmbH
• Changan Moble	• Microsoft	• TomTom
• Grab Taxi		

Source: The Economist, Schulte Research Estimates.

Baidu's Autonomous Driving Software

- Baidu is partnered with NVIDIA (Santa Clara, CA)
- Goal is to use Apollo while adopting NVIDIA's PX Platform for Baidu's self-driving car initiative
  - Combines data from multiple cameras
  - Lidar, radar, ultrasonic sensors
  - Uses NVIDIA's deep neural networks for detection and classification

Source: The Economist, NVIDIA Schulte Research Estimates.

Baidu's Autonomous Driving Software

- Algorithms are learning to fully understand the 360-degree environment around the car.
- In theory, a Beijing-trained self-driving car will travel well as its chaotic streets make Europe's or the US' seem simple.
- Microsoft is also part of the Apollo alliance.
  - Will provide a **global scale** for Apollo with Micrsoft Azure's Cloud.

Source: The Economist, NVIDIA Schulte Research Estimates.

Robotics

a. Baidu: Robotics

- Baidu, alongside AiNemo, has created Xiaoyo Zaijia (小魚), or "Little Fish".
- Has Amazon Alexa's talents PLUS a screen, camera, and the ability to move.
- Google and Amazon's similar products cater to single professionals. Baidu's product is more family oriented.
  - Helps parents with a busy schedule take care of children.

Source: Tech Wire Asia, The Verge, Schulte Research Estimates.

b. Baidu: Robotics software

- Little Fish's monitor turns to the user when he/she uses the hotword.
- Users can order meals, groceries, and medicine by voice, and are authorized to do so by a facial recognition program.
- Powered by Baidu's DeurOS .
- Available on JD.com for US $490.

| Enjoy music | Explore news | Get answers | Request on– demand services | Manage calendar | Control smart home | Make video calls |

Source: PR Newswire The Verge, Tech Wire Asia, Schulte Research Estimates.

Cloud

a. Baidu Cloud as infrastructure

- Though Baidu has its own cloud computing services, all AI-based projects will be using Microsoft Azure's cloud computing services instead.
  - Autonomous cars
  - Facial recognition
  - Speech recognition
  - Image recognition

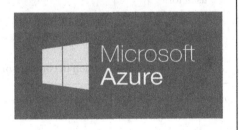

Source: SCMP, Schulte Research Estimates.

Summary: Baidu Pros

1. Strong AI-based partnerships and acquisitions in the past year
   - RavenTech
   - xPerception
   - Paypal
   - Kitt.AI
   - ZestFinance
2. Leading the autonomous driving industry, has Project Apollo, globally used API
   - Partnerships with Tier 1 and Tier 2 manufacturing companies (Continental AG, Robert Bosch).
3. PRC support. National Engineering Laboratory of Deep Learning Technology on Baidu Campus, joint venture between government and Baidu.
4. Strong 2Q2017 results show promising AI payoff.

Source: SCMP, Schulte Research Estimates.

Summary: Baidu Cons

1. Not much has come out of the Research Labs in the past year.
2. External acquisitions rather than internal development.
3. Cloud, the infrastructure for AI, is not its own cloud. Microsoft Azure is being used for:
   - Autonomous cars
   - Facial recognition
   - Speech recognition
   - Image Recognition
4. Behind on global expansion in comparison to Tencent, Alibaba.

Source: SCMP, Schulte Research Estimates.

Appendix 5

Zhong An

Zhong An Summary

Zhong An Ownership
Main ownership from Chinese tech giants Alibaba, Tencent, and Ping An

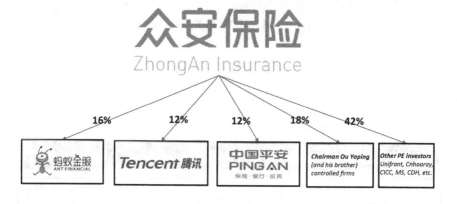

Zhong An: only 3 years old, the largest pure online, cloud-based insurance company globally.

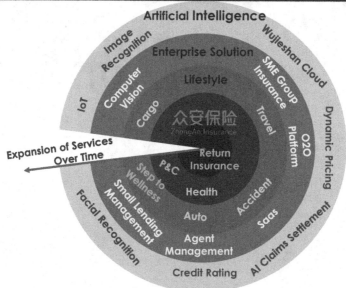

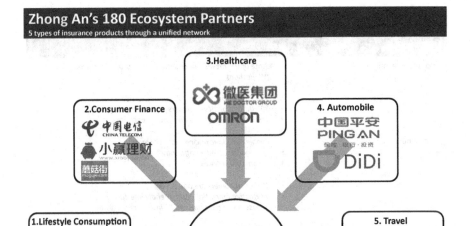

Zhong An Insurance — Business Model
Customer cross-selling for acquisition and retention with no agent or branch. Anytime, anywhere, any place.

Business Model Details

No branch offices. Entire value chain (distribution, underwriting, processing) is digitized and online.

Leverages cloud, Big Data analytics, user base + icon layout from Ping An, Alibaba, Tencent.

Sells insurance directly to customers. Small premiums and big volume.

Low fixed cost and highly scalable business model means unparalleled distribution capability.

Target customers include corporate and individual clients.

Benefits to Customers	Benefits to Zhong An
• Simple, affordable, relevant • Coverage of risks related to internet • Better claims experience • Anytime, anywhere, anyone	• Unparalleled distribution • Low fixed costs, scalable • Satisfying customer experience • Faster claims servicing

Source: Company Filings, The Digital Insurer, Schulte Research Estimates.

Zhong An Advantage: data driven and cloud based

Advantage	Description
Proprietary technology	• State of the art insurance system on its proprietary cloud (Wujieshan)
Extensive user data	• 492 mn customers and 7.2 bn policies sold • 800 data analytics engineers • Application of big data analytics in each step
Scalable business model	• Enables dynamic pricing and risk tracking • Personalized product offering • Automated services
Attract young	• 60% of customers are age 20–35 • Interactive marketing, user-friendly interface • Bypasses stagnant, unfriendly insurance websites
Ideal Usage	• Adjust frequency and timing of the popout notifications tuned to customers' daily routine • Scenario settings to induce purchases • Physical proximity to multiple Ant icons

Source: Company Filings, Schulte Research Estimates.

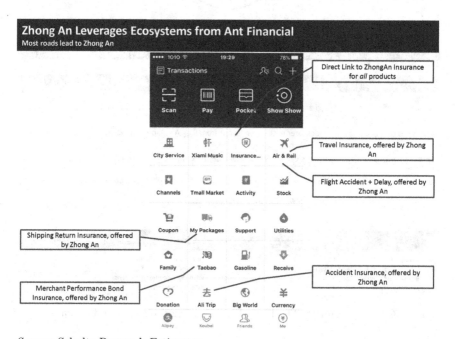

Source: Schulte Research Estimates.

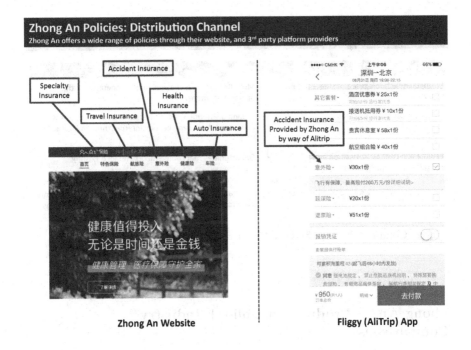

Zhong An Policies: Distribution Channel
Zhong An offers a wide range of policies through their website, and 3rd party platform providers

Zhong An Website

Fliggy (AliTrip) App

Zhong An Future Strategy: Internal
More customers, more products, more efficiency

Future Strategy	Description
Grow customer base	• Branding initiatives and targeted marketing
Maximize profitability/ expanding product mix	• Upgrade existing products
Drive operating efficiency	• Enhance risk management, internal control and asset management

Source: Company Filings, Schulte Research Estimates.

Zhong An Future Strategy: External
Provide the infrastructure for PRC insurance digitization (become the Ant Financial for insurance)

Future Strategy	Description
Strengthen technology leadership, big data analytics	• More R&D, more engineers, more cooperation with universities, more AI
Foster connective ecosytems	• Expand distribution capabilities through partnerships with new channels. Expand beyond internet through O2O and IOT.
Explore investment, acquisition and collaboration opportunities	• Opportunities that complement/ enhance existing operations
Monetize technology solutions in the future	• Zhongan Technology is a new venture: AI, Big Data Analytics, and Blockchain technology to provide enterprise solutions

Source: Company Filings, Schulte Research Estimates.

Zhong An — Product Portfolio + Industry Comparisons

Zhong An Product Portfolio: Lifestyle Consumption
Insurance policies that cover a range of sectors, from e-commerce to mobile phones.

Ecosystem (Sector)	Description
Lifestyle Consumption	Electronic products related to consumption. Issues such as product quality, delivery, security, etc.

Key Products	Description
Shipping Return Policy	For customers unsatisfied with the product they were delivered, they can return the good for no extra cost to them.
Merchant Performance Bond Insurance	Allow merchants of e-commerce sites to access consumer-related services without deposits; protection from potential repayment to consumers.
Phone Accident Policy	Repair services for newly purchased Xiaomi devices. Machine learning software is used to identify physical damage to devices from images uploaded by users.

Source: Company Filings, The Digital Insurer, Schulte Research Estimates.

Zhong An Product Portfolio: Consumer Finance
Policies and platforms which aim to connect users with financial institutions to allow for the free movement of capital.

Ecosystem (Sector)	Description
Consumer Finance	Solutions to protect customers against credit risks and connect funding providers, credit providers, and consumers.

Key Products	Description
Mashanghua Policy (mobile consumer lending)	An online credit platform which connects users with banks, funds, brokers, and other financial institutions to expand coverage of consumer finance
Orange Baitiao (credit guarantee)	Cooperating with China Telecom, Orange Baitiao allows users to apply for credit line through their mobile devices for certain transactions (mobile phone bills, partner merchandise, Baidu Takeout Delivery, etc.)
Baobei Open Platform	Credit assessment service and platform which connects financial institutions with individual users to provide lines of credit and funding.

Source: Company Filings, The Digital Insurer, Schulte Research Estimates.

Zhong An Product Portfolio: Health
Integration of intelligent devices and innovative technologies to provide modern health insurance policies.

Ecosystem (Sector)	Description
Health	Insurance for personal well-being, medical spending, and healthcare. Partnerships with hospitals and medical device manufacturers.

Key Products	Description
Personal Clinic Policy	Illness and disease insurance protections, and medical benefits. Customer's age ranges from 30 days to 60 years old, with renewal eligibility until 80 years of age. Any and all expenses exceeding the deductible incurred in Chinese hospitals with ordinary illnesses, malignant tumors, are fully reimbursed.
Walk to Wellness Policy	An internet-based health management plan which provides customized health protection. Customers' daily exercise is tracked using a pedometer and premiums are dynamically adjusted accordingly.
Diabetes Policy	An intelligent glucose meter is provided to measure daily glucose levels of users. Health plans and payment premiums are adjusted accordingly and customized depending on each user's health.

Source: Company Filings, The Digital Insurer, Schulte Research Estimates.

Zhong An Product Portfolio: Auto

Streamlined and efficient auto insurance policy which leverages on cooperation with insurance firm Ping An Insurance.

Ecosystem (Sector)	Description
Auto	Insurance against vehicle damage, personal injury, death, theft, and third party liabilities.

Key Products	Description
Baobiao Auto Insurance	Jointly launched with Ping An Insurance, Baobiao Auto Insurance provides a streamlined and efficient auto insurance policy. Compared with traditional auto insurance, Baobiao Auto Insurance connects online customers with offline insurance settlement providers and offers more convenient customer services.

Source: Company Filings, The Digital Insurer, Schulte Research Estimates.

Zhong An Product Portfolio: Travel

Cooperation with Ctrip to provide consumers with coverage for accidents or delays incurred through air travel.

Ecosystem (Sector)	Description
Travel	Travel insurance covering various risks arising from travel, such as accidents, delays, and cancellations. Major partnership with Ctrip.

Key Products	Description
Flight Accident Policy	Through cooperation with Ctrip, customers who purchase flight tickets on their website will be eligible to purchase Zhong An travel insurance. Any accident which occurs during flights are covered.
Flight Delay Policy	Also through cooperation with Ctrip, customers who experience extended flight delays over a certain length of time will be reimbursed. Same day flight delay is also covered under the Jijiubao scheme. Through advanced data analytics, coverages and delay times can be easily calculated.

Source: Company Filings, The Digital Insurer, Schulte Research Estimates.

Zhong An Product Portfolio
While focusing predominantly on Shipping return and lifestyle, Zhong An is shifting their focus towards Accidents, Travel, and Finance sectors.

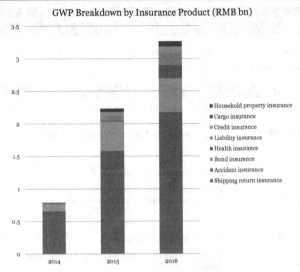

GWP Breakdown by Insurance Product (RMB bn)

- Household property insurance
- Cargo insurance
- Credit insurance
- Liability insurance
- Health insurance
- Bond insurance
- Accident insurance
- Shipping return insurance

Source: Company Filings, The Digital Insurer, Schulte Research Estimates.

Industry Overview
As the leading Insurtech company in the world, Zhong An's growth is far greater than its competitors.

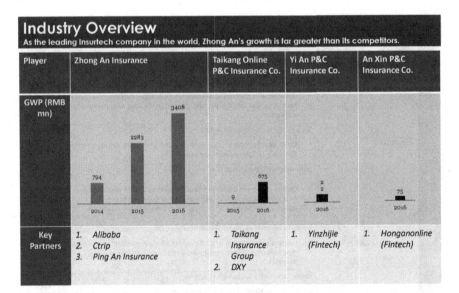

Player	Zhong An Insurance	Taikang Online P&C Insurance Co.	Yi An P&C Insurance Co.	An Xin P&C Insurance Co.
GWP (RMB mn)	2014: 794, 2015: 2283, 2016: 3408	2015: 9, 2016: 675	2016: 2	2016: 75
Key Partners	1. Alibaba 2. Ctrip 3. Ping An Insurance	1. Taikang Insurance Group 2. DXY	1. Yinzhijie (Fintech)	1. Honganonline (Fintech)

Source: HKEx IPO Filings.

Zhong An Technology

Zhong An Technology — S Series (Insurance/Financial Industry)
AI, Big Data, Blockchain Venture

A new venture which aims to leverage on new technologies: AI, Big Data Analytics, and Blockchain, to produce useful applications.

	Product	Description
S Series (Insurance and Financial Industry)	**E-commerce platforms**	Fast development of e-commerce platforms for insurance companies • Quick deployment of online sales channels, official home pages, mobile apps, e-commerce platforms, WeChat platforms, online marketplaces, etc.
	Insurance Core	Cloud-based insurance core systems • Offers capability to build online insurance services including policy distribution, payment services, claims handling, etc.
	Finance Core	Cloud-based finance core systems • Offers capability to build online mobile payment transactions, wealth management, etc.

Zhong An Technology – T Series (Cyber Security Services)
AI, Big Data, Blockchain Venture

	Product	Description
T Series (Security Services)	**Anlink**	Enterprise and cloud services for finance and health • Cyber security platform based on blockchain, AI, and other technologies. • Identity verification, digital assets circulation, clearing, data storage, copyright protection
	Ti-Capsule	Storage of important/sensitive data • Cloud-based storage of important data such as EMR and privacy information • All data is tamper-resistant and secure
	Ti-Sun	Secure digital ID generation • Unique and secure digital identities for users within the ecosystem • Smart identity contains all digitalized assets belonging to user • Enhances customer understanding and prevents money laundering
	Ti-GlassHouse	Digitalize off-chain assets into blockchain • Financial infrastructure which digitalizes off-chain assets into blockchain • Insurance products can be digitalized into blockchain for security and efficiency purposes
	Ti-Packet	Online, blockchain-based, signing system • Electronic, blockchain-based signing system • Allows signing parties to enter agreement by certified electronic signatures, which is authorized by blockchain • Agreements are non-amendable to protect interests of signing parties

Product	Description
X-Model	Risk management models
	• Tools for data analysis for customized risk management models
X-Decision	Transactional decision-making
	• Helps make transactional decisions, for loan applications based on data analysis provided by X-Data and X-Model
X-Man	Customer data analysis
	• Helps clients analyze customer data to create customer profiles
Smart Customer Service	AI Chatbot
	• Helps clients to analyze customer questions
	• Used in Personal Clinic Policy, handles 40,000 service incidents per week
Facial Recognition	Remote identity authentication
	• Allows clients to enhance secure identification authentication processes
X-Data	Data Uploading Service
	• Data uploading service to utilize vast amounts of data
	• Helps clients develop risk models and credit rating

Table title: **Zhong An Technology – X Series (Data services supported by AI)** AI, Big Data, Blockchain Venture — Row label: X Series (Data services)

Zhong An — Risks

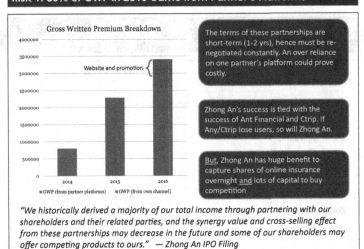

Risk 1: 86% of GWP in 2016 came from Partner's Platforms

Gross Written Premium Breakdown

Website and promotion

■ GWP (from partner platforms)   ■ GWP (from own channel)

The terms of these partnerships are short-term (1-2 yrs), hence must be re-negotiated constantly. An over reliance on one partner's platform could prove costly.

Zhong An's success is tied with the success of Ant Financial and Ctrip. If Any/Ctrip lose users, so will Zhong An.

But, Zhong An has huge benefit to capture shares of online insurance overnight and lots of capital to buy competition

"We historically derived a majority of our total income through partnering with our shareholders and their related parties, and the synergy value and cross-selling effect from these partnerships may decrease in the future and some of our shareholders may offer competing products to ours." — Zhong An IPO Filing

Source: HKEx IPO Filings.

Risk 2: Zhong An Operating Expense Breakdown:
Very high Platform fees!

Consulting and service fees paid to "ecosystem partners" (Alibaba, Ctrip, etc.) through their platforms is high + rising.

Zhong An Operating Expense Breakdown (% of net written premiums)	2014	2015	2016
Consulting Fees and Service Fees	13.3%	30.7%	33.9%
Handling Charges and Commissions	2.3%	5.2%	8.9%
Research and Development	3.1%	3.3%	6.7%

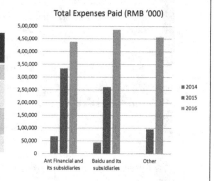

Total Expenses Paid (RMB '000)

"Ant Financial and Ping An Insurance are amongst our most important ecosystem partners. We would not be able to achieve our rapid growth without the support and resources provided by our shareholders." — Zhong An IPO Filing

Source: HKEx IPO Filings.

Risk 3: Product Diversification in Very Early Stages
Zhong An offers a wide range of insurance products. However, their products' success fluctuate greatly. This is a sign of inconsistent growth and risk!

Zhong An Offers **8** types of insurance:

1. *Shipping Return Insurance*
2. *Accident Insurance*
3. *Credit Guarantee*
4. *Health Insurance*
5. *Liability Insurance*
6. *Credit Insurance*
7. *Cargo Insurance*
8. *Household Property Insurance*

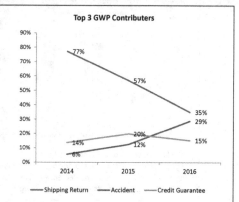

Top 3 GWP Contributers

While the majority of GWP has been attributed to Shipping Return Insurance, the contribution margin has been decreasing significantly over time.

Zhong An <u>needs</u> to find a product which can drive consistent growth for the future.

Source: HKEx IPO Filings.

Risk 4: Regulation, Regulation, Regulation
Insuretech is not well recognized in China. Zhong An is currently under careful examination. Future red tape?

Regulators in China tend to:

1. Allow sub-sectors to develop step by step. Stop. Examine. Proceed.

2. Closely examine and observe how the sub-sector develops and learn from it

3. Finally make a decision to impose strict or relaxed regulations

Zhong An is here right now! They have been granted a license on a special approval basis for the time being. However, anything can change.

As Zhong An is the first of its kind, it enjoys many **benefits** as well as **consequences**. Being a first mover means there is a large degree of uncertainty and risk involved in its future.

Source: HKEx IPO Filings.

Risk 5: Issue of Rapid Growth
Zhong An has experienced very rapid growth by partnering with Ant and Ctrip. This growth may not be sustainable.

"We may not be able to effectively manage our growth, control our expenses or implement our business strategies, in which case we may be unable to maintain high-quality services or adequately address competitive challenges..." — *Zhong An IPO Filing*

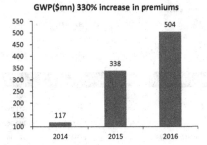

GWP($mn) 330% increase in premiums

As can be seen, partnerships with Ant Financial and Ctrip have been very effective in expanding Zhong An's business. However, rapid expansion requires even greater costs to rapidly develop infrastructure to ensure smooth business functions. Whether or not Zhong An has the management ability to deal with this is of great importance.

Source: HKEx IPO Filings.

Zhong An Financials

Overview: Transforming from a model of return insurance to a diversified model of pure online P&C insurance using an integrated platform.

- Sells all of its P&C insurance online along with handling claims.
- Ranked **Top 5** in KPMG's *Global Fintech 100* **2 years in a row.**

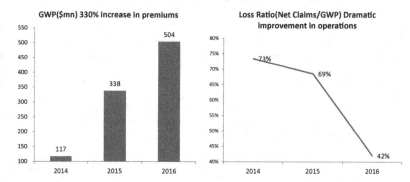

Source: ZhongAn, KPMG, Schulte Research Estimates.

Zhong An Balance Sheet
Debt-free and over-capitalized?

Zhong An Balance Sheet: Assets (RMB '000)	2014	2015	2016	2017 Est.
Investments classified as loans and receivables	408,299	1,207,896	1,707,648	2,424,860
Available-for-sale financial assets	368,130	3,556,804	3,670,260	5,138,364
Financial assets at fair value through profit /loss	121,486	1,321,398	1,599,230	2,270,907
Cash and cash equivalents	141,696	1,374,897	1,153,244	6,153,244
Acquisitions	0	0	0	4,000,000
Total Assets	**1,369,461**	**8,069,143**	**9,332,223**	20,033,357

Zhong An Balance Sheet: Liabilities (RMB '000)	2014	2015	2016	2017 Est.
Unearned Premium Reserves	87,459	441,579	601,256	853,784
Claim reserves	35,546	174,652	196,049	278,390
Investment contract liabilities (stopped by CIRC)	-	1,562	573,069	573,069
Securities sold under agreements to repurchase	140,000	1,600	282,674	339,208
Payables, premiums received in advance, and deferred tax liability	85,894	551,420	819,992	1,090,589
Total Liabilities	**348,899**	**1,170,825**	**2,473,251**	3,135,039
Total Equity	**1,020,562**	**6,898,318**	**6,858,972**	16,898,318

Source: HKEx IPO Filings.

Zhong An Income Statement (RMB '000)
The increase in GWP is partly offset by the increase in commissions to third parties. The reliance on third party channels (Alibaba) means small negotiation power.

	2014	2015	2016	YoY Growth	Est. 2017 YoY Growth	Est. 2017
Gross Written Premiums	794,097	2,283,042	3,408,048	49.3%	42%	4,839,428
Total Income	817,537	2,509,345	3,412,720	36.0%	42%	4,846,062
Net Claims Incurred	-522,903	-1,316,269	-1,355,293	3.0%	19%	-1,612,799
Loss Ratio (Net Claims/ GWP)	65%	57%	40%	-17%	-7%	33%
Consulting Fees and Service Fees*	-94,720	-589,899	-1,093,415	85.4%	27%	-1,388,637
Employee Benefits Expense	-42,021	-185,676	-302,547	62.9%	28%	-387,260
Employee Benefits (% of GWP)	5%	8%	9%	1%	-1%	8%
Handling Charges and commissions**	-16,154	-100,641	-287,109	185.3%	60%	-459,374
Research and Development Expense	-22,235	-63,925	-214,707	235.9%	150%	-536,768
R&D (% of GWP)	3%	3%	6%	3%	5%	11%
Other Expenses	-82,920	-193,342	-146,634	-24.2%	5%	-153,966
Operating Profit before Income tax	36,584	59,593	13,015	-78.2%	NA	307,258
Income Tax Expense	397	-15,336	-3,643	-76.2%	NA	-76,815
Income Tax Rate	NA	25.7%	28.0%	2.3%	-3%	25%
Net Profit for the year	36,981	44,257	9,372	-78.8%	NA	230,444

Notes: * Fees paid to third party platform providers (Ant, Ctrip, etc.) **Fees paid to travel agents and sales channels, mostly for travel related accident insurance.

Source: HKEx IPO Filings, Schulte Research Estimates.

Insurance Industry Valuation

Ticker	Name	Mkt Cap ($bn)	Total Asset ($bn)	ROA (%)	ROC (%)	ROE (%)	Asset/Equity	P/B	P/E	YTD Return
NA	Zhong An*	NA	3.0	1.2	1.3	1.4	1.2	NA	NA	NA
2318 HK EQUITY	PING AN	138	804	1.2	2.5	17.1	11.5	2.2	13.5	51.1
1299 HK EQUITY	AIA	92	185	2.6	11.9	13.6	5.2	2.4	18.1	38.0
PRU LN EQUITY	PRUDENTIAL	60	581	0.6	na	18.2	32.1	3.0	17.0	13.3
2628 HK EQUITY	CHINA LIFE	107	389	0.7	6.4	6.1	8.8	1.9	30.2	18.1
CS FP EQUITY	AXA	71	942	0.7	7.3	8.8	11.8	1.0	10.6	8.5
ALV GY EQUITY	ALLIANZ	96	932	0.8	8.3	11.2	12.6	1.3	11.2	21.7
AIG US EQUITY	AIG	57	498	-0.1	-0.3	-0.3	6.5	0.8	38.2	-2.5
MET US EQUITY	METLIFE	51	899	0.0	1.0	0.1	13.3	0.8	10.8	1.8
ZURN VX EQUITY	ZURICH INSURANCE	46	383	0.8	8.3	9.9	11.8	1.5	14.6	11.9
MFC CN Equity	Manulife	39	536	0.5	6.8	9.3	16.8	1.3	13.6	6.9
Total/Avg		756	6,149	0.7	5.2	8.6	12.0	1.6	17.8	16.9

Note: *Zhong An numbers are 2017 estimate.

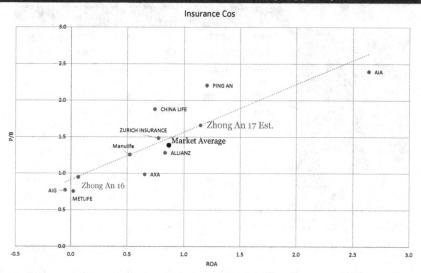

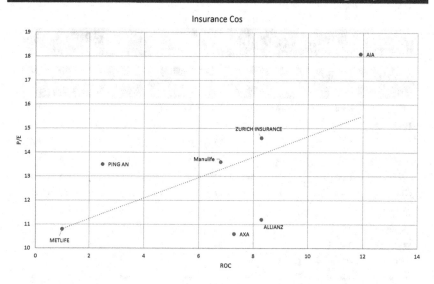

Appendix 6

Softbank

Softbank

 Schulte-Research

I. Softbank major portfolio companies: Alibaba stock price matters the most

Company Name	Category	Company Value (USD mn)	Softbank Ownership	Softbank Stake Value (USD mn)	Value Source
Alibaba	Finance	437,000	29.2%	127,604	Market Cap
Softbank (Domestic Telecom)	Telecom	57,265	100.0%	57,265	5.5x 18 EBITDA
ARM	Hardware	31,910	100.0%	31,910	Market Cap
Sprint	Telecom	31,855	83.4%	26,567	Market Cap
Yahoo! Japan	Search	26,000	36.4%	9,464	Market Cap
Didi Chuxing	Transportation	50,000	18.0%	9,000	CNBC
Nvidia	Hardware	99,490	4.9%	4,875	Market Cap
WeWork	Real Estate	20,000	22.0%	4,400	Forbes
OneWeb and Intelsat	Satellite	9,000	39.9%	3,591	Techcrunch
Fortress Investment Group	Finance	3,090	100.0%	3,090	Market Cap
Flipkart	ecommerce	11,600	21.6%	2,500	Techcrunch
Brightstar	Distribution	2,200	100.0%	2,200	Reuters
One97 Communications	Finance	7,000	20.0%	1,400	ET
Ola	Transportation	3,500	40.0%	1,400	Live Mint
SoFi	Finance	4,300	31.3%	1,344	Fortune
Grab	Transportation	7,000	14.3%	1,000	CNBC
Coupang	ecommerce	5,000	20.0%	1,000	Forbes
hike	Social Network	1,400	43.8%	613	Techcrunch
Improbable	Cloud	1,000	50.0%	500	Techcrunch
Guardant Health	Health	4,500	8.0%	360	Funderbeam
Kabbage	Finance	1,500	16.7%	250	Techcrunch
Nauto, Inc.	Transportation	1,000	10.0%	100	Seeking Alpha
Total		**815,610**	**na**	**290,432**	**na**

Note: USDJPY assumption 107.6.

Source: Techcrunch, Forbes, ET, CNBC, Live Mint, Crunchbase, Bloomberg, Schulte Research Estimate.

Softbank Portfolio Companies by Industry: FinTech is the future

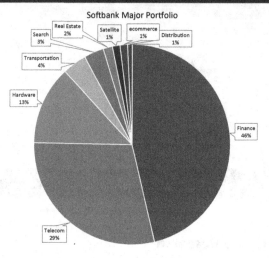

Softbank Major Portfolio

- Finance
- Telecom
- Hardware
- Transportation
- Search
- Real Estate
- Satellite
- ecommerce
- Distribution

Note: USDJPY assumption 107.6.

Source: Crunchbase, Bloomberg, Schulte Research Estimate.

Softbank Balance Sheet: Assets/Equity=5.4

Softbank B/S (USD Bn)			
Cash	12	Softbank debt	72
Receivables	19	Sprint debt	41
PP&E exec Sprint	19	Other debts	25
Sprint PP&E	18	Payables	14
Goodwill exec ARM	14	Derivative liabilities	4
ARM Goodwill	26	Deferred tax liabilities	18
Intangible Assets	65	Others	13
Investments	36	**Total Liabilities**	**188**
Others	23	Total Equity	43
Total Assets	**231**	**Total Equity and Liabilities**	**231**

Source: Company filing, Bloomberg, Schulte Research.

Softbank NAV

- Softbank portfolio total est. value: USD 290 bn
- Total Liabilities: USD 188 bn
- NAV est. = Portfolio Value – Total Liabilities = USD 102 bn

Source: Company filing, Bloomberg.

II. Softbank and Alibaba

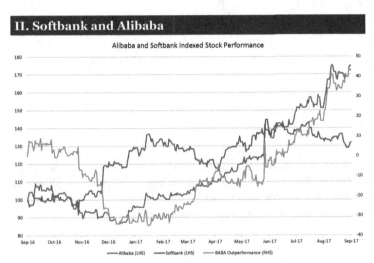

Source: Bloomberg, Schulte Research Estimates.

Alibaba and Softbank stock correlation: decreasing relationship

Correlation since BABA IPO (2014 Sep): 0.78
Correlation in 1 year: 0.60
Correlation YTD: 0.58

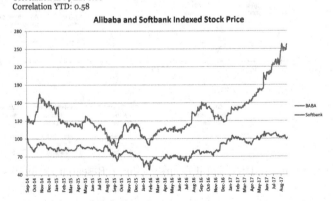

Source: Bloomberg, Schulte Research Estimates.

Softbank Market Cap Premium/Discount to NAV

Note: Historical NAV is adjusted based on historical price of Alibaba, Sprint, Yahoo! Japan and Nvidia, assuming all others the same.

Source: Bloomberg, Schulte Research Estimates.

III. Softbank Investment Timeline: speed up after setting up the vision fund

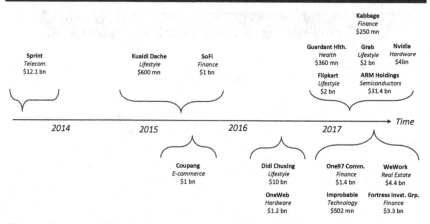

Source: Company Filings, The Digital Insurer, Schulte Research Estimates.

Major Holdings Within Softbank Vision Fund: ARM is the foundation of the mobile world

Company	Description
ARM Holdings **ARM**	**Key Facts** • Acquired by Softbank in 2016 for $41.4bn • World leader in semiconductor technology • Powers 90% of global mobile devices; used by 70% of global population **Reason for Investment** • Growing revenue and chips shipped per year • Developing chips for future devices (AI devices as well) • Powers Alibaba's smart speaker, Tmall Genie **Future Developments** • Create new IoT opportunities • Gain more share in long-term growth markets • Accelerate investment in new technology

Major Holdings Within Softbank Vision Fund: Sprint controls the data communication from telecom

Company	Description
Sprint Corporation **Sprint**	**Key Facts** • Major American telecommunications holding company • Provides wireless services and is an internet service provider • Acquired in 2012 for $12.1 bn **Reason for Investment** • Growing revenue, decreasing costs • Strengthened financial position since joining SoftBank Group • Since joining SoftBank: EBITDA doubled, Operating Income tripled, most improved US telcom **Future Developments** • Expected growth in EBITDA, Operating income, and Cash CAPEX • Signs of further growth in the future

Major Holdings Within Softbank - Didi Chuxing: a growing company which has defeated all local and global competitors already, poised for growth.	
Company	Description
Didi Chuxing	Key Facts
	• $10 bn investment into major Chinese ride-sharing company, Didi Chuxing
	• One of the largest Chinese ride-sharing companies, with 400mn users across 400 cities in China
	• Acquired Uber China in mid-2016
	• Major investors include Alibaba, Tencent, and Baidu
	Reason for Investment
	• The world's leading mobile transportation platform
	• 99% of Taxis, 95% of private drivers in China operate through Didi Chuxing
	Future Developments
	• Huge potential in a growing market, backed by large tech companies, Didi is on track for a successful future

Major Holdings Within Softbank Vision Fund – Wework is the initiative in real estate	
Company	Description
WeWork	Key Facts
	• American company which provides shared workspace services for entrepreneurs, freelancers, startups, small businesses, etc.
	• Designs and builds shared space and charges rental fees
	• 100,000+ members, locations in 23 US cities, 16 countries worldwide
	• $4.4 bn investment from Softbank
	Reason for Investment
	• Transform work style in Japan
	Future Developments
	• Global expansion

Major Holdings Within Softbank Vision Fund: the hardware bet for AI	
Company	**Description**
Nvidia 	**Key Facts** • American technology company which designs graphics processing units (GPUs), as well as system on a chip units (SoCs) • Focuses on gaming, professional visualization, data centers, and automobile markets • Also venturing into AI technology • $4 bn investment from Softbank in mid-2017 **Reason for Investment** • Overwhelming market position in GPUs for AI tech • Strong financial position with keen eye for future market development **Future Developments** • Expected to play a major role in supporting future AI development • Developing products for AI growth

Major Holdings Within Softbank Vision Fund: finance, transportation and ecommerce	
Company	**Description**
Fortress Investment Group 	**Key Facts** • Investment management firm operating out of the US • Manages approximately $70.2 bn (2016) • Acquired by Softbank for $3.3 bn
Grab 	**Key Facts** • Transportation App which offers ride-sharing, car-hailing, and rental service.
FlipKart 	**Key Facts** • India's largest e-commerce site, which offers a diverse range of products

Major Holdings Within Softbank Vision Fund: finance, ecommerce and cloud

Company	Description
SoFi	**Key Facts** • Personal financing firm which offers student loan refinancing, mortgage loans, personal loans, wealth management, and life insurance services • Leverages on new technology to provide services such as credit analysis
Coupang	**Key Facts** • South Korea's fastest-growing e-commerce site of all time • E-commerce site which sells a diverse range of products, similar to Amazon
Improbable	**Key Facts** • British tech company specializing in distributed computing and virtual world simulation • Advanced cloud computing platform used for developing modern video games • Uses AI and advanced cloud computing tech for research in smart cities

Major Holdings Within Softbank Vision Fund: finance, satellites and health

Company	Description
One97 Communication	**Key Facts** • India's premier one-stop portal for all mobile payment needs • Utilities, transportation, leisure, travel, e-commerce, PayTM offers a wide range of services
OneWeb	**Key Facts** • Produces low-cost, ultra-high performing satellites at high-volumes to deliver affordable Internet access globally • Objective is to "reach hundreds of millions of potential users residing in places without broadband access".
Roivant Sciences	**Key Facts** • Biotech company which in-licenses late-stage drug candidates and develops them through subsidiaries • Systematically reduces time and cost of drug development process • Partners with biopharmaceutical companies and academic institutions to ensure medicine is rapidly delivered and developed

Major Holdings Within Softbank Vision Fund: transportation, health, finance

Company	Description
Ola	**Key Facts** • India's most popular mobile app for transportation, integrates city transportation for customers and driver partners • Cab-hailing app with large growth and potential
Guardant Health	**Key Facts** • Leverages machine learning technology within their biopsy-free tumor sequencing test • Provides a service which is not only faster, but much more accurate • 10x better performance than other tests
Kabbage	**Key Facts** • Online financial technology which provides SME funding through an automated lending platform • Partnered with UPS to allow SMEs to share their shipping histories with Kabbage, allowing for better loans and rates

Appendix 7

Amazon AI

Artificial Intelligence is a renaissance — a golden age.

— Jeff Bezos, Amazon CEO

Summary. The Big Bang of Amazon

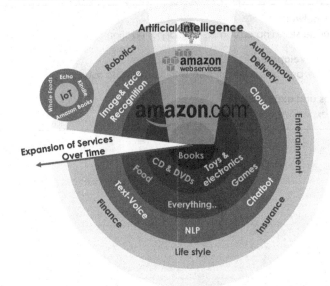

Amazon Pros

AWS: no.1 cloud globally with 33% mkt share globally
- Leading in data collecting through cloud storage and processing

First stop for US online shopper (29%) bypassing search engine (15%).
- Amazon is also leading the entrance point war for shopping

Complete ecosystem to cover all part of consumer's life
- Food, book, movie, music, habits, emotion, taste, etc.

Full information of 2mn SMEs on its marketplace
- Cash flow, working capital, inventory cycle, logistics, visited trends, etc.

Successful expansion into entertainment
- Emmys and Oscars winner

Online-offline connection
- The acquisition of Whole Foods provides a good O2O connection for IOT data

Vast intentional data — easy to process
- Most data are transaction related, easier to monetize

Amazon Cons

Weak in social network
- Behind in understanding customers' social connection

Slow in Fintech Development
- Fintech development is not matched with its vast data source

High valuations and expectations
- Relatively expensive comparing with peers

The virtuous circle of Amazon: Cloud based and AI driven

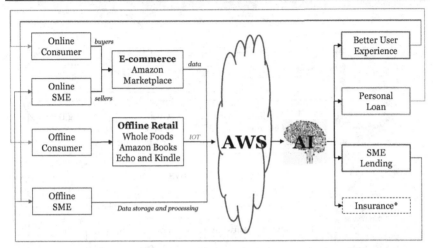

Note: *Amazon now only offers electronics damage insurance in UK.

Internal AI: efficiency, cost and user experience

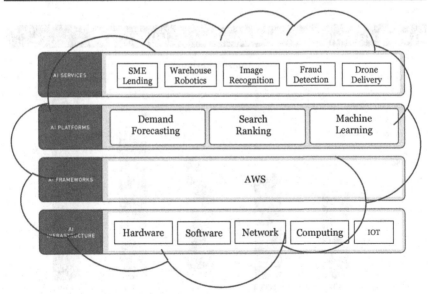

Source: Amazon company website, Schulte Research.

External AI: Amazon AWS based open AI strategy

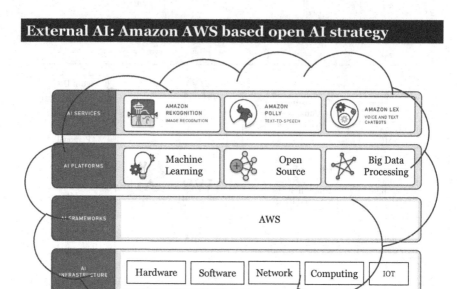

Source: Amazon company website, Schulte Research.

Amazon Revenue: the Cloud is the future

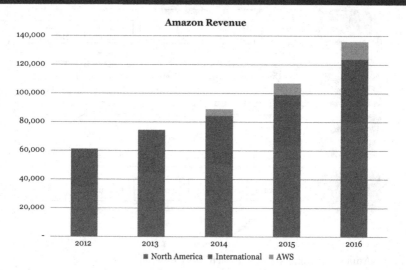

Note: Amazon started to separate out AWS revenue in 2014.

The war for data entrance

- The first step for any AI strategy is data.
- Amazon is leading the data entrance competition for US online shoppers.
- E-commerce internal search (Amazon, Alibaba) is competing with search engines (Google, Baidu) to get first hand info of customer.
- Shopping search data is easier to commercialize (ad.) than social network.

Where US online shoppers begin searching for products?
FIRST STOP: AMAZON

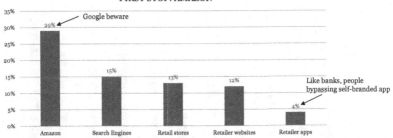

Note: Based on a survey of 5,189 online shoppers, Q1 2017 (Others=9%).

Source: Business Insider, UPS, comScore, Schulte Research.

Cloud: the war for data storage and processing

- If AI is the new electricity, then cloud is the electric dynamo.
- The value: service fees generated and live data pools sitting in providers' servers.
- Amazon is dominating the global cloud business (Microsoft and Google catching up quickly). Alibaba is leader in Asia.
- Global cloud market is expected to reach $400 bn in 2020(Gartner).

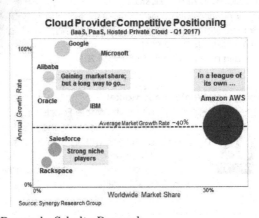

Source: Synergy Research, Schulte Research.

Amazon Cloud: AWS (Amazon Web Service)

- AWS is a $12 bn business with 59% CAGR in the past 2 years.
- 1 million global customers across all industries.
- No.1 with 33% market share in global cloud market.
- AWS provides data access to offline SMEs that are not on Amazon's marketplace.
- Wal-Mart asked its tech vendors to get off AWS to protect its own mkt share.

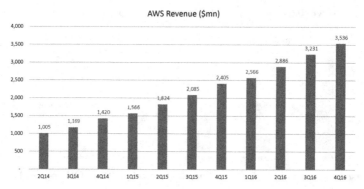

Source: Synergy Research, Schulte Research.

IoT: the war to digitalize the physical world: think of Echo as iphone + Alexa as Siri

Amazon Echo and Alexa Devices

- Smart speaker Echo launched in June 2015
- Sold ~5.2 mn Echo units in 2016
- 5,100 skills available in Alexa skill store
- Featured as family assistant to answer questions
- Echo is important step for its IoT strategy, digitalizing the physical world of its customers
- Echo was used in murder trial in November 2015

Whole Foods Market

- Amazon bought Whole Foods for $13.7 bn in June 2017
- Online-offline connection and in-store shopper data
- Camera tracking with image analysis of in-store shopper behavior
- Amazon can provide customized advertising, discount and award programs

AI Applications

Finance

a. Amazon Lending. 2 mn SMEs on its site. From data to credit. $3 bn in loans.

Introduction
- Lending to selected SMEs on Amazon platform
- Launched in 2011, $3 bn loans in total and $1 bn in past 1 year
- Instant loans; 1 to 12 months
- Rate range: 6%–17%
- Amazon is starting to offer personal loan (installment) in UK

Features
- Amazon has 2 mn SMEs on its marketplace
- SMEs pay Amazon to store, package and ship merchandise
- Loan funding from Amazon's balance sheet in one day
- Interest will be deducted from SMEs' Amazon account directly every 2 weeks
- SMEs' inventories in Amazon are viewed as collateral
- Amazon has nearly full information about the SME's cash flow

Source: Amazon company website, Bloomberg, Financial Times, CNBC, Schulte Research.

b. Insurance: Amazon Protect (only in UK)

amazonprotect

Introduction
- Damage and theft insurance for select electronic products
- Extended warranties to manufacturer's existing guarantee
- Manufacturer's coverage normally last only 2 years
- Launched in 2016
- UK only for now
- Amazon also offers third-party insurance and warranty extension plans

The Potential
- The potential to extend to other markets if it meets the regulation requirements
- Customized insurance is possible based on Amazon's data and technology advantage

Source: Amazon company website, venturebeat.com, Schulte Research.

Cognitive Services

Cognitive Services

a. Amazon Rekognition: Recognize, Search and Understand Images

Functions:
- Image and facial recognition — transaction verification
- Object and scene detection — where, when, who, why, what city
- Face comparison —are these the same people
- Emotion Detection — are you happy with my product

Features
- Easy to use — no deep learning knowledge required
- Batch analysis — image analysis in bulk for all photos stored in AWS
- Real time analysis — use video to analyze response to today's sale
- Continually improving – larger data sets give a better conclusion
- Low cost – $1 for 1,000 image processed (tier pricing)

Source: Amazon company website, Schulte Research.

Amazon Rekognition: snapshot

Searchable Image Library - REDFIN: property tags
- Property Agent Redfin uses Amazon Rekognition to automatically generate tags for hundreds of millions of listing property images
- Users are easier to search based on tags

REDFIN

Face-based user verification – C-SPAN: live TV tags
- TV firm C-SPAN uses Amazon Rekognition to tag who is speaking/on camera at what time down to the second
- Index twice as much content as before (3500 hours a year to 7,500 hours a year) with 97,000 entities in database

C-SPAN

Sentiment Analysis – Social Soup: brand matching
- Influencer marketing firm Social Soup uses Amazon Rekognition to match brand campaigns with right content creators
- Analysis of 30,000 influencer profiles

S⦿CIAL SOUP

Facial Recognition – SmugMug: content detection
- Online photo marketplace SmugMug uses Amazon Rekognition to help photographers manage their photo portfolio
- Automatically identify contents and people

SmugMug ☺

Source: Amazon company website, Schulte Research.

b. Amazon Polly: Text to Speech Powered by Deep Learning

Functions:
- Text in, life-like speech out
- Returns an mp3 audio stream
- Unlimited replay
- Lightning fast responses
- Speech in 24 languages

Features
- Natural sounding voices (47 voices)
- Store and redistribute speech
- Fast response
- Easy Integration
- Low cost

Applications
- Content Creation
- Education/E-learnig
- Mobile and desktop apps
- Customer contact centers
- IoT
- Language learning

Pricing
- Pay for what you use
- No setup cost and no minimum fee
- Fee based on number of characters of text to convert
- $4 per 1 million characters for speech request

Source: Amazon company website, Schulte Research.

Amazon Polly: snapshot

Content Creation – GoAnimate: animate voice generating
- Immediately give voice to animate characters for animate creator
- Multiple languages supported

GoAnimate

Education/E-learning – Wizkids: text reading
- Cloud-driven classroom of Wizkids uses Polly to read out highlighted text in real-time

Wizkids

Mobile and Desktop Apps - The Washington Post
- Audio versions of more than 1,200 daily stories
- High-quality voice with low cost

The Washington Post

Language Learning - Duolingo
- Speech for new language learner
- Accurate pronunciation as good as natural human

duolingo

Source: Amazon company website, Schulte Research.

c. Amazon Lex: build conversation bots

Functions:
- Automatic speech recognition
- Natural language understanding
- Same deep learning techs that power Amazon Alexa (Echo)
- You pay only what you use
- No upfront commitments or minimum fees

Features
- Easy to use
- Seamlessly deploy and scale
- Built-in integration with the AWS platform
- Cost effective

Pricing
- Pay as you use
- $0.004 per voice request to bot
- $0.00075 per text request

Source: Amazon company website, Schulte Research.

Amazon Lex: snapshot

Application Bots – Capital One
- Bank voice and text interaction customer service bots
- Seamless integration with other AWS services

Enterprise Productivity Bots – HubSpot: chatbot
- All-in-one chatbot
- Sophisticated natural language processing without having to code the algorithms by customer (HubSpot)

Internet of Things – NASA: robotic ambassador
- Robotic ambassador "Rov-E" to talk and answer students' questions about Mars supported by Amazon Lex
- NASA staff can navigate Rov-E via voice commands
- Connect and scale with NASA's Mars exploration data source

Source: Amazon company website, Schulte Research.

Autonomous Delivery

Autonomous Delivery

Autonomous Cars
- 12-people team to investigate the possibility of using autonomous driving technology to improve package delivery efficiency
- The team is more like a think-tank to leverage the autonomous driving tech in its supply chain
- The potential of autonomous driving can overcome human's 10-hour driving limit
- Amazon won a patent for "coordinating autonomous vehicles in a roadway"

Drone Delivery
- Large team in drone delivery research (2013)
- Plans to use drone delivery service named "Prime Air" by 2018
- Drone can work together with driverless vehicles to finish "last-mile delivery"
- In Dec 2016, 1st drone delivery testing in UK
- US regulation does *not* allow drone delivery

Robotics

Amazon Robotics to improve operating efficiency

Amazon Warehouse Robots
- Amazon spent $775 mn in 2012 to buy Kiva System and renamed it Amazon Robotics in 2015
- Kiva developed robots that can automate picking and packing process
- Kiva robots: 16 inches tall, 145 kg, 5 mph, weight limit 317 kg

Achievement
- Amazon has 45,000 robots across 20 centers, 3x growth in 2 years
- Use Fetch to increase efficiency: orders up 30% with no labor added by using automatic robots with path recognition from laser and camera

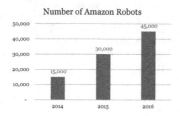

Number of Amazon Robots

Source: Amazon company website, Bloomberg, Financial Times, CNBC, youtube, Schulte Research.

Lifestyle

a. Books: "can a cover be judged by a book"?

a. Books

Arts & Photography • Biographies & Memoirs • Children's Books • Cookbooks, Food & Wine • History • Literature & Fiction • Mystery & Suspense • Romance • Sci-Fi & Fantasy • Teens & Young Adult

Channel
- Ecommerce marketplace
- Kindle
- Retail book store

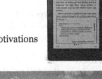

What information can Amazon get?
- Topics and highlights of books: interests, tastes and motivations
- Time of reading: habits
- Frequency of reading: condition changes
- $ amount spent on books: consumption power
- Language: culture and location

How can Amazon use the data?
- Advertising: recommendations
- Cross-selling based on your interests
- Credit rating
- Customized pricing

Source: Amazon company website, Schulte Research.

b. Food: monitor the whole food industry chain

Channel
- AmazonFresh: grocery delivery — fruits, vegetables, meats — fresh and frozen
- Whole Foods Market: 466 offline stores
- Amazon Restaurants: food delivery
- Amazon Launchpad: new products from startups

What information can Amazon get?
- Food categories: health condition
- Time of ordering: habits
- $ spent: consumption level
- Location: networks
- Food amount: social/family condition

How can Amazon use the data?
- Insurance
- Advertising
- Online-offline connection

The Future
- Whole food production process tracking: IoT sensors -> Cloud -> blockchain

Source: Amazon company website, Schulte Research.

c. Movie and TV: First tech firm to win Emmys and Oscars

Amazon Studios
- TV shows, films and comics; production, distribution; based on crowd-sourced feedback
- Launched in 2010
- Awards:
    a. In 2015, its original tv show "Transparent" made Amazon Studios 1st streaming media to win Emmys (five awards)
    b. 1st streaming media to win 3 Oscars ("Manchester by the Sea" and "The Salesman") in 2017

Amazon Storywriter
- Free, cloud-based screenwriting app
- Unlimited storage and auto-formatting
- Crowdsource the process of finding new scripts

Amazon Video
- Online on-demand video service
- Amazon collects data on customer preference on TV and films, which gives Amazon Studios the most important data to develop new products

Source: Quora, Amazon company website, Schulte Research.

d. Music: taste of music means a lot

Amazon Music Unlimited
- Access to millions of songs
- Echo and Alexa devices supported
- $7.99/month for prime members
- 9.99/month for non-prime members

What information can Amazon get?
- Types: character
- Time: habits
- Pattern: emotion change

Source: Quora, Amazon company website, Schulte Research.

Appendix 8

Google

Google

 —Schulte-Research Google

Introduction

Where we think AI belongs in the digital ecosystem

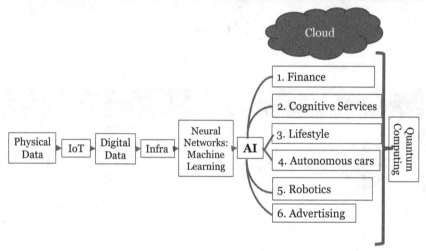

Source: Schulte Research Estimates.

Google AI:

"Google wants to be the best in search. To reach that goal Google wants to have the world's top AI research laboratory." — Larry Page

The Big Bang of Google

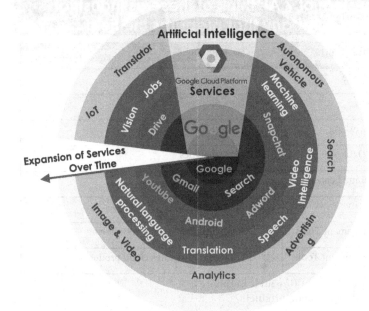

What makes a Google consumer profile

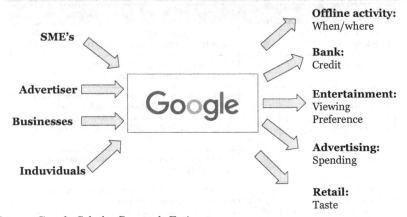

Source: Google Schulte Research Estimates
https://www.slideshare.net/abail019/customer-profile-for-google-inc

Google Cloud: Atmosphere to give life to ecosystem of 7 AI; think of these as industrial farming company that prepares the field

Internal Products		Description
1. Compute		Software development
2. Storage		Data Storage
3. Networking		Globally accessible anywhere
4. Big Data		Data analytics
5. Machine Learning		Cognitive services
6. Security		Encryption
7. Management Tools		Analytics performance monitoring
8. Development Tools		Links and integrates cloud products

Source: Google Schulte Research Estimates
https://cloud.google.com/products/

Google Cloud: The atmosphere to give life to the ecosystem of 7 AI: Think of this as seed products farmers develop into unique crops

External Products		Description
1. Cloud Machine Learning		GPU accelerated hardware allows Google to make predictions based on large data sets.
2. Cloud jobs		Integrates personal profile (location and qualifications) to attract employer adverts.
3. Cloud Video Intelligence		Video analytics: categorises videos within library and distinguishes attributes of each video.
4. Cloud Vision		Image recognition: identifies individual faces and objects.
5. Cloud Speech		Neural network models: recognise 80 languages that can convert speech to text
6. Cloud natural language		Extracts query-based information and conduct sentiment analysis from text.
7. Cloud Translation		Translates text, images and webpages.

Source: Google Schulte Research Estimates
https://cloud.google.com/products

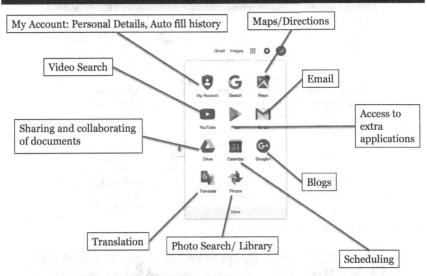

Source: Google Schulte Research Estimates.

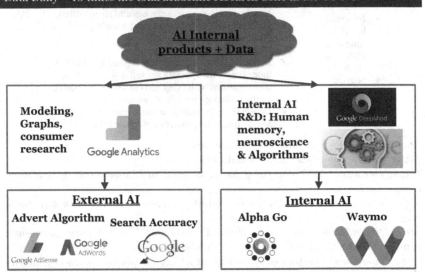

Source: Google Schulte Research Estimates.

Source: Google Schulte Research Estimates.

1. Location: Maps, street views + searches

Data Collected	How locational data will be used
• *Time+Place*: Where and when. • *Images*: location, time who's with you, clothing. • *Searches*: Most searched location.	**Advertising** • *Anticipatory psychology*: Google anticipates the adverts you are most likely to use based on: Nearby services, time, searches and store visited. • *Location*: Where is biggest sale? Should you eat now? What services do you need? **Personal life** • *Divorce/Cheating*: Change in schedule, spending more time with opposite sex (after work hours), conversations with other girl (usually ex), doesn't contact other girls while at home. • *Depression*: Average four times more use of phone, always home and analytics of music (more sad).

Source: Google Schulte Research Estimates.
https://privacy.google.com/your-data.html
https://www.mobileads.com/blog/location-based-mobile-advertising-small-business/
http://www.wired.co.uk/article/google-history-search-tracking-data-how-to-delete
https://www.forbes.com/sites/kashmirhill/2012/02/21/how-a-smartphone-app-can-detect-how-fit-or-fat-you-are/#72469daf2b63
https://news.northwestern.edu/stories/2015/07/your-phone-knows-if-youre-depressed

2. Searches: Sentiment analysis, anticipatory sickness, expected purchases & unconscious likes + dislikes

Data Collected

What and when did you search...
- Websites
- Videos
- Images
- Adverts clicked on
- When are you off Google?

Search engine manipulation effect (SEME)

- 25–80% sway in national voting
- 2016 US election: Estimated 2.3–10 million people swayed to Clinton, sentiment of searches of Clinton were more positive than Bing & Yahoo.

Google knows everything

- *Sickness*: 35% of US do this: Searching for symptoms; information on past illnesses are stored.
- *Pregnancy*: Symptoms searched up & shopping habits (increase purchases in natural lotions, scent free soap & minerals).
- *Abortion*: how many children, demographic $ age.
- *Cyberstalking*: Who is stalking you? Correlation between time spent on your profile and mental state. Less accurate than data available to Facebook.

Source: Google Schulte Research Estimate.
https://privacy.google.com/your-data.html
http://money.cnn.com/2017/04/05/technology/online-privacy-faq/index.html
http://techaeris.com/2016/06/10/google-manipulating-results-favor-hillary-clinton/
http://observer.com/2015/05/these-ten-google-medial-innovations-may-dramatically-improve-your-health/
http://www.bbc.co.uk/newsbeat/article/32379961/cyber-stalking-when-looking-at-other-people-online-becomes-a-problem
http://www.teenvogue.com/story/teen-pregnancy-prevention-facts
https://www.plannedparenthood.org/learn/abortion/considering-abortion

3. Android (part 1): Phone operating system

Data Collected	Firebase

1) Android Auto ⚠
  • Time & habits
  • Phone use while driving
2) Google Play ▶
  • Apps use: Where, when, how long, what
  • Music taste
3) Calendar 31
  • Contacts, calendar
4) Android TVandroidtv
  • Watching habits: what & when
5) Android wear (watch)android wear
  • Heart rate, activeness, calories burnt
	Analytics tool that records the statistics and performance of a mobile apps.
	
	Analytics ⚙
	Visual way of displaying data collected on apps
	
	Firebase Cloud Messaging ⚙
	Message notifications

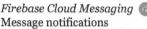

Source: Google Schulte Research Estimates.
https://privacy.google.com/how-ads-work.html
http://www.webmd.com/mental-health/recognizing-suicidal-behavior#1
https://www.forbes.com/sites/rent/2015/06/26/why-do-people-spend-money-t
hey-dont-have/#43226a8a1cc1
https://www.quora.com/How-does-Google-know-my-credit-card-statement-bill
-amount
http://www.bbc.com/news/technology-36783521
https://www.theverge.com/2016/2/25/11112366/deepmind-health-launches-me
dical-apps-group

3. Android (part 1): Health, Emotions & money

Tapping into consumers lives
Medical issues • Streams: Medical app that collects data on risk of developing kidney problems. • Searching for yellow colored skin as a symptom for rashes can mean jaundice.
Psychological issues • Suicidality :Phone use at night indicates sleep problems, searches for tool to die, more likely with terminal illness, family history of suicide.
Financial Issues • Lower demographic spend more on useless things, discounts, stress (psychological state), increased alcohol purchases, post midnight ATM withdrawal and gambling habits.

Source: Google Schulte Research Estimates.
https://privacy.google.com/how-ads-work.html
http://www.webmd.com/mental-health/recognizing-suicidal-behavior#1
https://www.forbes.com/sites/rent/2015/06/26/why-do-people-spend-money-t
hey-dont-have/#43226a8a1cc1
https://www.quora.com/How-does-Google-know-my-credit-card-statement-bill
-amount
http://www.bbc.com/news/technology-36783521
https://www.theverge.com/2016/2/25/11112366/deepmind-health-launches-me
dical-apps-group

4. Social Media (part 1): Shopping

Data Collected	Prediction using data

Gmail
• Who
• Topic/ message
• Attachments
• Images
• When
• 3rd party links
	• *What I buy:*
	  ○ Ecommerce receipts
	  ○ Promotion codes/ coupons/ QRF codes
	  ○ Subscriptions
	  ○ Advertisements clicked on
	• *Where and when am I buying:*
	  ○ Web based and retail store accounts
	• *When I buy:*
	  ○ Season
	  ○ Section of month
	• *What am I doing when I buy?*
	  ○ Image and video analysis

Source: Google Schulte Research Estimates.
https://privacy.google.com/your-data.html

4. Social Media (part 2)

Google can predict illegal activity

- *Guns*: Bailey Dwayne Case, use of gmail chats to show threats and order records of guns in prosecution case.

- *Drugs*: The probability of a substance addiction: Financial problems, antisocial, association with other addicts, stopping old hobbies (change in schedule).

- *Criminals*: Correlation of previous violation/ violence record, unemployed, alcoholic/drug addict, social life/support, family problem, searches for murder tools.

Source: Google Schulte Research Estimates.
https://privacy.google.com/your-data.html
http://www.technewsworld.com/story/84649.html
https://www.google.com/transparencyreport/userdatarequests/legalprocess/
http://www.newstatesman.com/science-tech/internet/2016/08/how-your-googl
e-searches-can-be-used-against-you-court
http://time.com/3068396/crime-predictor/#

Data Collector 5: Snapchat — 44% ownership

Data Collected

- *Personal information*: personal details, credit card, photos & messages sent.
- *Usage*: Photos sent, phone photos, contacts book, web searches
- *Third party information*: partners, government request; third party sources.

Data statistics

- 166 mn users a day
- 9000 photos uploaded a second

Functions

1) *Geofilters*
- Add filters to picture depending on location eg: High School will have effects specific for that.

2) *Snapchat ads*
- Advertising customised by snapchat effects

3) *Snapcodes*
- Information transferred through code scanning

Source: Google Schulte Research Estimates.
https://www.snap.com/en-US/privacy/privacy-policy/
https://techcrunch.com/2017/05/10/snapchat-user-count/

Data Collector 3: Google cloud platform

Data Collected	Privacy
Open Cloud • Software developer (Compute) • Storage • Globally accessible (Networking) • Data analysis (Big Data) • Cognitive services (Machine Learning) • Security • Analytics review (Management tools) • Links other cloud products (Development tools)	• *Government sharing:* NSA gets information; About 10 000 accounts are shared per year (all information Google has on them), however Google does not always give information out. • *Shares information:* Shares information except medical facts, racial or ethnic origins, political or religious beliefs or sexuality. Shares to third party apps or businesses.

Source: Google Schulte Research Estimate.
https://cloud.google.com/products/
http://www.dailymail.co.uk/sciencetech/article-2551277/Technology-giants-reveal-ordered-turn-information-Government.html
https://www.google.com/policies/privacy/

The heart of Google's AI

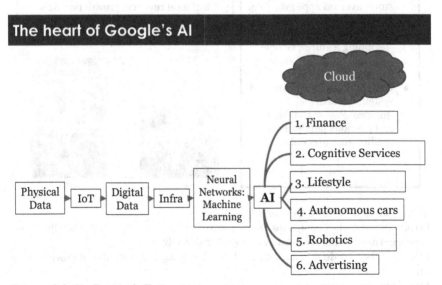

Source: Schulte Research Estimates.

1. Financial Services

A.   **E-commerce**

Google Play

B.   **Credit Rating**

Android Pay

Source: Google Schulte Research Estimates.

1.a. E-commerce, Google Play

AI in E-commerce	Contribution to Revenue
• *Google play*: Online app store; over 1m apps available. • *Google Assistant*: Retail stores pay cuber rent "shopping feature"; Google Assistant tells users store location. • *UPI (fund transfer)*: Android system features • *Adword*: Past search history and previous purchases will predict what users like.	Google Play Revenue: $3.3 bn 3.6% of total revenue, growth rate 82%

Q4 2016 Net Revenue, App Store and Google Play, Worldwide

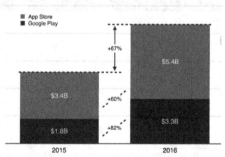

Source: Sensor Tower

Source: Google Schulte Research Estimates.
https://www.forbes.com/sites/jiawertz/2017/02/26/how-artificial-intelligence-can-benefit-e-commerce-businesses/#6852fd6854b2
https://www.recode.net/2017/5/23/15681596/google-assistant-ecommerce-revenue
http://www.gsmarena.com/google_play_app_revenue_up_82_in_q4_2016_60_growth_for_the_app_store-news-22759.php
https://abc.xyz/investor/pdf/20161231_alphabet_10K.pdf

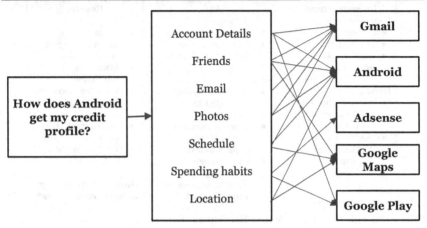

Source: Google Schulte Research Estimates.

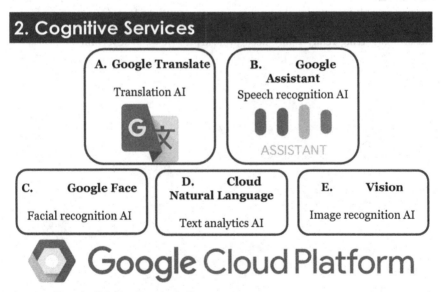

Source: Google Schulte Research Estimates.

2a. Google Translate - Translates 128,000 bibles/day

Infrastructure	AI	Results
Recurrent neural networks: • Predicts dialect • Uses syntax meaning, not direct translation *Technique:* • Word Embedding (Gives a number to a word) • Each sentence is a mathematical sequence	• Adapts to dialect • Translates broader meaning of sentence *AI links* • Speech recognition → App to translate spoken • Image recognition → Words on pictures to text • Image recognition → Writing down characters instead of using keyboard	• 103 languages: Direction languages are translated • Inaccurate translation: Reported inaccurate translations helps machines learn • Offline translation available • Works in noisy environment **Human accuracy: 77%** **Google accuracy: 72%**

100 Billion words translated a day

Source: Google Schulte Research Estimates.
https://www.quora.com/How-does-Google-translate-work-Do-they-have-datab
ase-for-all-words-of-a-particular-language
https://play.google.com/store/apps/details?id=com.google.android.apps.transla
te\&hl=en
http://www.k-international.com/blog/google-translate-facts/
https://www.washingtonpost.com/news/innovations/wp/2016/10/03/google-tr
anslate-is-getting-really-really-accurate/?utm_term=.998bc4e1bfaf

2.b. Google Assistant — Speech Recognition

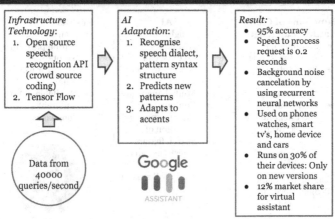

Infrastructure Technology:
1. Open source speech recognition API (crowd source coding)
2. Tensor Flow

Data from 40000 queries/second

AI Adaptation:
1. Recognise speech dialect, pattern syntax structure
2. Predicts new patterns
3. Adapts to accents

Google
ASSISTANT

Result:
- 95% accuracy
- Speed to process request is 0.2 seconds
- Background noise cancelation by using recurrent neural networks
- Used on phones watches, smart tv's, home device and cars
- Runs on 30% of their devices: Only on new versions
- 12% market share for virtual assistant

Source: Google Schulte Research Estimates.
https://9to5google.com/2017/06/01/google-speech-recognition-humans/
https://www.cnet.com/how-to/the-difference-between-google-now-and-google-assistant/
http://www.greenbot.com/article/2986737/google-apps/google-says-its-voice-search-is-faster-hears-you-better-in-noisy-environments.html
http://www.techradar.com/news/google-assistant-here-are-the-phones-and-devices-with-googles-ai-helper

2.c. Google Face — Facial Detection

Infrastructure	AI capabilities	Results
• *Detects landmark*: Eyes, nose, mouth • *Detects classification:* Characteristic ie; eyes opened or closed	• Facial detection • Emotions present • Facial tracking Can't do • Facial recognition • Emotional depth • Indentify: Age, ethnicity, gender	Facial detection accuracy: 99.96% Identify people with up to 36 degree turn Useful for: 1. Police 2. Advertising 3. Businesses 4. Research

2 bn consumer profiles

Source: Google Schulte Research Estimates.
https://developers.google.com/vision/face-detection-concepts
https://www.kairos.com/blog/face-recognition-kairos-vs-microsoft-vs-google-vs-amazon-vs-opencv
https://www.fastcompany.com/1768963/how-googles-new-face-recognition-tech-could-change-webs-future

2.d. Cloud Natural Language — Text Analytics

Functions of Cloud Natural Language
- Sentimental Analysis: Rating
- Parse Intent: Meaning of sentences
- Open source API and integrate into own software

Extract information: People, place, time, event

AI
- Insights from customers: reviews, call centers and social media
- Multi-Language
- Content classification and assortion of data
- Search feature: Questions can be asked about a text and it will answer

Results
- Can analyse 740bn pages in 1 second
- 8% error word rate
- Available in 80 different languages

Source: Google Schulte Research Estimates.
https://cloud.google.com/natural-language/
https://cloud.google.com/blog/big-data/2017/01/new-features-in-the-google-cloud-natural-language-api-thanks-to-your-feedback
https://www.quora.com/How-well-does-Google-Search-understand-natural-language-at-this-point-in-time

2.e. Vision — Image Recognition

Image recognition

- 10% of Google's traffic is on Google Images, Google developed image recognition for better search results. Can detect what is happening in the picture with 93.9% accuracy.

- Google advertising tracks behavior patterns and locational data by image recognition

- Google Image recognition average annual growth rate 19.5%

- Vision can solve CAPTCHA (use of images to display text to prevent bots) 99.8% of the time.

Image Recognition applications

- Google Search engine can search for images on your own computer by using key words

- Safesearch: Detects inappropriate content and filters those images out

- Extract text from images

- Detect (from thousands of categories) objects in a picture such as type of flower.

Source: Google Schulte Research Estimates.
https://www.theverge.com/2014/4/16/5621538/google-algorithm-can-solve-recaptcha-almost-every-time
https://www.dpreview.com/news/7658223665/google-algorithm-can-describe-image-contents-with-93-9-accuracy

Case study: Deepmind division

<u>Deepmind</u>

- Alphabet Inc's division for AI development that specializes in Neural networks. This was acquired for US $625 million.

Intends to expand in:

- Machine memory.

- Deep Reinforced learning using unsupervised learning, AI can learn how a system works using the pixels as its only raw data input.

- Partner with Hospitals in order to use neural networks detect diseases before they arrive.

- Reduce operations costs. Deep learning AI have reduced the energy usage of Google's servers by 40%.

Source: Google Schulte Research Estimates.

3. Lifestyle Services

Source: Google Schulte Research Estimates.

3.a. Music

Music — YouTube

Scale of data:
- 1.3 bn Total users on YouTube
- 300hrs of videos uploaded per min
- Global scale: 80% of videos viewed outside US; total 3.25 bn hrs watched on YouTube every month
- 9% of SME's in the US use YouTube

Financials:
- 6% of ad revenue comes from YouTube.
- Annual cost of running YouTube is $6.4 bn

AI in Music

Recommendations for new videos
- What people watch next after a video
- Consumer credit profile
- Comments, likes & shares

Advertising
- Video analytics determine genre of ad shown
- Considers credit profile

Analytics of Music
- Deep learning neural networks classify genre of a song

Source: Google Schulte Research Estimates.
https://fortunelords.com/youtube-statistics/
http://variety.com/2016/digital/features/media-companies-youtube-ai-1201912832/

4. Autonomous car: Waymo

a. **Hardware**

Velodyne
LiDAR™

b. **Software**

WAYMO

Source: Google Schulte Research Estimates.

4.a. Hardware, will it succeed?

LiDAR Systems

- Detection of:
  - o Vision
  - o Light detection
  - o Radars
  - o Ranging radars
- Use 3 LiDAR systems: short, medium long
  - o Solved short range blind spot problem
- Karficks solution reduced cost of LiDAR by 90%

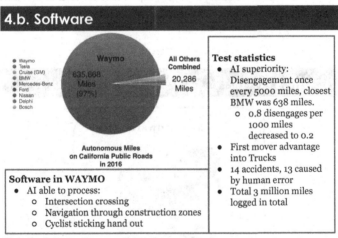

Navigant Research 2017 Automated Driving Systems Leaderboard

Source: Los Angeles times

Waymo have the best technology but do not have the capacity nor production deals to commercialise.

Source: Google Schulte Research Estimates.
http://www.latimes.com/business/autos/la-fi-hy-driverless-rankings-20170404-story.html
http://fortune.com/2017/01/08/waymo-detroit-future

4.b. Software

- Waymo
- Tesla
- Cruise (GM)
- BMW
- Mercedes-Benz
- Ford
- Nissan
- Delphi
- Bosch

Waymo 635,868 Miles (97%)

All Others Combined 20,286 Miles

Autonomous Miles on California Public Roads in 2016

Test statistics
- AI superiority: Disengagement once every 5000 miles, closest BMW was 638 miles.
  - o 0.8 disengages per 1000 miles decreased to 0.2
- First mover advantage into Trucks
- 14 accidents, 13 caused by human error
- Total 3 million miles logged in total

Software in WAYMO
- AI able to process:
  - o Intersection crossing
  - o Navigation through construction zones
  - o Cyclist sticking hand out

Source: Google Schulte Research Estimates, Forbes.
https://www.forbes.com/sites/chunkamui/2017/02/08/waymo-is-crushing-it/#4af4bee5aa9f
https://www.androidheadlines.com/2017/06/waymo-beginning-test-autonomous-software-trucks.html
https://www.theverge.com/2017/2/1/14474790/google-waymo-self-driving-car-disengagement-dmv-california
https://9to5google.com/2017/05/09/waymo-miles-3-million-may/

Industry outlook: Autonomous Cars

Waymo — Google Car

- Google looks to build the worlds first completely automated car
- Currently behind only on production deals and capacity
- On average invested 30 million per year project to build it
- Expected to launch the product at the year 2020–2025
- Projected 12 million units of completely driverless cars to be sold by 2035
- Boston Consulting group projects industry to be worth 42 billion by 2025
- Opportunity: 15% of cars on the road projected to be fully autonomous

Source: Google Schulte Research Estimates.
https://www.forbes.com/sites/brookecrothers/2015/11/12/google-is-leader-in-r
evolutionary-self-driving-cars-says-ihs/#5ffebc5a39b7
https://www.wired.com/insights/2013/08/top-5-market-trends-driverless-cars-
will-rev-up-in-the-future/

5. Robotics

Source: Google Schulte Research Estimates.

5.a. Hardware

ROBOTIC ARMS: Boston Dynamics Division
Google has partnered with Johnson and Johnson to create a robotic
arm that can assist in medical surgery, robotic dog and human.

Robotic arms current progress:

Unsupervised adaptation	Adverse objects	Deep visual foresight	Grasp detection	Q-learning	Hand-eye coordination
Robots can adapt to tasks without instructions from a human.	Robots can develop memory on dangerous objects	A model that lets the robot do tasks in environments without supervision in order to collect data	Ability to grab and move objects from the right angle	Robot arms takes optimal action, this has been tested to pick up objects for safety.	Robots able to detect objects and pick them up
i.e: moving objects out of the way	i.e: learning sharp objects		i.e: angle to pick up blocks	i.e: not knocking down other objects	

Source: Google Schulte Research Estimates.

6. Advertisement

Source: Google Schulte Research Estimates.

6.a. Search engine (part 1) — AI and Data used in advertising algorithms

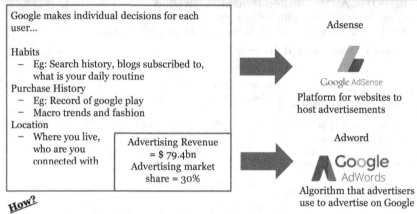

Google makes individual decisions for each user...

Habits
  – Eg: Search history, blogs subscribed to, what is your daily routine
Purchase History
  – Eg: Record of google play
  – Macro trends and fashion
Location
  – Where you live, who are you connected with

Advertising Revenue
= $ 79.4bn
Advertising market
share = 30%

Adsense

Google AdSense
Platform for websites to host advertisements

Adword

Google AdWords
Algorithm that advertisers use to advertise on Google

How?

Google uses AI to predict who is likely to click: CTR (% that an ad gets clicked on) Average of 1.91% on search network and 0.35% on display network.

Source: Google Schulte Research Estimates.
https://www.wired.com/2009/03/google-ad-annou/
http://expandedramblings.com/index.php/google-advertising-statistics/

6.a. Search engine (part 2) — Google Ad Revenue

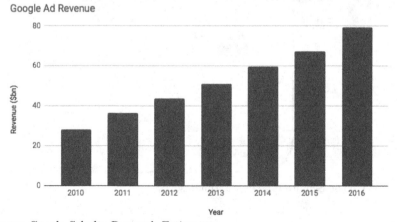

Google Ad Revenue

Source: Google Schulte Research Estimates.
https://www.statista.com/statistics/266249/advertising-revenue-of-google/

6.a. Search engine (part 3) — Average cost per click within industries

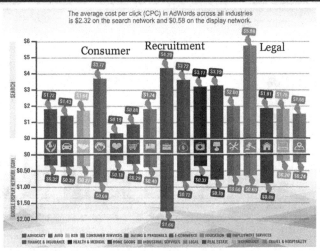

Source: Google Schulte Research Estimates.
http://www.wordstream.com/blog/ws/2016/02/29/google-adwords-industry-be
nchmarks

6.a. Search engine (part 4) — Which industries are most likely to be clicked

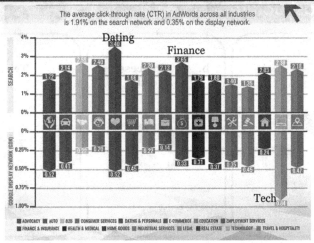

Source: Google Schulte Research Estimates.
http://www.wordstream.com/blog/ws/2016/02/29/google-adwords-industry-be
nchmarks

6.a. Search engine (part 5) — Google's top 10 customers; Amazon doubles 2nd place

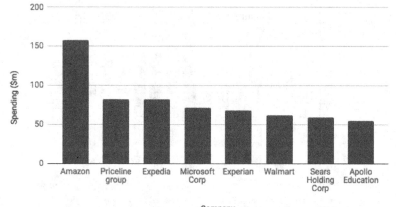

Top 9 customers of Adword

Source: Google Schulte Research Estimates.
https://www.hawkeyeppc.com/single-post/2016/09/15/The-Top-10-Most-Inter esting-Unique-Statistical-Facts-about-Google-AdWords

7. Infrastructure

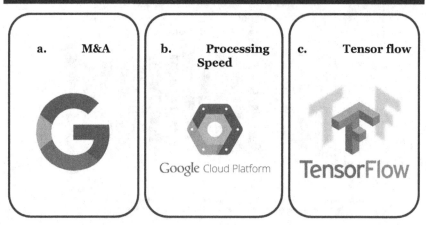

| a. M&A | b. Processing Speed | c. Tensor flow |

Source: Google Schulte Research Estimates.

7.a. M&A

Company Acquired & Date	Function	Ownership	Projects/ Divisions
Image Recognition			
DNN research (2013)	Image Search Engine	100%	Google Search Engine
Moodstocks (2016)	Image enhancement	100%	Image enhancement
JetPac (2014)	Deep learning	100%	Google Maps
Speech recognition			
API.AI (2016)	Speech/ Intent recognition	100%	Chatbots
Dark Blue labs (2014)	Language query	100%	Google Translate
Neural System			
Deep Mind (2014)	Cognitive thinking	100%	Google Now/ Waymo
Emu (2014)	Schedules/ Appointments	100%	Google Calendar
Timeful (2015)	Machine learning - Schedules	100%	Gmail
Data Mining & Big Data			
Granata (2015)	Scenario Analysis	100%	Adword/ Adsense

Source: Google Schulte Research Estimates.

7.b. Processing Speed: 4% Marketshare

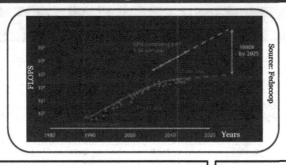

1 FLOPS:
1 instruction
executed per
second.

Speed	TPU: Tensor Processing Unit
• TPU2 speed at 45 teraflops • As Moore's law slows down, GPU/TPU computing is taking over • TPU2's speed was doubled from TPU	• TPU2 upgraded from TPU • Operations cost rise: Power usage of 40 Watt's quadrupled to 160

Source: Google Schulte Research Estimates.
https://www.fedscoop.com/era-ai-computing/
https://www.nextplatform.com/2017/05/22/hood-googles-tpu2-machine-learning-clusters/

7.c. Tensor Flow

Tensor flow is a combination of...

| Python | C++ | Java | Go |

Tensor flow using AI for Research:

Banking	**Media**	**Business Apps**	**Pharma**	**Translation**
Using AI to enhance search result accuracy according to consumer credit profile	Programs such as vision that use image recognition	AI that automatically replies to emails.	Neural network that identifies possible drug candidates	Optical character recognition allows for real time instant translation

Source: Google Schulte Research Estimates.

Financials: Top 3 Assets + Liabilities

	Assets			
Year:	**2016**	**2015**	**2014**	**2013**
Cash	12,918.00	16,549.00	18,347.00	18,898.00
Marketable Securities	73,415.00	56,517.00	46,048.00	39,819.00
PPE	34,234.00	29,016.00	23,883.00	16,524.00

	Liabilities			
Year:	**2016**	**2015**	**2014**	**2013**
Accrued Post Retirement Compensation	3,976.00	3,539.00	3,069.00	2,502.00
Long Term Debt	3,935.00	1,995.00	3,228.00	2,236.00
Accrued Expenses	2,942.00	2,329.00	1,952.00	1,729.00

Source: Google Schulte Research Estimates.

Appendix 9

Apple

Schulte-Research

"AI is Horizontal in nature, running across all products." — Tim Cook

The Big Bang of Apple

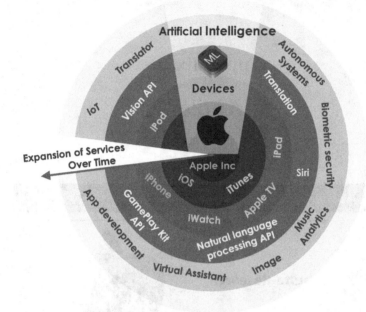

How does Apple make its consumer profile?

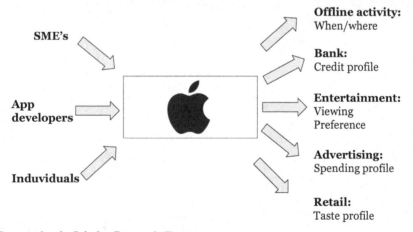

Source: Apple Schulte Research Estimates.

How does Apple get its data?

Itunes:
800 mn users

Cloud:
782 mn users

Devices:
1 bn active devices

Apps Store:
2.2 mn apps available

Source: Apple Schulte Research Estimates.

What do they do with their data?

Understand customer behavior: Partnered with IBM for data analytics

Improving and producing new Apple products

Improving and producing new Apple products

Improving mass media Advertising

Source: Apple Schulte Research Estimates.

Apple's scale of data

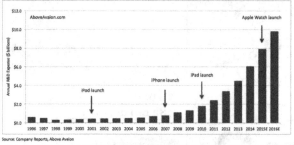

Expenditure on R&D

US $10 billion spent on Research and development

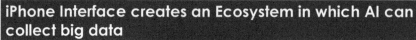

2017: 120-150 MILLION

2016: 123.5 MILLION

2015: 122.8 MILLION

iPhones sold
As of mid 2016, there was a cumulative total of 1 billion iPhones active.

Source: Apple Schulte Research Estimates, Forbes.

iPhone Interface creates an Ecosystem in which AI can collect big data

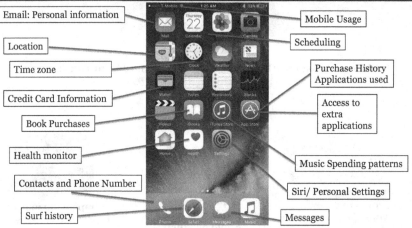

Source: Apple Schulte Research Estimates.

Apple's Internal AI development

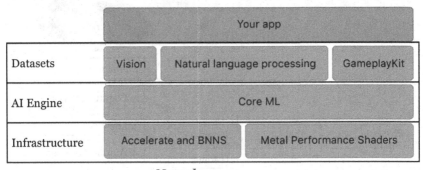

Datasets	Vision	Natural language processing	GameplayKit
AI Engine	Core ML		
Infrastructure	Accelerate and BNNS	Metal Performance Shaders	

Neural Network   **GPU chips**

Source: Apple Schulte Research Estimates.
https://developer.apple.com/documentation/coreml

Apple's external AI development; powered by Apple's Neural Chip engine

Products	Description	How does it make money
Vision	<ul><li>Facial detection & recognition</li><li>Image analysis (also reads text)</li><li>Barcode detection</li><li>Image alignment</li><li>Tracking</li></ul>	<ul><li>Security</li><li>Business licensing<ul><li>R&D</li><li>Marketing</li></ul></li><li>Healthcare</li></ul>
Natural language processing	<ul><li>Language identification</li><li>Tokenization</li><li>Lemmatization</li><li>Part of speech</li><li>Identifying nouns</li></ul>	<ul><li>Translation</li><li>Business licensing</li><li>Music analytics</li></ul>
GameplayKit	Gaming and app development. Uses AI to determine computer bot behavior.	<ul><li>Appstore</li><li>Game development</li></ul>

Source: Apple Schulte Research Estimates.
https://developer.apple.com/

The heart of Apple's AI

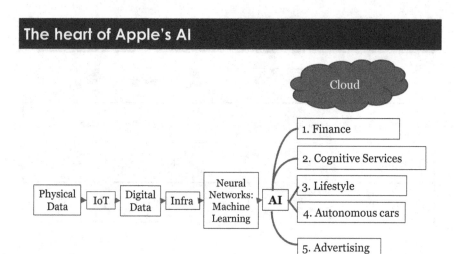

Source: Schulte Research Estimates.

1. Financial Services

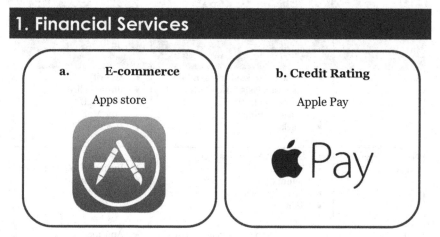

Source: Apple Schulte Research Estimates.

1a. Ecommerce

<table>
<tr><td>

AI in E-commerce

- *Appstore*: Online app store, over 2.2 million apps.
- *Genius Bar*: Online booking system for device repair. Collects data on problems with devices and consumer profiles for Applecare.
- *Apple store*: Integrated into android

Contribution to Revenue

Total E-commerce sales: $16.8 bn
7.8% of total revenue, year end 2016

</td><td>

Ikea Partnership: Bringing augmented reality to shopping

App: Augmented reality to envision what objects look like in a room. Upgraded from "Place in your room".

Ecommerce through augmented reality through a Camera:

1. Allows users to try and see how IKEA furniture looks in a room
2. Measurements see if it would fit
3. Click buy & have it shipped

</td></tr>
</table>

Source: Apple Schulte Research Estimates.
https://www.dezeen.com/2017/06/21/ikea-apple-collaboration-ar-shopping-app/
https://www.statista.com/statistics/263795/number-of-available-apps-in-the-apple-app-store/

1b. Credit profile: 85 million users on Apple Pay

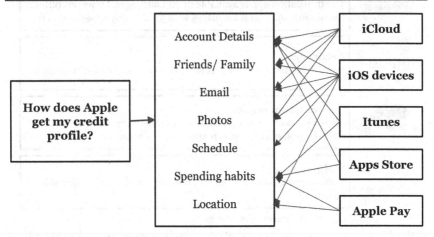

Source: Apple Schulte Research Estimates.

2. Cognitive Services

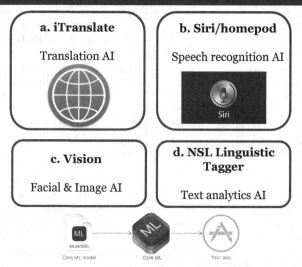

Source: Apple Schulte Research Estimates.

2a. Translation

Languages:
The translation app is available in 90 Languages. However, offline translation only has 8 languages with 3$ monthly subscription fee.

Accuracy:
Voice activated translation is not effective in noise environments.

Conversation mode:
Allows people to converse in different languages with the conversation option. Speech-speech translation.

Webpage:
Translates websites into 38 different languages, only available in Safari.

Source: Apple Schulte Research Estimates.
http://www.telegraph.co.uk/travel/advice/The-five-best-translation-apps-for-t ravellers/

2b. Speech recognition

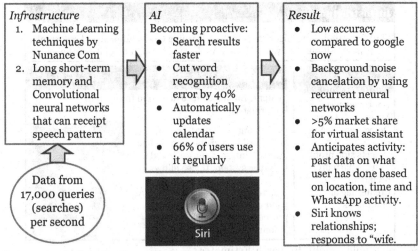

Infrastructure
1. Machine Learning techniques by Nunance Com
2. Long short-term memory and Convolutional neural networks that can receipt speech pattern

Data from 17,000 queries (searches) per second

AI
Becoming proactive:
- Search results faster
- Cut word recognition error by 40%
- Automatically updates calendar
- 66% of users use it regularly

Siri

Result
- Low accuracy compared to google now
- Background noise cancelation by using recurrent neural networks
- >5% market share for virtual assistant
- Anticipates activity: past data on what user has done based on location, time and WhatsApp activity.
- Siri knows relationships; responds to "wife.

Source: Apple Schulte Research Estimates.

https://www.bloomberg.com/news/videos/2015-06-08/siri-how-many-questions-do-you-answer-per-minute-

https://9to5mac.com/2016/06/06/siri-embarrassment-poll/

http://www.businessinsider.com/amazon-echo-google-home-microsoft-cortana-apple-siri-2017-1

http://www.iphonefaq.org/archives/972621

http://www.pocket-lint.com/news/129922-apple-homekit-and-home-app-what-are-they-and-how-do-they-work

http://www.cnbc.com/2017/05/20/siri-vs-google-assistant-on-iphone.html

2c. Vision, Facial Recognition (part 1)

Infrastructure

Shift from traditional algorithms to deep learning (convolutional and recurrent). Identifies positions of: Eyes & Mouth. Apple can identify 1 bn of its consumers.

AI	Results
Improved accuracy. Misses less faces. Detects smaller faces. Detects side profile. Detects partially occluded. Creates facial landmarks.	Patent: Enhanced face detection from depth information Depth data of a user's face is collected by a visible light image sensor, infrared sensor & a 3D capture system. It also allows for the inclusion of UV sensor, scanning laser & ultrasound. This allows high accuracy of facial recognition, thus a form of biometric security. 

Source: Apple Schulte Research Estimates.
http://www.telegraph.co.uk/travel/advice/The-five-best-translation-apps-for-travellers/
https://developer.apple.com/machine-learning/
http://appleinsider.com/articles/17/03/07/apple-patents-method-of-detecting-faces-using-depth-information-bolsters-iphone-8-rumors

2c. Vision, Image Recognition (part 2)

Image Recognition method

| Requests | → | Request Handler | → | Results |

Classification/ Detection:
- Active Vision
- Can do optimization

What can it detect
- Text
- Rectangles
- Bar code
- Objects/ people

Tracking:
- Square locked to object/ person
- Cannot do optimization

Image registration
- Can align two images together, even if one is at a different angle

Unlike Google, Apple's Vision API is hosted on device and not on the cloud, making it free and private, draw back is it is not as fast as Google.

Source: Apple Schulte Research Estimates.
https://developer.apple.com/machine-learning/

2d. Natural Language Processing

Input	AI	Results
Social media	Language identification: detects language	Available in 52 languages
Article	Tokenization: Break down into meaning of word, sentence and paragraph	90% accuracy
Document		Part of speech tagging power: increased from 50,000 to 80,000 tokens per second
Blog	Part of speech tagging: nouns, verbs, adjectives etc	
Report		Name of entity recognition: Increased from 40,000 to 65000 tokens per second.
Reviews	Lemmatization: Root of verbs, important for unbounded vocab such as Russian	
Posts	Name of entity recognition: Recognising Common nouns	

Source: Apple Schulte Research Estimates.
https://developer.apple.com/machine-learning/
https://willowtreeapps.com/ideas/apples-natural-language-processing-nlp-api

3. Lifestyle Services

a.   Music

iTunes

Source: Apple Schulte Research Estimates.

3a. Music

Apple Music:

- Mobile Music Market share: 18%, revenue $4.8 bn
- Digital downloads (iTunes): 75% Global market share, revenue $1.6 bn
- Total 40.7 mn subscriptions
- Total songs 10 mn songs

Revenue:

- $2.5 bn revenue (Half of Spotify)

AI in Music:

Recommendations for new songs based on...
- Genre of playlist's and songs
- Apple credit profile
- Billboard/ common interests of songs

Analytics of Music
- Deep learning neural networks classify genre of a song

Source: Apple Schulte Research Estimates.
https://www.statista.com/statistics/475978/online-music-services-popular-young-people-usa/
https://www.macrumors.com/2017/03/30/us-music-industry-revenue-growth-streaming-music/
https://www.imore.com/apple-music
http://appleinsider.com/articles/13/06/20/apples-itunes-accounts-for-75-of-global-digital-music-market-worth-69b-a-year
https://www.digitalmusicnews.com/2017/03/30/apple-music-spotify-music/
http://www.investors.com/news/technology/spotify-apple-music-lead-in-streaming-tunes-but-its-early-days/

4. Autonomous car: Project Titan

1. Hardware	2. Software

Source: Apple Schulte Research Estimates.

4a. Hardware

Apple is not making its own hardware, but has started to make deals with other companies...

Contemporary Amperex Technology	Magna

Solved Apple's bottleneck: Electric batteries. CATL tripled Lithion-ion batteries production making it second largest consumer of car batteries behind Tesla.	Apple has been trying to negotiate with Magna to manufacture their Apple Car. Apple has been already rejected Daimler and BMW to be manufacturing partners.

Source: Apple Schulte Research Estimates.

http://www.businessinsider.com/everything-we-know-about-apples-project-titan-electric-car-2016-8?op=0#/#but-apple-has-reportedly-had-a-difficult-time-getting-an-automaker-to-help-develop-the-car-with-bmw-and-daimler-rejecting-the-tech-giants-offers-15

https://9to5mac.com/2017/07/20/apple-electric-car-batteries-china-battery-maker/

4b. Software

Project Titan car kit	Autonomous System: CarPlay
Entertainment: iTunes/Apple Music Assistance/Voice: Siri Navigation/Local: Apple Maps Image Processing/Recognition (Autonomous Driving): iPhone Camera Security locking of the car: Touch ID Third-party Software: App Store (for software updates)	Project Titan team are developing the car system rather than the car itself. The CarPlay kit will be licensed to manufacturers to earn money. *AI dependent on IoT*: Utilise IoT; image recognition and sensors are inputted into navigation system that will drive the car. Everything will be controlled on an iPad.

Source: Apple Schulte Research Estimates.

http://loupventures.com/apples-dream-car-hardware-software/

https://www.pcmag.com/commentary/348158/apples-project-titan-is-not-about-building-a-car

5. Advertisement

Source: Apple Schulte Research Estimates.

5a. Targeting Consumers: The gold from their data mine (part 1)

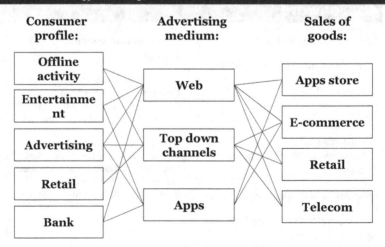

Source: Apple Schulte Research Estimates.

5b. Retail Stock management in the US: Big data used in predicting Retail sales (part 2)

Individual
1. Location
2. Demographic
3. Previous phone
4. Pre-order data

Past Trends
1. % of sales in Apple Stores in the past
2. Online demand
3. Sales through other channels such as telecom (30% in total)
4. Price modelling of different devices specifications (memory, ram, size, etc.)

66% of revenue comes from iPhone sales

USA

Market share growth (retail stores) of 6.4%, totalling 37.1%

Global

Apple is the most efficient retail company, they have made retail sales of 5.5k per square foot

Source: Apple Schulte Research Estimates.
https://www.wired.com/2009/03/google-ad-annou/
https://iterativepath.wordpress.com/2015/09/25/10-problems-with-foursquare-big-data-prediction-of-iphone-6s-sales

5c. Results: regional Market share (part 3)

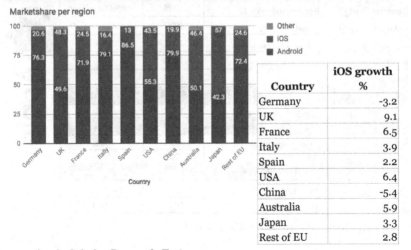

Marketshare per region

| | iOS growth |
Country	%
Germany	-3.2
UK	9.1
France	6.5
Italy	3.9
Spain	2.2
USA	6.4
China	-5.4
Australia	5.9
Japan	3.3
Rest of EU	2.8

Source: Apple Schulte Research Estimates.
https://9to5mac.com/2017/01/11/ios-market-share-kantar/

6. Infrastructure

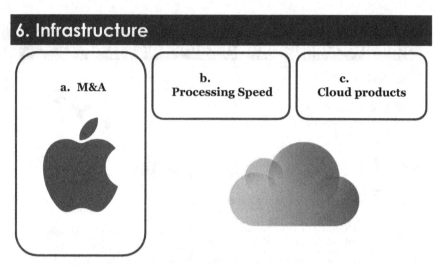

Source: Apple Schulte Research Estimates.

6a. Apple Acquisitions and Projects

Company Acquired & Date	Function	Ownership	Projects/ Divisions
Facial and image Recognition			
Emotient (2016)	Sentimental analysis	100%	Facial Detection
Polar Rose (2010)	Facial Detection	100%	Biometric security
Faceshift (2015)	Sentimental Detection	100%	
RealFace (2017)	3D Depth modeling	100%	
Perceptio (2015)	Deep Learning	100%	
Speech recognition			
Novauris (2014)	Speech Recognition	100%	Siri
Virtual Assistant			
Volaquil (2015)	Voice Recognition	100%	Siri
Cue (2013)	Scheduler	100%	
Data Mining & Big Data			
Tuplejump (2016)	Data Analytics	100%	Marketing
Turi (2016)	Data Analytics	100%	R&D

Source: Apple Schulte Research Estimates.

6b. Processing Speed

Speed test for 1gb of data

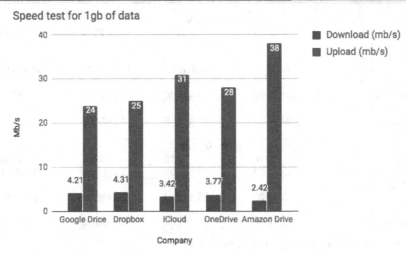

Source: Apple Schulte Research Estimates.

6c. Cloud Products

Products		Description	Subscription
Backup & Restore		Users can reload old data previously stored.	Free
Back to my Mac		Log in remotely to your mac	Free
Email		Stores emails	Free
Find my Friends		Share your location with other iOS users	Free
Find my iPhone		Track your iPhone's location	Free
iCloud Keychain		Security for stored passwords, credit cards	Free
iTunes Match		Scans CD's and records for iTunes music use	Annual Fee
iWork		Creating and editing of documents	One off Payment
Photostream		Stores 1000 photos	Monthly fee
iCloud photo Lib		Stores all photos but at lower definitions	Free
Storage		Storage of any documents	5gb Free
iCloud Drive		Host of operating systems	5gb Free

Source: Apple Schulte Research Estimates.

Financials: Top 3 Assets + Liabilities

Assets				
Year	2016	2015	2014	2013
Long Term marketable securities	170,430.00	164,065.00	164,065.00	106,215.00
Short Term investments	46,671.00	20,484.00	11,233.00	26,287.00
Cash + Cash equiv	20,484.00	21,120.00	13,844.00	14,259.00

Liabilities				
Year	2016	2015	2014	2013
Long Term debt	75,427.00	53,329.00	28,987.00	16,960.00
Accounts Payable	37,294.00	35,490.00	30,196.00	22,367.00
Accrued expenses	22,027.00	25,181.00	18,453.00	13,856.00

Source: Apple Schulte Research Estimates.

Appendix 10

Facebook

Schulte-Research **facebook.**

"AI is closer to being able to do more powerful things than most people expect — driving cars, curing diseases, discovering planets, understanding media. Those will each have a great impact on the world, but we're still figuring out what real intelligence is."

— Mark Zuckerberg

The Big Bang of Facebook

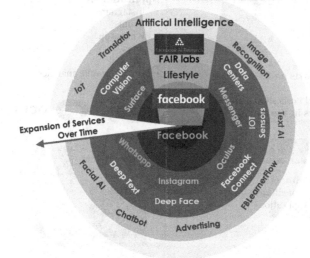

How does Facebook make its consumer profile? Advantage is access to unintentional data

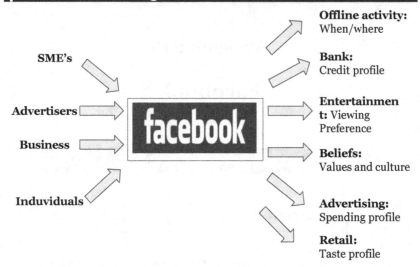

Source: Apple Schulte Research Estimates.

How does Facebook get its data? It lacks the ability to collect Ecommerce data

Source: Facebook Schulte Research Estimates.

What do they do with their data? 2.5 Billion pieces of content Daily + 105 TB every 30 minute

Facebook Storage: Cloud locker + Data Centers

Facebook insights has joint with Google Analytics

Security:
Algorithms designed to use past data to detect unusual behavior

Advertising:
Algorithms and direct marketing

Improve products and services:
Provide suggestions and inform you about nearby services

Source: Facebook Schulte Research Estimates.

Facebook's scale of data: On level with Google's consumer profiles

Active Quarterly Users

Content shared (billions per month)

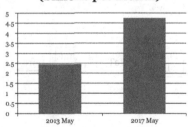

Currently Q2 2017, 2 billion Users active

Contents shared on Facebook has nearly doubled since 2013

Source: Facebook Schulte Research Estimates.
https://zephoria.com/top-15-valuable-facebook-statistics/

Facebook Interface creates an Ecosystem in which AI can collect big data

Source: Facebook Schulte Research Estimates.

The heart of Facebook's AI

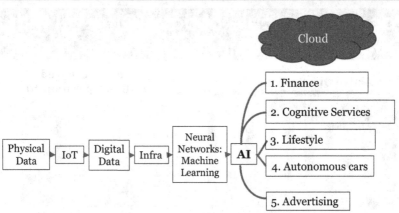

Source: Schulte Research Estimates.

1. Financial Services

a. E-commerce	b. Credit Rating
Shopify - Facebook Shop	Friend-to-friend pay

Source: Facebook Schulte Research Estimates.

1a. Ecommerce, shopify. Facebook does not have its own E-commerce platform, it should.

What is it?	How does Shopify use AI?

• Facebook has partnered with Shopify to set up E-commerce platform
• Facebook hosts the deals while Shopify handles transactions

	• *Facebook recommendations*: Based on Facebook's credit profile, AI will recommend shops to users.
	  ○ Previous purchases and what other people have purchased next
	  ○ Likes and shares of genres
	• *Shopify's Virtual marketing assistant Kit*: Connects merchants with Facebook's advertising platform, has made 900 million unique connections in total
	  ○ Chatbot handles: Page ads, emails (direct marketing and thank you) and updates
	  ○ Open API so merchants can edit messages

Source: Facebook Schulte Research Estimates.
https://www.shopify.com/facebook
https://venturebeat.com/2016/07/12/shopify-owned-kits-artificial-intelligence-marketing-bot-now-on-facebook-messenger/

1a. Credit rating; Facebook's own transaction based platform does not make money

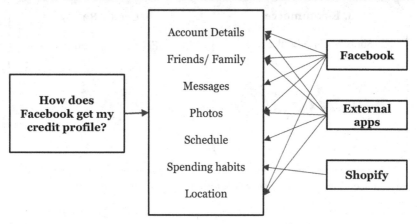

Source: Facebook Schulte Research Estimates.

2. Cognitive Services; no speech AI

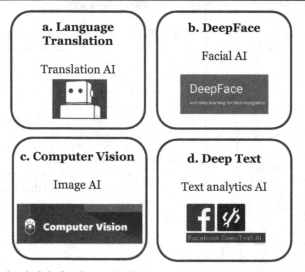

Source: Facebook Schulte Research Estimates.

2a. Translation: a unique neural network and more efficient way to translate

Infrastructure:	AI:	Results:
• *High parallel GPU*: Unifying of GPU's • *Convolutional Neural Networks*: Unlike RNN, it doesn't go left to right, it computes the meaning of each word simultaneously. This utilizes the GPU parallelism more efficiently. CNN also processing information hierarchy which makes it better at catching complex relationships.	• *High parallel GPU*: Unifying of GPU's • *Convolutional Neural Networks*: Unlike RNN, it doesn't go left to right, it computes the meaning of each word simultaneously. This utilizes the GPU parallelism more efficiently. CNN also processing information hierarchy which makes it better at catching complex relationships.	• 2 billion texts are translated daily, in 40 different languages and 1800 directions • Bing uses FAIR's translation • 800 million people use this; 40% of its users

Source: Facebook Schulte Research Estimates.
https://code.facebook.com/posts/1978007565818999/a-novel-approach-to-neural-machine-translation/
https://techcrunch.com/2016/05/23/facebook-translation/
https://techcrunch.com/2017/06/13/facebook-messenger-translator/

2b. DeepFace; Facebook can identify ¹/₃ of the world's population

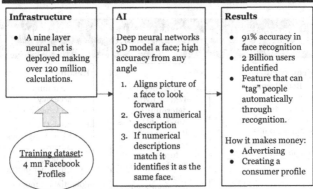

Infrastructure	AI	Results
• A nine layer neural net is deployed making over 120 million calculations. Training dataset: 4 mn Facebook Profiles	Deep neural networks 3D model a face; high accuracy from any angle 1. Aligns picture of a face to look forward 2. Gives a numerical description 3. If numerical descriptions match it identifies it as the same face.	• 91% accuracy in face recognition • 2 Billion users identified • Feature that can "tag" people automatically through recognition. How it makes money: • Advertising • Creating a consumer profile

Source: Facebook Schulte Research Estimates.
http://www.computervisiononline.com/blog/deepface-facebooks-face-verification-algorithm
https://www.robots.ox.ac.uk/~vgg/publications/2015/Parkhi15/parkhi15.pdf
https://www.cs.toronto.edu/~ranzato/publications/taigman_cvpr14.pdf
http://www.investopedia.com/news/could-facebook-google-survive-face-recognition-lawsuit-fb-goog/
http://www.npr.org/sections/alltechconsidered/2016/05/18/477819617/facebooks-facial-recognition-software-is-different-from-the-fbis-heres-why
http://theconversation.com/facial-recognition-is-increasingly-common-but-how-does-it-work-61354

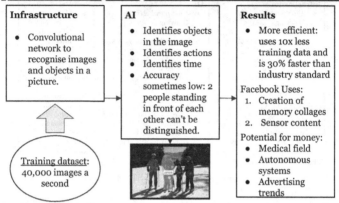

Source: Facebook Schulte Research Estimates.
http://www.socialmediatoday.com/social-networks/facebooks-evolving-image-r
ecognition-technology-offers-new-opportunities
https://www.fastcompany.com/40428910/facebooks-image-recognition-tech-is-
teaching-40000-images-a-second-to-understand-context
https://research.fb.com/category/computer-vision/

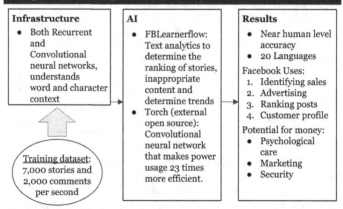

Source: Facebook Schulte Research Estimates.
https://code.facebook.com/posts/181565595577955/introducing-deeptext-faceb
ook-s-text-understanding-engine/
https://techcrunch.com/2016/06/01/facebook-deep-text/
https://research.fb.com/fair-open-sources-deep-learning-modules-for-torch/

3. Lifestyle Services

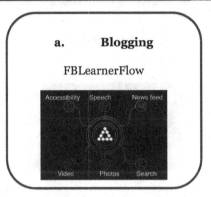

Source: Facebook Schulte Research Estimates.

3a. FBLearnerFlow: The ranking system that allows Facebook to determine and predict what we would want to see

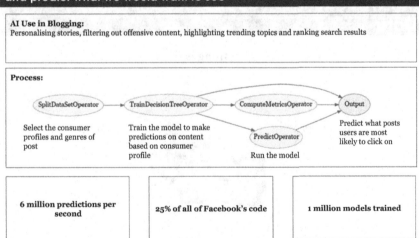

Source: Facebook Schulte Research Estimates.
https://code.facebook.com/posts/1072626246134461/introducing-fblearner-flow-facebook-s-ai-backbone/

4. Advertisement

Source: Facebook Schulte Research Estimates.

4a. What defines Facebook's unintentional Data

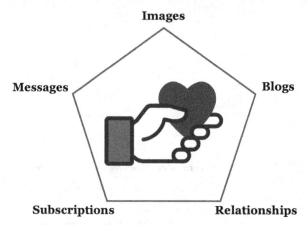

Source: Facebook Schulte Research Estimates.

4b. Advertising: what Facebook's unintentional data says: consumer data can be used in SME lending, credit analysis and HRtech

1. Demographic	2. Personal Life	3. Work life
1. Location	17. Anniversary within 30 days	33. Employer
2. Age	18. Travel from family or	34. Industry
3. Generation	hometown	35. Job title
4. Gender	19. Friends with someone	36. Office type
5. Language	who's newly married or	37. Interests
6. Education level	engaged, moved, birthday	38. Motorcycles owners
7. Field of study	20. Long-distance relationship	39. Plan to buy a car
8. School	21. New relationships	40. Who bought auto parts
9. Ethnic affinity	22. New jobs	41. Likely to need auto parts or services
10. Income and net	23. Newly engaged	42. Style and brand of car you drive
worth	24. Newly married	43. Year car was bought
11. Home ownership/	25. Recently moved	44. Age of car
type	26. Birthdays soon	45. How much money on next car
12. Home value	27. Parents	46. Where user is likely to buy next car
13. Property size	28. Expectant parents	47. # of Employees your company has
14. Square ft. of home	29. Mothers	48. Who owns SMEs
15. Year home was built	30. Likely to engage in politics	49. Work in management or exec
16. Household items	31. Conservatives and liberals	
	32. Relationship status	

Source: Facebook Schulte Research Estimates.
https://www.washingtonpost.com/news/the-intersect/wp/2016/08/19/98-perso
nal-data-points-that-facebook-uses-to-target-ads-to-you/?utm_term=.5ba221d8
4233

4c. Advertising: Facebook's unintentional can give indications of intentional (Cont'd)

4. Spending habits	5. What users buy
50. Donated to charity	66. Active credit card users
51. Operating system	67. Credit card type
52. Who plays canvas games	68. Debit card users
53. Who owns a gaming console	69. Balance on their credit card
54. Who have created an FB event	70. Users who listen to the radio
55. Who has used FB Payments	71. Preference in TV shows
56. Who are top spenders	72. Users who use a mobile device
57. Users who administer FB page	73. Internet connection type
58. Users who uploaded photos	74. Purchases of smartphone/tablet
59. Internet browser	75. Internet usage
60. Email service	76. Users who use coupons
61. Early/late adopters of	77. Clothing purchases
technology	78. Household shopping habits
62. Expats	79. Users who are "heavy"drinkers
63. Users credit union, bank	80. Users who buy groceries
64. Users who investor	81. Users who buy beauty products
65. Number of credit lines	82. Users who buy medications

Source: Facebook Schulte Research Estimates.
https://www.washingtonpost.com/news/the-intersect/wp/2016/08/19/98-perso
nal-data-points-that-facebook-uses-to-target-ads-to-you/?utm_term=.5ba221d8
4233

4d. Advertisement revenue segment growth; decline in payments

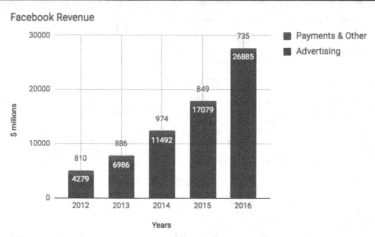

Facebook Revenue

Source: Facebook Schulte Research Estimates.

4e. Click through rate

Source: Facebook Schulte Research Estimates.
http://www.wordstream.com/blog/ws/2017/02/28/facebook-advertising-bench marks

4f. Cost per clicking on an advert

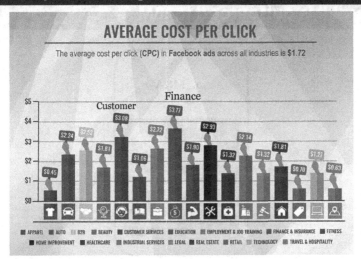

Source: Facebook Schulte Research Estimates.
http://www.wordstream.com/blog/ws/2017/02/28/facebook-advertising-bench
marks

4g. Cost of showing media (video and images)

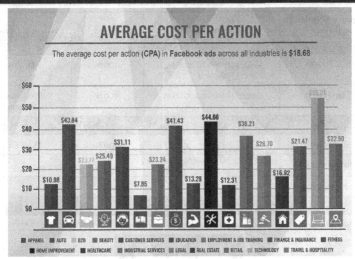

Source: Facebook Schulte Research Estimates.
http://www.wordstream.com/blog/ws/2017/02/28/facebook-advertising-bench
marks

5. Infrastructure: Facebook does not have a cloud! It is outsourced to IBM

a. M&A

Source: Facebook Schulte Research Estimates.

Facebook Acquisitions and Projects

Company Acquired and Date	Function	Ownership	Projects/ Divisions
Speech Recognition			
Wit.Ai (2015)	Voice activated Command system	100%	Oculus
Jibbigo (2014)	Speech based translation	100%	Voice Command System
Facial recognition and Image Recognition			
Face.com (2012)	Facial Recognition	100%	Deep Face
Masquerade (2016)	Facial Recognition using 3D depth modeling	100%	
Zurich Eye (2016)	3d Depth Modeling	100%	
Virtual Reality			
Oculus Rift (2014)	Gaming Virtual Reality	100%	Oculus

Source: Facebook Schulte Research Estimates.

Financials: Top 3 Assets + Liabilities

Assets				
Year	2016	2015	2014	2013
Marketable Securities	20,546.00	13,527.00	6,884.00	8,126.00
Cash + Cash equiv	8,903.00	4,907.00	4,315.00	3,323.00
PPE	8,591.00	5,687.00	13,390.00	13,070.00

Liabilities				
Year	2016	2015	2014	2013
Accrued expenses	2,203.00	1,449.00	866	555
Accounts payable	302.00	196.00	176.00	87.00
Platform Partners Payable	280.00	217.00	202.00	181.00

Source: Facebook Schulte Research Estimates.

Appendix 11

Microsoft AI

Microsoft AI

I. Summary — The Big Bang of Microsoft

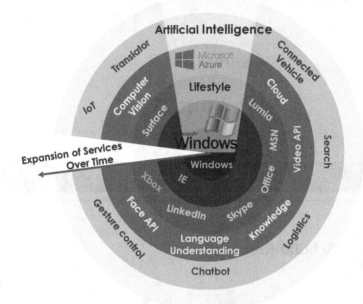

Microsoft Pros

Strong in corporate/business solutions
- Most applications and data are business faced

Fastest growing cloud
- Azure is the fastest growing cloud business (97% YOY)

Leader for desktop
- Windows is still dominating desktop operating system (92% mkt share)

High barriers to entry for MS Office
MS Office provides high barriers to entry for business solutions

Strong R&D in cognitive service

Unique data access from LinkedIn, Skype, and Xbox
- LinkedIn provides unique professional network data
- Skype provides vast video and voice data
- Xbox provides home IoT data

Microsoft Cons

Get lost in mobile era
- People move from using Windows/Mac OS to Android/IOS
- Windows mobile system only gets 1% mkt share.

Weak in consumer data access
- Most applications are towards business solutions
- Lacks consumer transaction data

No Fintech
- Weak in transforming data and technology into finance products

Microsoft data source

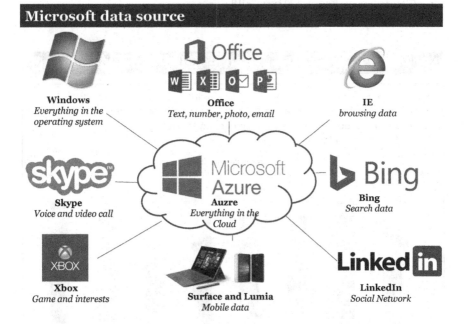

Windows
Everything in the operating system

Office
Text, number, photo, email

IE
browsing data

Skype
Voice and video call

Auzre
Everything in the Cloud

Bing
Search data

Xbox
Game and interests

Surface and Lumia
Mobile data

LinkedIn
Social Network

Microsoft by the numbers: From Windows/Office to the Cloud

Operating System and Applications
- More than 400 mn devices are running Windows 10
- 1.2 bn people in 140 countries and 107 languages use MS Office
- Office has been downloaded 340 mn time in smartphones
- Outlook.com has more than 400 mn active users
- 669,000 apps in Windows Store
- Skype: 3 bn minutes of calls each day
- Skype Translator can translate voice calls in 9 different languages
- LinkedIn: 510 mn users and 9 mn companies

The Cloud
- 80% of Fortune 500 is on Microsoft Cloud
- 40% of Azure revenue comes from startups and software SME
- Cloud revenue target: $20 bn in 2018

Microsoft Revenue Breakdown: Azure is the new engine

- Cloud business — Azure and Office 365 — is the main driver for growth
- Azure in the category "Intelligent Cloud", no detailed breakdown
- Azure achieved 97% YOY growth, according to Microsoft
- 43% YOY growth for Office 365

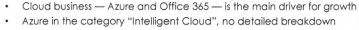

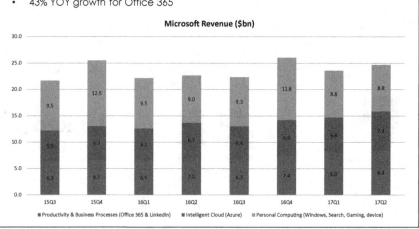

Source: Microsoft website, Schulte Research.

Can Microsoft keep its advantage in the mobile era?

The War of Operating System
- Windows is the dominant leader in desktop operating system (92%).
- The concern is the world is **moving from PC to mobile, from Windows & Mac OS to IOS and Android**. Microsoft is weak in mobile operating system (1%).
- Windows is still the first choice for most business due to the legacy advantage.
- Microsoft has more advantage in corporates (business) than consumer (lifestyle).

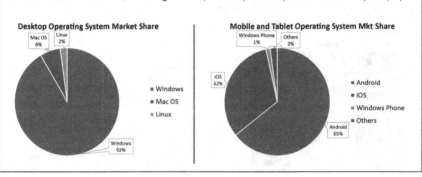

Source: Microsoft website, Schulte Research.

II. Microsoft AI Applications

1. Cognitive Services

Vision   Speech   Language   Knowledge

a. Computer Vision API: image recognition

Analyze an image
- Provides descriptions, tags, confidence level
- Identifies image types and color

Read text in images
- Detects text in an image, extract the words
- Users can take photos of text instead of typing to save time
- Handwritten text recognition supported

Example

Input	Output

Description
{ "tags": ["train", "platform", "station", "building", "indoor", "subway", "track", "walking", "waiting", "pulling", "board", "people", "man", "luggage", "standing", "holding", "large", "woman", "yellow", "suitcase"], "captions": [{ "text": "people waiting at a train station", "confidence": 0.8331026 }] }

Tags
[{ "name": "train", "confidence": 0.9975446 }, { "name": "platform", "confidence": 0.995543063 }, { "name": "station", "confidence": 0.9798007 }, { "name": "indoor", "confidence": 0.927719653 }, { "name": "subway", "confidence": 0.838939846 }, { "name": "pulling", "confidence": 0.431715637 }]

Source: Microsoft website, Schulte Research.

Computer Vision API cont'd

Recognize celebrities and landmarks
- Recognizes 200k celebrities (business, politics, sports and entertainment)
- Recognizes 9,000 landmarks globally
- Continuously evolving feature

Example

Input	Output

Categories
{"name": "building_street",
 "score": 0.66796875,
 "detail": {"celebrities": null,
 "landmarks": [{"name": "Bank of China Tower",
 "confidence": 0.9997583}

Tags
"outdoor", "building", "city", "tall", "large", "view", "tower", "front", "street", "traffic", "bridge", "sign", "top", "standing", "pole", "stop", "clock", "man", "flying", "riding", "air", "bus", "people", "group"

Captions
[{ "text": "a tall building in a city", "confidence": 0.96250556379848}

Source: Microsoft website, Schulte Research.

Face API

Face verification
- Identifies if two faces belong to the same person
- Wide application for customer verification and document check

Face detection
- Detects the position of faces
- Returns features (age, emotion, gender, pose, smile, hair) for each face

Emotion detection
- Returns emotion for each face with confidence level
- Anger, contempt, disgust, fear, happiness, neutral, sadness, surprise

Applications
- Face identification
- Similar face search
- Face grouping

Source: Microsoft website, Schulte Research.

b. Video API

Stabilize shaky videos
- Stabilize and smooth shaky videos

Face detection and tracking
- Detects up to 64 faces with face location
- Returns features like gender, age, description

Motion detection
- Detect the time motion occurred
- Wide applications in security monitoring

Generate video thumbnails
- Automatic thumbnail summary for preview

Analyze videos in real time
- Scenario detection with tags
- Emotion detection in real-time video
- Key IoT process to connect offline to online with cameras

Source: Microsoft website, Schulte Research.

Video Indexer: Unlock video insights

Features
- *Easy to use*: upload your video and go
- AI automatically extracts insights from the video
- Make video contents easier to be discovered (search)
- Detects key topics, sentiments, people, etc.
- *Wide applications*: education, conference, news, etc.

Source: Microsoft website, Schulte Research.

c. Speech Recognition

Translator Speech API
- Ten languages supported
- Available for mobile, desktop and web applications
- *Functions*: partial transcriptions as you speak, partial text translations, final transcriptions, final text translation, or audio text-to-speech translation

Speaker Verification
- Voice verification to detect if it is the same person
- Wide applications for customer verification and security checks

Speaker Identification
- Identify who is speaking
- Input the voice of the speaker and AI can detects that voice against others

Text-speech converter
- *Speech to text*: real-time speech recognition to text
- *Text to speech*: "talk back" to customers with text inputs

Source: Microsoft website, Schulte Research.

d. Language Understanding

Text Analytics API

- Extract and analyze information from text
- Sentiment analysis with score between 0 to 1
- 15 languages supported
- Automatic key content extraction
- Topic detection

Example

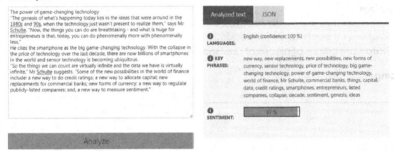

Source: Microsoft website, Schulte Research.

Language Understanding cont'd

Bing Spell Check API

- Spell check includes word breaks, slang, names, homonyms, brands, etc.

Translator Text API

- 60+ supported languages across mobile, desktop, and web
- Crowd-source translation improvement (improved with user feedback)

Language Understanding Intelligent Service (LUIS)

- A sets of APIs to understand human comment
- LUIS improves with more data input and user feedback

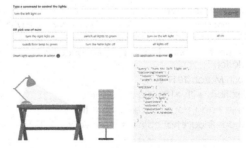

Source: Microsoft website, Schulte Research.

e. Knowledge: Recommendations API

Recommendations API
- Azure Machine learning uses customer historical data and live online activities to offer recommendation:
  - Frequently bought together
  - Learn from click patterns to increase discoverability of catalog
  - Customized recommendations from history

Use Case
Allrecipes
- Allrecipes uses the recommendations API to provide personalized recipe solutions for each family for cooking
- Data from billions of pieces of user-shared experiences/reviews

Pricing

	FREE	STANDARD S1	STANDARD S2	STANDARD S3	STANDARD S4
Price (per month)	$0	$75.02	$500.03	$2,500.15	$4,999.99
Maximum Transactions included in tier	10,000	100,000	1,000,000	10,000,000	50,000,000
Overage Rate (per 1,000 transactions)	N/A	$0.75	$0.50	$0.25	$0.10

Source: Microsoft website, Schulte Research.

f. Bot Platform: Zo — the chatbot with emotion

| China 小冰 2014 | Japan りんな 2015 | → | US Zo Today |

"There are two sides to conversational AI — the task-completion or productivity side and the emotional side. You need both to truly realize the promise of AI."

Microsoft Journey with Chatbot
- Microsoft chatbot started in May **2014** in China (**Xiaoice (小冰)** — 40 mn users). Average 23 back and forth with users.
- In July **2015**, launched **Rinna** in Japan. Regular conversations with 20% of Japan population.
- Launched **Zo** in March **2017** in US.
  - i.   Conversations with >100,000 people in US
  - ii. 5,000 users had 1 hour conversation with Zo
  - iii. Longest continual chat conversation record: 1,229 turns, lasting 10 hours

Source: Microsoft website, Schulte Research.

Open Bot framework

Introduction
- Open source builder SDKs (software development kit)
- 67,000 developers are using Microsoft's bot framework and cognitive services
- Channels: Bing, Cortana, Skype

Use Cases
- The Bank of Kochi in Japan is developing a receptionist bot
- Rockwell Automation, a bot to automate productions
- The Dept. of Human Services in Australia, a bot to improve customer engagement

Source: Microsoft website, Schulte Research.

2. Microsoft Azure Labs

Project Prague
Gesture based controls

Project Cuzco
Event associated with Wikipedia entries

Project Nanjing
Isochrones calculations

Project Abu Dhabi
Distance matrix

Project Johannesburg
Route logistics

Project Wollongong
Location insights

Project Prague: gesture-based controls

Project Prague is an easy-to-use SDK (software development kit) for hand gestures understanding and controls
- Ability to define desired hand poses
- AI will detect when user does that gesture
- Action can be assigned to response to different gesture
- The potential to totally change the way we send comment to machine (computer, mobile, TV, etc.)

Source: Microsoft website, Schulte Research.

Project Nanjing, Abu Dhabi, Johannesburg, Wollongong

Project Nanjing
- An API to calculate time and distance-based recommendations for enterprise route optimization
- Example: identify the meeting point between two locations

Project Abu Dhabi
- API to create distance matrices
- Histogram of travel times considering time-windows and traffic conditions

Project Johannesburg
- Truck-routing service for professions
- Considers speed limits, weight, length, height, bridge limits, materials, etc.

Project Wollongong
- An API to score the attractiveness of a location, based on amenity nearby
- User can set time or distance limit to reach customized criteria (cinemas, bars, theaters, parks, etc.)

Source: Microsoft website, Schulte Research.

3. Autonomous Driving: Connected Vehicle Platform

Microsoft Prediction
- **Connected**: 90% of new cars will be connected by 2020
- **Shared**: 10% will give up ownership for on-demand access by 2020
- **Autonomous**: 15% of cars will be self-driving
- **Personalized**: 31% of customers desire more personalized experience

Microsoft Strategy: instead of making real vehicle, Microsoft is building a connected vehicle platform.

Functions:
1. Telematics and predictive services
2. Productivity and digital life
3. Connected advanced driver assistance systems
4. Advanced navigation
5. Customer insights and engagement

Source: Microsoft website, Schulte Research.

Microsoft Connected Vehicle Platform

Telematics and predictive services

Using IoT data to improve user experience: predictive maintenance alerts, "find my car", warranty and recall issues

Productivity and digital life

Help user keep productivity and reduce distraction for safety in driving by using voice controlled tools like Skype, Cortana, Bing, Office 365

Connected advanced driver assistance systems

Provide real-time road and environment information to increase driving safety and performance

Advanced navigation

Personalized and dynamic navigation service from a combination of various location sources and user preference

Customer insights and engagement

Improve customer relationship management to increase brand loyalty

Source: Microsoft website, Schulte Research.

4. Lifestyle — Xbox and Skype

Xbox: the entertainment center at home

- Top 2 most popular game console globally
- Xbox 360: 78 mn sold
- Xbox one: ~30 mn sold
- Xbox live Monthly Active Users: 55 mn
- $9.4 bn revenue from Xbox in 2016

⊗ XBOX LIVE

Xbox Live Active Users
(in millions)

Sep-14 Dec-14 Mar-15 Jun-15 Sep-15 Dec-15 Mar-16 Jun-16 Sep-16 Dec-16
Source: Microsoft Earnings Reports   GEEKWIRE.COM

Skype: vast voice and video call data

- Launched in 2003, bought by Microsoft in 2011
- Monthly Active Users: 300 mn
- 1 bn Skype mobile app downloaded
- 3 bn minutes of calls each day
- 2 tn minutes Skype video calls
- Skype Translator can translate voice calls in nine different languages

Source: Microsoft website, Schulte Research.

Lifestyle cont'd. LinkedIn: The key social network data

Introduction

- Microsoft acquired LinkedIn in December 2016 for $26 bn
- 510 million users in 200 countries
- 10 million active job posts
- 9 million companies
- LinkedIn contributed $1.1 bn revenue to Microsoft in Q4 FY2017

The Implication

- LinkedIn provides the valuable social network data that Microsoft lacks
- The professional network data fits Microsoft's advantage in corporate (business) solutions
- Professional and employment data is easier to commercialize into finance products since it is directly linked to income condition

5. Search: Will Bing catch up in mobile search through AI?

- Bing is the no.2 search engine for desktop due to the default setting in Windows IE
- Google enlarge its no.1 advantage in mobile market due to Android system
- Bing has only 0.9% mkt share in mobile market (from 7% desktop market share)
- Bing open AI API is the initiative to improve user experience and get back market share

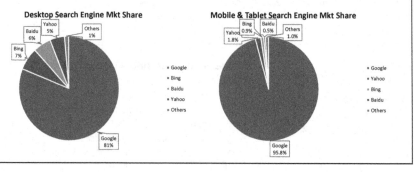

Source: Netmarketshare, Schulte Research.

Printed in the United States
By Bookmasters